Student Solutions Manual for Tussy and Gustafson's

ELEMENTARY ALGEBRA

Pat Averbeck

Oregon State University

Brooks/Cole Publishing Company

I(T)P® *An International Thomson Publishing Company*

Pacific Grove • Albany • Belmont • Bonn • Boston • Cincinnati • Detroit • Johannesburg • London
Madrid • Melbourne • Mexico City • New York • Paris • Singapore • Tokyo • Toronto • Washington

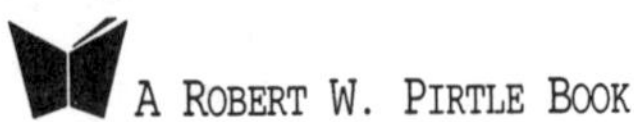

A ROBERT W. PIRTLE BOOK

Sponsoring Editor: *Linda Row*
Marketing Representative: *Jennifer Huber*
Editorial Assistant: *Peggi Rodgers*
Production: *Dorothy Bell*
Cover Design: *Vernon T. Boes*
Cover Illustration: *Joe Miles*
Printing and Binding: *Patterson Printing*

For more information, contact:

BROOKS/COLE PUBLISHING COMPANY
511 Forest Lodge Road
Pacific Grove, CA 93950
USA

International Thomson Publishing Europe
Berkshire House 168-173
High Holborn
London WC1V 7AA
England

Thomas Nelson Australia
102 Dodds Street
South Melbourne, 3205
Victoria, Australia

Nelson Canada
1120 Birchmount Road
Scarborough, Ontario
Canada M1K 5G4

International Thomson Editores
Seneca 53
Col. Polanco
11560 México, D. F., México

International Thomson Publishing GmbH
Königswinterer Strasse 418
53227 Bonn
Germany

International Thomson Publishing Asia
60 Albert Street
#15-01 Albert Complex
Singapore 189969

International Thomson Publishing Japan
Hirakawacho Kyowa Building, 3F
2-2-1 Hirakawacho
Chiyoda-ku, Tokyo 102
Japan

Printed in the United States of America

10 9 8 7 6 5 4 3 2

ISBN 0-534-35698-2

Table of Contents

STUDY SET Section 1.1

VOCABULARY

1. sum

3. product

5. Variables

7. expressions

9. formula

11. horizontal; 1

CONCEPTS

13. equation

15. algebraic expression

17. algebraic expression

19. equation

21. a) multiplication; subtraction
b) x

23. a) addition; subtraction
b) m

25. a) For 0°:
$$90° - 0° = 90°$$
For 30°:
$$90° - 30° = 60°$$
For 45°:
$$90° - 45° = 45°$$
For 60°:
$$90° - 60° = 30°$$
For 90°:
$$90° - 90° = 0°$$
b)

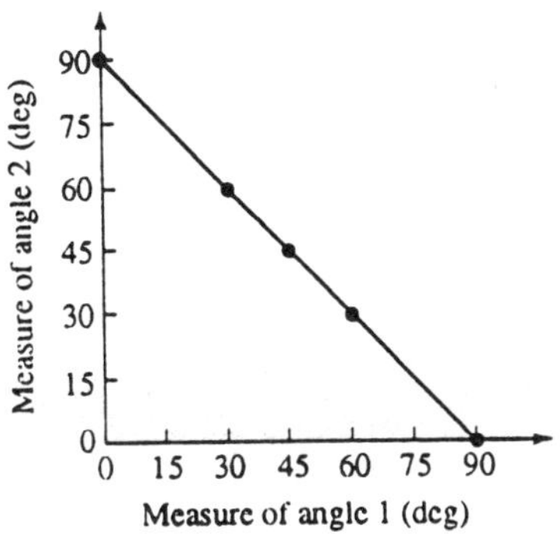

27. a) Because the vertical line starts at the 15 on the horizontal line, we can determine how much 15-year-old machinery is worth. By looking at the horizontal dotted line, we can see that the machinery is worth $450.
b) Because the graph goes down as we trace it to the right, the value decreases.

NOTATION

29. 5•6 or 5(6). Note: putting the 5 and 6 next to each other would look like the number 56.

31. 34•75 or 34(75)

33. $4x$. Note: we can put numbers and letters together because they won't be confused as another number.

35. $3rt$

37. lw

39. Prt

41. $\frac{32}{x}$

43. $\frac{90}{30}$

PRACTICE

45. the product of 18 and 24

47. the difference of 11 and 9

49. the product of 2 and x

51. the quotient of 66 and 11

53. Replace sale price with s and discount with d to get $s = 100 - d$.

55. Replace age of dog with d and dog's equivalent human age with h to get $7d = h$.

57. Replace sand with s and cement with c to get $s = 3c$.

59. Replace weight of truck with w and weight of engine with e to get $w = e + 1,200$.

61. Replace profit with p and revenue with r to get $p = r - 600$.

63. Replace laps with l and miles with m to get $\frac{l}{4} = m$.

65. Evaluate $g = 4$:

$$f = 16 + g$$
$$f = 16 + (4)$$
$$f = 20$$

Evaluate $g = 12$:

$$f = 16 + g$$
$$f = 16 + (12)$$
$$f = 28$$

Evaluate $g = 34$:

$$f = 16 + g$$
$$f = 16 + (34)$$
$$f = 50$$

67. Evaluate $d = 200$:

$$t = 1,500 - d$$
$$t = 1,500 - (200)$$
$$t = 1,300$$

Evaluate $d = 300$:

$$t = 1,500 - d$$
$$t = 1,500 - (300)$$
$$t = 1,200$$

Evaluate $d = 400$:

$$t = 1,500 - d$$
$$t = 1,500 - (400)$$
$$t = 1,100$$

69. The pattern appears to be to divide the first column by 100 to get the second column. Let s represent the number of schools and a represent acres. $s = \frac{a}{100}$. The number of schools is the quotient of the number of acres and 100.

71. The pattern appears to be to subtract 10 from the first column to get the second column. Let o represent output voltage and i represent input voltage. $o = i - 10$. The output voltage is 10 less than the input voltage.

APPLICATIONS

73. The number of legs is four times the number of chairs. Let l represent the number of legs and c represent the number of chairs. $l = 4c$.
The number of arms is two times the number of chairs. Let a represent the number of legs. $a = 2c$.
The number of seats is the same as the number of chairs. Let S represent the number of legs. $S = c$.
The number of backs is same as the number of chairs. Let b represent the number of legs. $b = c$.
The number of arm pads is two times the number of chairs. Let p represent the number of legs. $p = 2c$
The number of screws is twenty times the number of chairs. Let s represent the number of legs. $s = 20c$.

75. The number of left-side doors is the sum of the number of cars and 500. Let d represent left-side doors and x represent the number of cars. $d = x + 500$.

77.

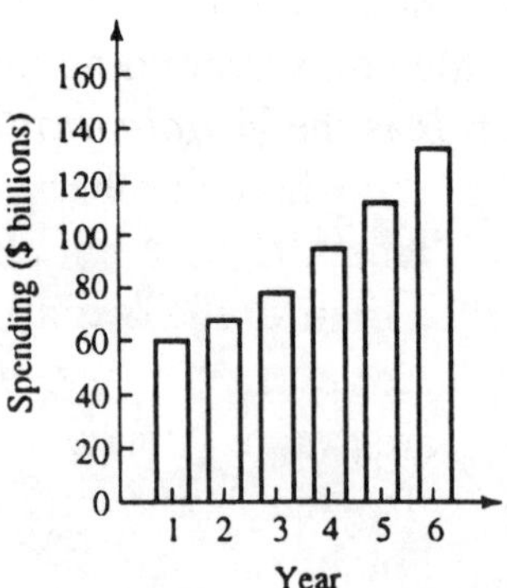

STUDY SET Section 1.2

VOCABULARY

1. whole

3. negative; positive

5. rational; integer

7. integers

9. irrational

11. absolute value

CONCEPTS

13. $\frac{6}{1}, \frac{-9}{1}, \frac{-7}{8}, \frac{7}{2}, \frac{-3}{10}, \frac{283}{100}$

15. Using a number line, start at 5 and count 8 places to the right to get 13. Then at 5 count 8 places to the left to get -3.

17. a) Because a is to the left of b on the number line: $a < b$.
b) Because b is to the right of a on the number line: $b > a$.
c) Because b is to the right of 0 on the number line: $b > 0$.
Because a is to the left of 0 on the number line: $a < 0$.

19. The strategy is to work from the inside box outwards. The first level of numbers are the natural numbers, label the innermost box *Natural numbers*. Since all natural numbers are whole numbers, the box surrounding the natural numbers box is the *Whole numbers* box. All of the whole numbers are integers, so label the next box *Integers*. The next box, half of the overall box, is labelled *Rational numbers* because all integers are rational numbers. The other half is labeled *Irrational numbers*. Finally, all of the above numbers are real numbers, so label the big box *Real numbers*.

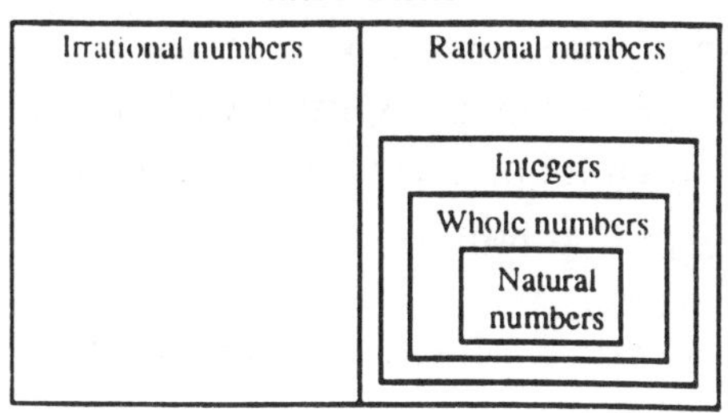

NOTATION

21. square root

23. is approximately equal to

25. numerator; denominator

27. 4π means $4 \cdot \pi \approx 12.6$

29. $-2 > -3$

31. Because $\sqrt{10} \approx 3.2$ and $|3.4| = 3.4$, $|3.4| > \sqrt{10}$.

33. Because $-|-1.1| = 1.1$, $-|-1.1| < -1$.

35. $\frac{-5}{8} < \frac{-3}{8}$

37. Because $|-\frac{15}{2}| = |-7.5|$, and $|-7.5| = 7.5$, $|-\frac{15}{2}| = 7.5$.

39. $\frac{99}{100} = .99$

41. Because $0.3 = 0.3000....$, $0.333... > 0.3000...$

43. Because $|-\frac{15}{16}| = \frac{15}{16}$, $1 > |-\frac{15}{16}|$.

PRACTICE

45. There are not natural numbers. 0 is the only whole number. 0 and -50 are integers. Every number but $\sqrt{2}$ are rational numbers. $\sqrt{2}$ is the only irrational number. All of the numbers are real numbers.

47. a) true
b) false
c) false
d) true

49. $-5 > -6$

51.

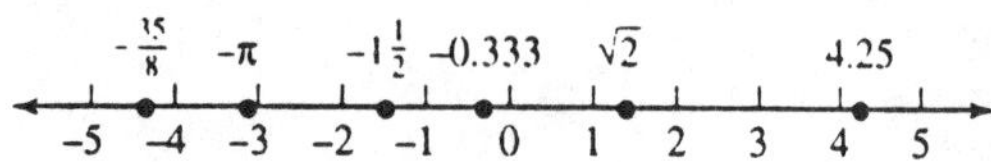

53. 2.236

55. 9.950

57. 4.472

59. 15.708

61. -5

63. $-(-\frac{7}{8}) = \frac{7}{8}$

APPLICATIONS

65. Only 750 and 5,000 are natural, whole, and integer numbers. All of the numbes are rational. None are irrational numbers. All are real numbers.

67. '90: \$.03; '91: $-$ \$0.8; '92: \$0.9; '93: $-$ \$2.3; '94: \$3.8; '95: \$2.0; '96: \$3.8.

69. Evalaute for $d = 26$:
$$C = \pi \bullet 26$$
$$C \approx 81.7 \text{ in.}$$

71.

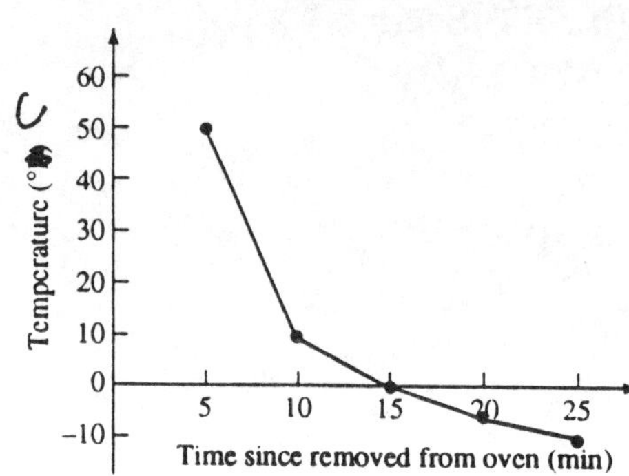

73. For the motorcycle, the speed is $|-30| = 30$ mph. For the car, the speed is $|55| = 55$ mph.

75. The difference between the markings of the timeline is 500 years.

77. The $B.C/A.D$ is zero point on the line. Since $B.C.$ is to the left of this point, $B.C.$ is negative. Since $A.D.$ is to the right of this point, $A.D.$ is positive.

REVIEW

83. The difference of 7 and 5 is 2.

85. The quotient of 30 and 15 is 2.

87. Evaluate for $g = 10$:
$$T = 15g = 15(10) = 150$$
Evaluate for $g = 12$:
$$T = 15g = 15(12) = 180$$
Evaluate for $g = 15$:
$$T = 15g = 15(15) = 225$$

STUDY SET Section 1.3

VOCABULARY

1. factors

3. squared; cubed

5. mean; average

CONCEPTS

7. a) addition; multiplication
b) $4 + 5 \cdot 6 = 9 \cdot 6 = 54$
or $4 + 5 \cdot 6 = 4 + 30 = 34$
c) 34 is the correct answer because multiplication is done before addition. If addition was in parentheses, then addition would be done before multiplication.

9. a) Divide the sum of the scores by the number of scores.
b) average $= \frac{75+81+47+53}{4} = \frac{256}{4} = 64$

11. a) addition, power, and multiplicaton
b) power, multiplication, and addition

13. In the numerator, multiplication is done first. In denominator, subtraction is done first because it is in parentheses.

NOTATION

15. 12^6

17. $50 + 6 \cdot 3^2 = 50 + 6 \cdot 9$
$= 50 + 54$
$= 104$

19. $19 - 2[(1+2) \cdot 3] = 19 - 2[(3) \cdot 3]$
$= 19 - 2[9]$
$= 19 - 18$
$= 1$

PRACTICE

21. Let d = distance and t = time. The square of time $= t^2$. Thus, $d = t^2$.

23. Evaluate $s = 1$:
$V = (1)^2 = 1$
Evaluate $s = 2$:
$V = (2)^2 = 4$
Evaluate $s = 3$:
$V = (3)^2 = 9$
Evaluate $s = 4$:
$V = (4)^2 = 16$

25. b^3

27. $10^2 k^3$

29. $4\pi r^2$

31. $6x^2y^3$

33. $12 - 2 \cdot 3 = 12 - 6 = 6$

35. $15 + (30 - 4) = 15 + 26 = 41$

37. $100 - 8(10) + 60 = 100 - 80 + 60$
$= 20 + 60$
$= 80$

39. $22 - (15 - 3) = 22 - 12 = 10$

41. $2(9) - 2(5) = 18 - 10 = 8$

43. $5^2 + 13^2 = 25 + 169 = 194$

45. $3 \cdot 8^2 = 3 \cdot 64 = 192$

47. $8 \cdot 5 - 4 \div 2 = 40 - 4 \div 2$
$= 40 - 8$
$= 32$

49. $14 + 3(7 - 5) = 14 + 3(2)$
$= 14 + 6$
$= 20$

51. $4 + 2[16 - 5(3)] = 4 + 2[16 - 15]$
$= 4 + 2[1]$
$= 4 + 2$
$= 6$

53. $19 - (45 - 41)^2 = 19 - (4)^2$
$= 19 - 16$
$= 3$

55. $2 + 3 \cdot 2^2 \cdot 4 = 2 + 3 \cdot 4 \cdot 4$
$= 2 + 12 \cdot 4$
$= 2 + 48$
$= 50$

57. $3(4)(5)(6) = 12(5)(6)$
$= 60(6)$
$= 360$

59. $5[9(2) - 2(8)] = 5[18 - 16]$
$= 5[2]$
$= 10$

61. $75 - 3 \cdot 1^2 = 75 - 3 \cdot 1$
$= 75 \cdot 3$
$= 72$

63. $5(150 - 3^3) = 5(150 - 27)$
$= 5(123)$
$= 615$

65. $(4 + 2 \cdot 3)^4 = (4 + 6)^4$
$= (10)^4$
$= 10,000$

67. $3(2)^5(2)^2 = 3(32)(4)$
$= 96(4)$
$= 384$

69. $6(\frac{25}{5}) - \frac{36}{9} + 1 = 6(5) - 4 + 1$
$= 30 - 4 + 1$
$= 26 + 1$
$= 27$

71. $\frac{5(68-32)}{9} = \frac{5(36)}{9}$
$= \frac{180}{9}$
$= 20$

73. $\frac{(6-5)^4+21}{27-4^2} = \frac{(1)^4+21}{27-4^2}$
$= \frac{1+21}{27-16}$
$= \frac{22}{11}$
$= 2$

75. $\frac{13^2-5^2}{3(9-5)} = \frac{169-25}{3(4)}$
$= \frac{144}{12}$
$= 12$

77. $\frac{8^2-10}{2(3)(4)-5(3)} = \frac{64-10}{2(3)(4)-5(3)}$
$= \frac{64-10}{6(4)-15}$
$= \frac{64-10}{24-15}$
$= \frac{54}{9}$
$= 6$

79. $5^7 - (45 \cdot 489) = 78,125 - (45 \cdot 489)$
$= 78,125 - 22,005$
$= 56,120$

81. $54^3 - 16^4 + 19(3)$
$= 157,464 - 65,536 + 19(3)$
$= 157,464 - 65,536 + 57$
$= 91,928 + 57$
$= 91,985$

APPLICATIONS

83. Total prize money = grand prize
+ total of 1st prizes + total of 2nd prizes
+ total of 3rd prizes
$= 1(2,500) + 4(500) + 35(150)$
$+ 85(25)$
$= 2,500 + 2,000 + 5,250 + 2125$
$= \$11,875$

85. average award (in thousands)
$= \frac{300+200+175+300+200+175+150}{8}$
$= \frac{1,650}{8}$
$= 206.25$
Average award was $206,250.

87. Value D $= \frac{6(\text{Value C})-(\text{Value A}+\text{Value B})}{12}$

Row 1:

$$\frac{6(8)-(20+4)}{12} = \frac{6(8)-(24)}{12}$$
$$= \frac{48-24}{12}$$
$$= \frac{24}{12}$$
$$= 2$$

Row 2:

$$\frac{6(16)-(9+3)}{12} = \frac{6(16)-(12)}{12}$$
$$= \frac{96-12}{12}$$
$$= \frac{84}{12}$$
$$= 7$$

Row 3:

$$\frac{6(11)-(1+5)}{12} = \frac{6(11)-(6)}{12}$$
$$= \frac{66-6}{12}$$
$$= \frac{60}{12}$$
$$= 5$$

89. a) Week 1: 5 can be written as 5^1.
Week 2: 25 can be written as 5^2.
Week 3: 125 can be written as 5^3.
Week 4: 625 can be written as 5^4.
b) Notice the pattern that the exponent matches the week. so, for the 5th week, the exponent should be 5: $5^5 = 3125$.

REVIEW

95. $|-5| = 5$

97. True

99. Starting at -3, move 6 numbers to the right to get 3. Then, at -3 move 6 numbes to the left to get -9.

STUDY SET Section 1.4

VOCABULARY

1. evaluate

3. subtraction

CONCEPTS

5. Any combination of 6, 20, and x with mathematical operations will work as long as no equal signs are used.
$\frac{6-x}{20}$, $6-\frac{20}{x}$, $6+20x$, ...

7. Instead of being $3(4)-6$, we would get $34-6$. The difference is multiplying 3 by 4 or the number 34.

9. a) Because we want to the weight of the van and do not know the weight of the car, let weight of car $= x$.
Key phrase: less than
Translation: subtract
Key word: twice
Translation: multiply by 2.
van = 2•car − 500
van $= 2x - 500$
b) van $= 2(2,000) - 500$
$= 4,000 - 500$
$= 3,500$ lb.

11. a) The top of the second column usually is the algebraic expression: $8x - x^2$.
b) The values of x are usually in the first column and under the x: 3, 4, 5.
c) $8x - x^2 = 8(4) - (4)^2$
$= 8(4) - 16$
$= 32 - 16$
$= 16$

NOTATION

13. $9a - a^2 = 9(5) - (5)^2$
$= 9(5) - 25$
$= 45 - 25$
$= 20$

PRACTICE

15. Key word: sum
Translation: add
The phrase translates to $l + 15$.

17. Key word: product
Translation: multiply
The phrase translates to $50x$.

19. Key word: ratio
Translation: division
The phrase translates to $\frac{w}{l}$.

21. Key word: increased
Translation: add
The phrase translates to $P + p$.

23. Key word: square
Translation: power 2
Key word: minus
Translation: subtract
The phrase translates to $k^2 - 2,005$.

25. Key word: reduced
Translation: subtract
The phrase translates to $J - 500$.

27. Key phrase: split equal ways
Translation: division
The phrase translates to $\frac{1,000}{n}$.

29. Key phrase: more than
Translation: add
The phrase translates to $p + 90$.

31. Key phrase: total
Translation: add
The phrase translates to $35 + h + 300$

33. Key phrase: fewer than
Translation: subtract
The phrase translates to $p - 680$.

35. Key word: product
Translation: multiply
Key word: decreased
Translation: subtract
The phrase translates to $4d - 15$.

37. Key word: twice
Translation: multiply by 2
Key word: sum
Translation: add
Because multiplying 2 to the sum, need to use parentheses to indicate adding first. The phrase translates to $2(200+t)$.

39. Key word: difference
Translation: subtract
Because it said absolute value of the difference, need to put the subtraction inside the absolute value signs
The phrase translates to $|a-2|$.

For 41 and 43, the answers will vary depending on which phrase is used from the lists in the book.

41. 7 less than a number

43. the product of 7 and a number increased by 4.

45. Each hour has 60 minutes; this indicates multiplication.
a) $60(5)=300$ minutes
b) $60(h)$

47. Each yard has 3 feet:
a) To find number of feet, multiply: $3y$.
b) To find number of yards, divide: $\frac{f}{3}$

49. 2 inches are added to the skirt of length x. This phrase translates to $x+2$.

51. Key word: shrink
Translation: subtract
The phrase translates to $36-x$.

53. The amount earned is the product of the number of hours and the amount of pay.
Key word: product
Translation: multiply
The phrase translates to
a) $\$8x$
b) $\$40x$

55. The total cost of tickets is the product of 5 and the number of tickets. The number of tickets is the sum of the family and the neighbors. The phrase translates to $\$5(x+2)$.

57. $6x=6(7)=42$

59. $3(t-6)=3(8-6)$
$=3(2)$
$=6$

61. Evaluate for $g=0$:
$g^2-7g+1=(0)^2-7(0)+1$
$=0-7(0)+1$
$=0-0+1$
$=0+1=1$
Evaluate for $g=7$:
$g^2-7g+1=(7)^2-7(7)+1$
$=49-7(7)+1$
$=49-49+1$
$=0+1=1$
Evaluate for $g=10$:
$g^2-7g+1=(10)^2-7(10)+1$
$=100-7(10)+1$
$=100-70+1$
$=30+1=31$

63. Evaluate for $s=1$:
$\frac{5s+36}{s}=\frac{5(1)+36}{(1)}$
$=\frac{5+36}{1}$
$=\frac{41}{1}$
$=41$
Evaluate for $s=6$:
$\frac{5s+36}{s}=\frac{5(6)+36}{(6)}$
$=\frac{30+36}{6}$
$=\frac{66}{6}$
$=11$
Evaluate for $s=12$:
$\frac{5s+36}{s}=\frac{5(12)+36}{(12)}$
$=\frac{60+36}{12}$
$=\frac{96}{12}$
$=8$

65. Evaluate for $x = 100$:

$$\begin{aligned} 2x - \tfrac{x}{2} &= 2(100) - \tfrac{(100)}{2} \\ &= 2(100) - 50 \\ &= 200 - 50 \\ &= 150 \end{aligned}$$

Evaluate for $x = 300$:

$$\begin{aligned} 2x - \tfrac{x}{2} &= 2(300) - \tfrac{(300)}{2} \\ &= 2(300) - 150 \\ &= 600 - 150 \\ &= 450 \end{aligned}$$

67. Evaluate for $a = 4$:

$$\begin{aligned} 3a^2 + 1 &= 3(4)^2 + 1 \\ &= 3(16) + 1 \\ &= 48 + 1 \\ &= 49 \end{aligned}$$

Evaluate for $a = 8$:

$$\begin{aligned} 3a^2 + 1 &= 3(8)^2 + 1 \\ &= 3(64) + 1 \\ &= 192 + 1 \\ &= 193 \end{aligned}$$

69. Evaluate for $a = 4 \,\&\, b = 12$:

$$\begin{aligned} a^2 + b^2 &= (5)^2 + (12)^2 \\ &= 25 + 144 \\ &= 169 \end{aligned}$$

71. Evaluate for $s = 23 \,\&\, t = 21$:

$$\begin{aligned} \tfrac{s+t}{s-t} &= \tfrac{23+21}{23-21} \\ &= \tfrac{44}{2} \\ &= 22 \end{aligned}$$

73. Evaluate for $h = 5,\ b = 7,\ \&\, c = 9$:

$$\begin{aligned} \tfrac{h(b+c)}{2} &= \tfrac{5(7+9)}{2} \\ &= \tfrac{5(16)}{2} \\ &= \tfrac{80}{2} \\ &= 40 \end{aligned}$$

75. Evaluate for $r = 4.5$:

$$\begin{aligned} \tfrac{4\pi r^3}{3} &= \tfrac{4\pi(4.5)^3}{3} \\ &\approx 381.7 \end{aligned}$$

APPLICATIONS

77. Let $n =$ the number of candidates entering the program. The number in each squad is the original group reduced by 6, then split into 8 equal groups. The phrase is translated to the number in each squad $= \frac{n-6}{8}$.

79. Let $a =$ number of apartments occupied before rent was reduced. The situation can be interpreted as the number of apartments now occupied is double a less 3.
Key word: double.
Translation: multiply by 2.
Key word: less.
Translation: subtract.
The phrase can be translated to number of apartments $= 2a - 3$.

81. Evaluate for $t = 0$:

$$\begin{aligned} 64t - 16t^2 &= 64(0) - 16(0)^2 \\ &= 64(0) - 16(0) \\ &= 0 - 0 \\ &= 0 \end{aligned}$$

Evaluate for $t = 1$:

$$\begin{aligned} 64t - 16t^2 &= 64(1) - 16(1)^2 \\ &= 64(1) - 16(1) \\ &= 64 - 16 \\ &= 48 \end{aligned}$$

Evaluate for $t = 2$:

$$\begin{aligned} 64t - 16t^2 &= 64(2) - 16(2)^2 \\ &= 64(2) - 16(4) \\ &= 128 - 64 \\ &= 64 \end{aligned}$$

Evaluate for $t = 3$:

$$\begin{aligned} 64t - 16t^2 &= 64(3) - 16(3)^2 \\ &= 64(3) - 16(9) \\ &= 192 - 144 \\ &= 48 \end{aligned}$$

Evaluate for $t = 4$:

$$\begin{aligned} 64t - 16t^2 &= 64(4) - 16(4)^2 \\ &= 64(4) - 16(16) \\ &= 256 - 256 \\ &= 0 \end{aligned}$$

83. a) The computer formula translates to
$2BC + 2BD + 2CD$

Row 1: $B = 12$, $C = 6$, & $D = 6$

$$\begin{aligned}2BC + 2BD + 2CD &= 2(12)(6) + 2(12)(6) + 2(6)(6)\\ &= 24(6) + 24(6) + 12(6)\\ &= 144 + 144 + 72\\ &= 288 + 72\\ &= 360\end{aligned}$$

Row 2: $B = 18$, $C = 12$, & $D = 12$

$$\begin{aligned}2BC + 2BD + 2CD &= 2(18)(12) + 2(18)(12) + 2(12)(12)\\ &= 36(12) + 36(12) + 24(12)\\ &= 432 + 432 + 288\\ &= 864 + 288\\ &= 1,152\end{aligned}$$

Row 3: $B = 18$, $C = 24$, & $D = 18$

$$\begin{aligned}2BC + 2BD + 2CD &= 2(18)(24) + 2(18)(18) + 2(24)(18)\\ &= 36(24) + 36(18) + 48(18)\\ &= 864 + 648 + 864\\ &= 1,512 + 864\\ &= 2,376\end{aligned}$$

b) Because each term, $2BC$, $2BD$, and $2CD$ have two variables multiplying and each variable has inch as the unit, the unit of the result is squared inches, in^2. Adding does not change the unit and the 2 does not have a unit.

REVIEW

89. {...-3, -2, -1, 0, 1, 2, 3...}

91. $|-\frac{2}{3}| = \frac{2}{3}$

93. $c{\bullet}c{\bullet}c{\bullet}c = c^4$.

95. $$\begin{aligned}\text{average} &= \tfrac{84+93+72}{3}\\ &= \tfrac{249}{3}\\ &= 83\end{aligned}$$

STUDY SET Section 1.5

VOCABULARY

1. equation

3. check; true

5. equivalent

7. isolate

CONCEPTS

9. a) subtraction of 8; addition of 8
b) addition of 8; subtraction of 8
c) division by 8; multiplication by 8
d) multiplication by 8; division by 8

11. a) $x + 6$
b) Neither, because it depends on what we put in for x. When x is replaced with a number, then we can determine whether the equation is true.
c) To determine if $x = 5$ is a solution, we substitute 5 for x in the equation:
$x + 6 = 12$
$(5) + 6 = 12$
$11 = 12$
5 is not a solution.
d) To determine whether $x = 6$ is a solution, we substitute 6 for x.
$x + 6 = 12$
$(6) + 6 = 12$
$12 = 12$
6 is a solution.

NOTATION

13. $x + 15 = 45$
$x + 15 - 15 = 45 - 15$
$x = 30$

15. 12 is 40% of what number.
$12 = 40\% \cdot x$
$12 = 0.40 \cdot x$
$\frac{12}{0.40} = \frac{0.40x}{0.40}$
$30 = x$

17. a) $35\% = \frac{35}{100} = 0.35$
b) $3.5\% = \frac{3.5}{100} = 0.035$
c) $350\% = \frac{350}{100} = 3.50$
d) $\frac{1}{2}\% = 0.5\% = \frac{0.5}{100} = 0.005$

PRACTICE

19. $x + 12 = 18$
$(6) + 12 = 18$
$18 = 18$
Yes, 6 is a solution.

21. $2b + 3 = 15$
$2(5) + 3 = 15$
$10 + 3 = 15$
$13 = 15$
No, 5 is not a solution.

23. $0.5x = 2.9$
$0.5(5) = 2.9$
$2.5 = 2.9$
No, 5 is not a solution.

25. $33 - \frac{x}{2} = 30$
$33 - \frac{(6)}{2} = 30$
$33 - 3 = 30$
$30 = 30$
Yes, 6 is a solution.

27. $|c - 8| = 10$
$|(20) - 8| = 10$
$|12| = 10$
$12 = 10$
No, 20 is not a solution.

29. $3x - 2 = 4x - 5$
$3(12) - 2 = 4(12) - 5$
$36 - 2 = 48 - 5$
$34 = 43$
No, 12 is not a solution.

31. $x^2 - 5x + 6 = 0$
$(3)^2 - 5(3) + 6 = 0$
$9 - 5(3) + 6 = 0$
$9 - 15 + 6 = 0$
$-6 + 6 = 0$
$0 = 0$
Yes, 3 is a solution.

33. $\frac{2}{a+1}+5=\frac{12}{a+1}$
$\frac{2}{(1)+1}+5=\frac{12}{(1)+1}$
$\frac{2}{2}+5=\frac{12}{2}$
$1+5=6$
$6=6$
Yes, 1 is a solution.

35. $\sqrt{x-5}+1=15$
$\sqrt{(201)-5}+1=15$
$\sqrt{196}+1=15$
$14+1=15$
$15=15$
Yes, 201 is a solution.

37. $x+7=10$
$x+7-7=10-7$
$x=3$
check: $(3)+7=10$
$10=10$

39. $a-5=66$
$a-5+5=66+5$
$a=71$
check: $(71)-5=66$
$66=66$

41. $0=n-9$
$0+9=n-9+9$
$9=n$
check: $0=(9)-9$
$0=0$

43. $9+p=90$
$9-9+p=90-9$
$p=81$
check: $9+(81)=90$
$90=90$

45. $9+p=9$
$9-9+p=9-9$
$p=0$
check: $9+(0)=9$
$9=9$

47. $203+f=442$
$203-203+f=442-203$
$f=239$
check: $203+(239)=442$
$442=442$

49. $4x=16$
$\frac{4x}{4}=\frac{16}{4}$
$x=4$
check: $4(4)=16$
$16=16$

51. $369=9c$
$\frac{369}{9}=\frac{9c}{9}$
$41=c$
check: $369=9(41)$
$369=369$

53. $4f=0$
$\frac{4f}{4}=\frac{0}{4}$
$f=0$
check: $4(0)=0$
$0=0$

55. $23b=23$
$\frac{23b}{23}=\frac{23}{23}$
$b=1$
check: $23(1)=23$
$23=23$

57. $\frac{x}{15}=3$
$15(\frac{x}{15})=15(3)$
$x=45$
check: $\frac{(45)}{15}=3$
$3=3$

59. $\frac{l}{24}=2$
$24(\frac{l}{24})=24(2)$
$l=48$
check: $\frac{(48)}{24}=2$
$2=2$

61. $35 = \frac{y}{4}$
$4(35) = 4(\frac{y}{4})$
$140 = y$
check: $35 = \frac{(140)}{4}$
$35 = 35$

63. $0 = \frac{y}{11}$
$11(0) = 11(\frac{y}{11})$
$0 = y$
check: $0 = \frac{(0)}{11}$
$0 = 0$

65. What number is 48% of 650?
$x = 48\%•650$
$x = 0.48•650$
$x = 312$

67. 78 is what percent of 300?
$78 = x•300$
$\frac{78}{300} = \frac{x•300}{300}$
$0.26 = x$
$0.26 = 26\%$

69. 78 is 25% of what number?
$78 = 25\%•x$
$78 = 0.25•x$
$\frac{78}{0.25} = \frac{0.25x}{0.25}$
$312 = x$

71. What number is 92.4% of 50?
$x = 92.4\%•50$
$x = 0.924•50$
$x = 46.2$

73. 0.42 is what percent of 16.8
$0.42 = x•16.8$
$\frac{0.42}{16.8} = \frac{x•16.8}{16.8}$
$0.025 = x$
$0.025 = 2.5\%$

75. 128.1 is 8.75% of what number?
$128.1 = 8.75\%•x$
$128.1 = 0.0875•x$
$\frac{128.1}{0.0875} = \frac{0.0875x}{0.0875}$
$1,464 = x$

APPLICATIONS

77. a) Area of rectangle $= lw$
$= 2\,ft•3\,ft$
$= 6\,ft^2$
b) Area of red disc $= \pi r^2$
$= \pi(0.625\,ft)^2$
$\approx 1.2\,ft^2$
c) The area of the disc is what percent of the area of the flag?
1.2 is what percent of 6?
$1.2 = x•6$
$\frac{1.2}{6} = \frac{x•6}{6}$
$0.2 = x$
$0.2 = 20\%$

79. 180 is what percent of 10,000,000
$180 = x•10,000,000$
$\frac{180}{10,000,000} = \frac{x•10,000,000}{10,000,000}$
$0.000018 = x$
$0.000018 = 0.0018\%$

81. Net interest is 15% of $1,519 billion.
$x = 15\%•1,519$
$x = 0.15•1,519$
$x = 227.85$
Net interest is $227.85 billion

83. Tip is 15% of $75.18.
$x = 15\%•75.18$
$x = 0.15•75.18$
$x = 11.277$
$x \approx 12$
The tip was $12.

85. The decrease is \$1,050 − \$925 = \$125
\$125 is what percent of \$1,050
$125 = x\bullet 1,050$
$\frac{125}{1,050} = \frac{x\bullet 1,050}{1,050}$
$0.1190476 \approx x$
$0.12 \approx x$
$0.12 = 12\%$

87. \$63 is 44% of what amount?
$63 = 44\%\bullet x$
$63 = 0.44\bullet x$
$\frac{63}{0.44} = \frac{0.44x}{0.44}$
$143 \approx x$
The total amount given to charity was \$143 billion

89. \$11.30 is 20% of what amount?
$11.30 = 20\%\bullet x$
$11.30 = 0.20\bullet x$
$\frac{11.30}{0.20} = \frac{0.20x}{0.20}$
$56.50 = x$
The medical bill was \$56.50.

91. a) The decline in U.S. exports occurred between 1994 and 1995. The amount of decrease was $51 - 46 = 5$ billion.
5 billion is what percent of 51 billion.
$5 = x\bullet 51$
$\frac{5}{51} = \frac{x\bullet 51}{51}$
$0.098 \approx x$
$0.100 \approx x$
$0.100 = 10\%$

b) The biggest increase occurred between 1995 and 1996. The amount of increase was $57 - 46 = 11$ billion.
11 billion is what percent of 46 billion?
$11 = x\bullet 46$
$\frac{11}{46} = \frac{x\bullet 46}{46}$
$0.239 \approx x$
$0.24 \approx x$
$0.24 = 24\%$

93. The amount of increase was
$211,500 - 700 = 210,800$.
210,800 is what percent of 700?
$210,800 = x\bullet 700$
$\frac{210,800}{700} = \frac{x\bullet 700}{700}$
$301.14 \approx x$
$301.14 = 30,114\%$

95. The number of wins is what percent of the total games played during the season.
Total games = wins + loses
Chicago Bulls:
Total games $= 72 + 10 = 82$.
72 is what percent of 82?
$72 = x\bullet 82$
$\frac{72}{82} = \frac{x\bullet 82}{82}$
$0.88 \approx x$
$0.88 = 88\%$
Miami Dolphins:
Total games $= 14 + 0 = 14$.
14 is what percent of 14?
$14 = x\bullet 14$
$\frac{14}{14} = \frac{x\bullet 14}{14}$
$1.00 = x$
$1.00 = 100\%$
Chicago Cubs:
Total games $= 116 + 36 = 152$.
116 is what percent of 152?
$116 = x\bullet 152$
$\frac{116}{152} = \frac{x\bullet 152}{152}$
$0.76 \approx x$
$0.76 = 76\%$

REVIEW

101. $9 - 3x = 9 - 3(3)$
$= 9 - 9$
$= 0$

103. $45 - x$

105. $3\pi \approx 9.424777961$
≈ 9.4

STUDY SET Section 1.6
VOCABULARY

1. variable

3. equation

CONCEPTS

5. Analyze the problem, Form an equation, Solve the equation, State the result, and Check the result.

7. $x =$ length of the running portion of the triathlon.
The total length of the triathlon is the sum of the lengths of the swimming, biking, and running portions.
$16 = 1 + 10 + x$

9. $x =$ number of pages in the preface.
Total number of pages is the sum of the pages in the table of contents, preface, text, and index.
$430 = 4 + x + 400 + 12$

11. $x =$ amount of interest earned in savings.
The total amount of interest earned is the sum of the interest in savings and checking.
$678 = x + 73$

PRACTICE

13. Key phrase: disbursed equally
Translation: division.

15. Key word: melted
Translation: subtraction

17. Key phrase: reclaimed
Translation: addition

19. Key word: eroded
Translation: subtraction.

21. Key phrase: sectioned off equally
Translation: division

23. Key word: cut
Translation: subtraction.

25. Key phrase: diluted to twice their size
Translation: multiplication.

27. the total length of the water line is the sum of the original length and the length of the extension.
$1,525 = 1,000 + x$

29. The number of patients seen is the difference between the number of scheduled appointments and the number of no-shows.
$8 = x - 4$

31. Key phrase: cut into 12-inch pieces
Translation: division
$\frac{x}{12} = 5$

33. The total cost is the product of the number of postage stamps and the value.
$15x = 480$

35. **Analyze:** Need to find the old unit requirement. We know that the unit requirement was reduced by 6. The new unit requirement is 28.
Form: Let $x =$ the old unit requirement.
Key Word: reduced
Translation: subtract
Old reduced by 6 is new.
$x - 6 = 28$
Solve: $x - 6 = 28$
$x - 6 + 6 = 28 + 6$
$x = 34$
State: The old unit requirement was 34 units.
Check: If we reduce the old unit requirement of 34 by 6, we have $34 - 6 = 28$. The answer checks.

APPLICATIONS

37. **Analyze:** Americans eat 47 pints of ice cream per year and eat 20 pints more than the Canadians. We want to find the number of pints that the Canadians eat.
Form: Let c = number of pints Canadians eat.
Key word: more   Translation: add
Americans = Canadians + 20
$47 = c + 20$
Solve: $47 = c + 20$
$47 - 20 = c + 20 - 20$
$27 = c$
State: The Canadians per capita consume 27 pints of ice cream yearly.
Check: $47 = (27) + 20$
$47 = 47$

39. **Analyze:** Billy Jean King won 40 Grand Slam tennis tournaments. She won 14 less than Martina Navratilova. We want to find how many tournaments did Martina Navratilova win.
Form: Let w = number of wins for Martina Navratilova.
Key Word: less   Translation: subtract
King's wins = Navratilova win's − 14
$40 = w - 14$
Solve: $40 = w - 14$
$40 + 14 = w - 14 + 14$
$54 = w$
State: Martina Navratilova won 54 Grand Slam tennis tournaments.
Check: $40 = (54) - 14$
$40 = 40$

41. **Analyze:** Subway costs $564 million to construct. The cost was 3 times the estimate. We want to find the estimate.
Form: Let e = estimate to construct subway.
Key phrase: 3 times
Translation: multiply by 3.
Cost = 3•estimate
$564 = 3 \cdot e$
Solve: $564 = 3e$
$\frac{564}{3} = \frac{3e}{3}$
$188 = e$
State: The estimate for constructing the subway was $188 million.
Check: $564 = 3(188)$
$564 = 564$

43. **Analyze:** Withdrawal was $35.00 and the new balance was $287.00. The new balance is the old balance less the withdrawal. We want to find the old balance.
Form: Let b = old balance.
Key word: less
Translation: subtract
new balance = old balance − withdrawal
$287 = b - 35$
Solve: $287 = b - 35$
$287 + 35 = b - 35 + 35$
$322 = b$
State: The balance before the withdrawal was $322.
Check: $287 = (322) - 35$
$287 = 287$

45. **Analyze:** The interview is split into 3 equal segments. Each segment is 9 minutes long. We want to find the length of the interview.
Form: Let i = length of inteview.
Key phrase: split into 3 equal
Translation: divide by 3
segment length = $\frac{\text{interview length}}{3}$
$9 = \frac{i}{3}$
Solve: $9 = \frac{i}{3}$
$3(9) = 3(\frac{i}{3})$
$27 = i$
State: The interview was 27 minutes long.
Check: $9 = \frac{(27)}{3}$
$9 = 9$

47. Analyze:

Year	*States*
$1800 - 1950$	15
$1851 - 1900$	14
$1901 - 1950$	3
$1950+$	2

There are 50 states in the Union. The number of states in the Union before 1800 is the total less the number of states added after 1800.

Form: Let $s =$ number of states before1800

Key word: less

Translation: subtract

States before 1800

$=$ total states $-$ states added after 1800

$s = 50 - (15 + 14 + 3 + 2)$

Solve: $s = 50 - (15 + 14 + 3 + 2)$

$s = 50 - (34)$

$s = 16$

State: There were 16 states in the Union before 1800.

Check: $180 = 50 - (15 + 14 + 3 + 2)$

$180 = 50 - (34)$

$180 = 180$

49. Analyze:

Body Part	Number of Bones
legs	60
trunk	31
neck	26
head	29

There are 206 bones in the body. The number of remaining bones is the total number of bones less the number of bones in the chart.

Form: Let $b =$ number of remaining bones.

Key word: less

Translation: subtract

Remaining bones

$=$ total bones $-$ chart total

$b = 206 - (60 + 31 + 26 + 29)$

Solve: $b = 206 - (60 + 31 + 26 + 29)$

$b = 206 - (146)$

$b = 60$

State: There are 60 bones not listed.

Check:

$60 = 206 - (60 + 31 + 26 + 29)$

$60 = 206 - (146)$

$60 = 60$

51. Analyze: The triangle made between the three locations has a perimeter of $3,075$ miles. the Melbourne/Puerto Rico is $1,100$ miles. The Puerto Rico/Bermuda side is $1,000$ miles. We want to find the length of the Bermuda/Melbourne side.

Form: Let $BM =$ length of the Bermuda/Melbourne side.

The formula for perimeter of a triangle is:

$P = a + b + c$

$3,075 = 1,100 + 1,000 + BM$

Solve:

$3,075 = 1,100 + 1,000 + BM$

$3,075 = 2,100 + BM$

$3,075 - 2,100 = 2,100 - 2,100 + BM$

$975 = BM$

State: The imaginary line between Bermuda and Melbourne is 975 miles long.

Check: $3,075 = 1,100 + 1,000 + 975$

$3,075 = 3,075$

53. Analyze: The space craft is made up of 5 sections. The first, second, third sections are 138, 98, and 46 ft tall, respectively. The escape tower is 28 ft tall. The total hieght of the space craft is 364 ft. We need to find the height of the Lunar Module.

Form: Let $h =$ height of Lunar Module.

Total height $=$ sum of the five sections.

Key word: sum Translation: add

Total height $=$ 1st $+$ 2nd $+$ 3rd

$+$ Lunar Module $+$ tower

$364 = 138 + 98 + 46 + h + 28$

Solve: $364 = 138 + 98 + 46 + h + 28$

$364 = 138 + 98 + 46 + 28 + h$

$364 = 310 + h$

$364 - 310 = 310 - 310 + h$

$54 = h$

State: The Lunar Module was 54 ft tall.
Check:

$364 = 138 + 98 + 46 + (54) + 28$
$364 = 364$

55. **Analyze:** The sum of the measures of the angles in a stop sign is 1080°. Since the angles ina stop sign are the same, the total measure is split evenly among 8 angles. We want to find the measure of one angle.
Form: Let a = measure of the angle.
Key phrase: split evenly among 8
Translation: divide by 8

$\text{measure of the angle} = \frac{\text{total measure}}{8}$

$a = \frac{1080}{8}$

Solve: $a = \frac{1080}{8} = 135$
State: The measure of the angle is 135°.
Check: $135 = \frac{1080}{8}$
$135 = 135$

REVIEW

61. 25 is what percent of 85?
$25 = x \text{•} 85$
$\frac{25}{85} = \frac{x \text{•} 85}{85}$
$0.29 \approx x$
$0.29 = 29\%$

63. What is 35% of 7,800?
$x = 35\% \text{•} 7,800$
$x = 0.35 \text{•} 7,800$
$x = 2,730$

65. $x - 12 = 20$
$(34) - 12 = 20$
$22 = 20$
No, 34 is not a solution.

67. Evaluate $x = 4$:
$2x^2 - 3x = 2(4)^2 - (4)$
$= 2(16) - 4$
$= 32 - 4$
$= 28$

CHAPTER 1 REVIEW

1. The production of wide screen TV's is increasing.

3. a) The difference of 15 and 3 is 12.
b) The sum of 15 and 3 is 18.
c) The quotient of 15 and 3 is 5.
d). The product of 15 and 3 is 45.

5. a) $8 \bullet b = 8b$
b) $x \bullet y = xy$
c) $2 \bullet l \bullet w \bullet = 2lw$
d) $P \bullet r \bullet t = Prt$

7. Let T = number of tellers and
a = number of accounts.
number of tellers
$= \frac{\text{number of bank accounts}}{350}$
$T = \frac{a}{350}$

9. Each number in the second column is 5 times the first column. Let c = number of children and f = total fees. $f = 50c$.

11.

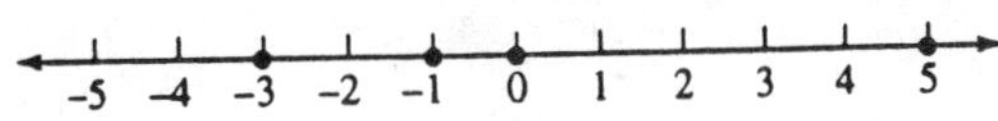

13. a) $0 < 5$
b) $6 > 4$
c) $-12 > -13$
d) $-3 < 2$

15.

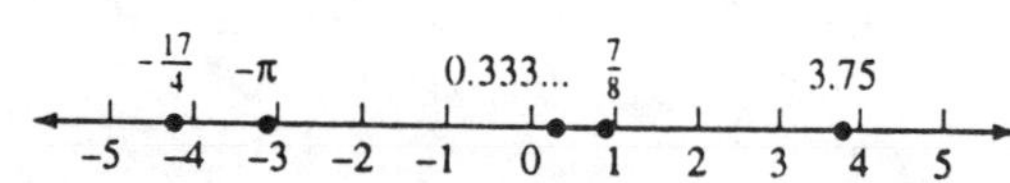

17. a) false
b) true
c) true
d) true

19. a) -10
b) $-(-3) = 3$
c) $-(-\frac{9}{16}) = \frac{9}{16}$
d) 0

21. a) 8^5
b) 2^3
c) $5^3 \bullet 9^2$
d) a^4
e) $9\pi r^2$
f) x^3y^4
h) 1^6

23. Evaluate $s = 15$:
$A = s^2 = (15)^2 = 225 \text{ in}^2$

25. There are 4 operations. The order is (1) power, (2) multiplication, (3) subtraction, and (4) addition.

27. parentheses; brackets; fraction bar.

29. Average $= \frac{19{,}981+21{,}456+20{,}599+22{,}500}{4}$
$= \frac{84{,}536}{4}$
$= \$21,134$

31. a) Key word: more Translation: add
$h + 15$

b) Key word: less
Translation: subtract
$s - 15$

c) Key word: times
Translation: multiply
$\frac{1}{2}t$

d) Key word: product
Translation: multiply
$6x$

33. x = time to complete test.
$3x$ = time spent standing in line.

35. a) The number of years is the product of 10 and the number of decades.
Key word: product
Translation: multiply
$10d$

b) The number of dozen donuts is the ratio of the total number of donuts and 12.
Key word: ratio Translation: divide
$\frac{x}{12}$

c) The age of the patio is the difference between the age of the house and 5.
Key word: difference
Translation: subtract
$x - 5$

37. Evaluate $x = 0$:
$$\begin{aligned}20x - x^3 &= 20(0) - (0)^3\\ &= 20(0) - 0\\ &= 0 - 0\\ &= 0\end{aligned}$$
Evaluate $x = 1$:
$$\begin{aligned}20x - x^3 &= 20(1) - (1)^3\\ &= 20(1) - 1\\ &= 20 - 1\\ &= 19\end{aligned}$$
Evaluate $x = 4$:
$$\begin{aligned}20x - x^3 &= 20(4) - (4)^3\\ &= 20(4) - 64\\ &= 80 - 64\\ &= 24\end{aligned}$$

39. a) Evaluate $x = 6$:
$6x = 6(6) = 36$

b) Evaluate $x = 4$:
$$\begin{aligned}7x^2 - \tfrac{x}{2} &= 7(4)^2 - \tfrac{(4)}{2}\\ &= 7(16) - \tfrac{(4)}{2}\\ &= 112 - \tfrac{(4)}{2}\\ &= 112 - 2\\ &= 110\end{aligned}$$

c) Evaluate $b = 10,\ a = 3,\ c = 5$:
$$\begin{aligned}b^2 - 4ac &= (10)^2 - 4(3)(5)\\ &= 100 - 4(3)(5)\\ &= 100 - 60\\ &= 40\end{aligned}$$

d) Evaluate $c = 9$:
$$\begin{aligned}2(24 - 2c)^3 &= 2(24 - 2(9))^3\\ &= 2(24 - 18)^3\\ &= 2(6)^3\\ &= 2(216)\\ &= 432\end{aligned}$$

e) Evaluate $r = 7,\ h = 5$:
$$\begin{aligned}3r^2h + 1 &= 3(7)^2(5) + 1\\ &= 3(49)(5) + 1\\ &= 735 + 1\\ &= 736\end{aligned}$$

f) Evaluate $x = 19,\ y = 17$
$$\begin{aligned}\tfrac{x+y}{x-y} &= \tfrac{(19)+(17)}{(19)-(17)}\\ &= \tfrac{36}{2}\\ &= 18\end{aligned}$$

41. a) $x - 34 = 50$
$(84) - 34 = 50$
$50 = 50$
Yes, 84 is a solution.

b) $5y + 2 = 12$
$5(3) + 2 = 12$
$15 + 2 = 12$
$17 = 12$
No, 3 is not a solution.

c) $\frac{x}{5} = 6$
$\frac{(30)}{5} = 6$
$6 = 6$
Yes, 30 is a solution.

d) $a^2 - a - 1 = 0$
$(2)^2 - (2) - 1 = 0$
$4 - 2 - 1 = 0$
$2 = 0$
No, 2 is not a solution.

e) $5b - 2 = 3b + 3$

$5(3) - 2 = 3(3) + 3$

$15 - 2 = 9 + 3$

$13 = 12$

No, 3 is not a solution.

f) $\frac{2}{y+1} = \frac{12}{y+1} - 5$

$\frac{2}{(1)+1} = \frac{12}{(1)+1} - 5$

$\frac{2}{2} = \frac{12}{2} - 5$

$1 = 6 - 5$

$1 = 1$

Yes, 1 is a solution.

43. a) $x - 9 = 21$

$x - 9 + 9 = 21 + 9$

$x = 30$

check: $(30) - 9 = 21$

$21 = 21$

b) $y + 15 = 32$

$y + 15 - 15 = 32 - 15$

$y = 17$

check: $(17) + 15 = 32$

$32 = 32$

c) $4 = v - 1$

$4 + 1 = v - 1 + 1$

$5 = v$

check: $4 = (5) - 1$

$4 = 4$

d) $100 = 7 + x$

$100 - 7 = 7 - 7 + x$

$93 = x$

check: $100 = 7 + (93)$

$100 = 100$

e) $2x = 40$

$\frac{2x}{2} = \frac{40}{2}$

$x = 20$

check: $2(20) = 40$

$40 = 40$

f) $120 = 15c$

$\frac{120}{15} = \frac{15c}{15}$

$8 = c$

check: $120 = 15(8)$

$120 = 120$

g) $x - 11 = 0$

$x - 11 + 11 = 0 + 11$

$x = 11$

check: $(11) - 11 = 0$

$0 = 0$

h) $p + 3 = 3$

$p + 3 - 3 = 3 - 3$

$p = 0$

check: $(0) + 3 = 3$

$3 = 3$

i) $\frac{t}{8} = 12$

$8(\frac{t}{8}) = 8(12)$

$t = 96$

check: $\frac{(96)}{8} = 12$

$12 = 12$

j) $3 = \frac{q}{26}$

$26(3) = (\frac{q}{26})$

$78 = q$

check: $3 = \frac{(78)}{26}$

$3 = 3$

k) $6b = 0$

$\frac{6b}{6} = \frac{0}{6}$

$b = 0$

check: $6(0) = 0$

$0 = 0$

l) $\frac{x}{14} = 0$

$14(\frac{x}{14}) = 14(0)$

$x = 0$

check: $\frac{(0)}{14} = 0$

$0 = 0$

45. Increase is 3.5% of \$764.
increase $= 3.5\% \bullet 764$
$= 0.035 \bullet 764$
$= \$26.74$

47. \$625 is what percent of $\$1,890$.
$625 = x \bullet 1,890$
$\frac{625}{1,890} = \frac{x \bullet 1,890}{1,890}$
$0.33 \approx x$
$0.33 = 33\%$

49. Increase $= 100 - 6 = 94$.
94 is what percent of 6?
$94 = x \bullet 6$
$\frac{94}{6} = \frac{x \bullet 6}{6}$
$15.67 \approx x$
$15.67 = 1,567\%$

51. **Analyze:** Each worker has 75 clients. There are 20 workers. We want the number of clients.
Form: Let $c =$ number of clients.
clients $= 75 \bullet$workers.
$c = 75 \bullet 20$
Solve: $c = 75 \bullet 20$
$c = 1,500$
State: Theare 1,500 clients.
Check: $(1,500) = 75 \bullet 20$
$1,500 = 1,500$

53. **Analyze:** There are 54 total cards. There are 2 jokers, 4 aces, and 12 face cards. The remainder of the deck are number cards. We want to find how many number cards there are.
Form: Let $c =$ number cards.
total $=$ jokers $+$ aces $+$ face $+$ number
$54 = 2 + 4 + 12 + c$
Solve: $54 = 2 + 4 + 12 + c$
$54 = 18 + c$
$54 - 18 = 18 - 18 + c$
$36 = c$
State: There are 36 number cards.
Check: $54 = 2 + 4 + 2 + (36)$
$54 = 54$

STUDY SET Section 2.1

VOCABULARY

1. positive

3. zero

5. commutative

CONCEPTS

7.

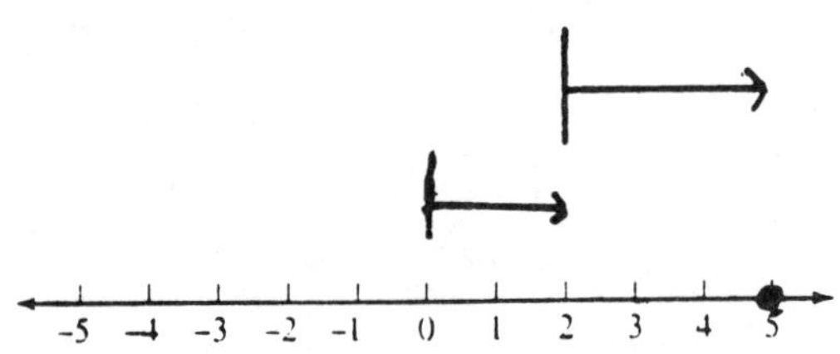

9.

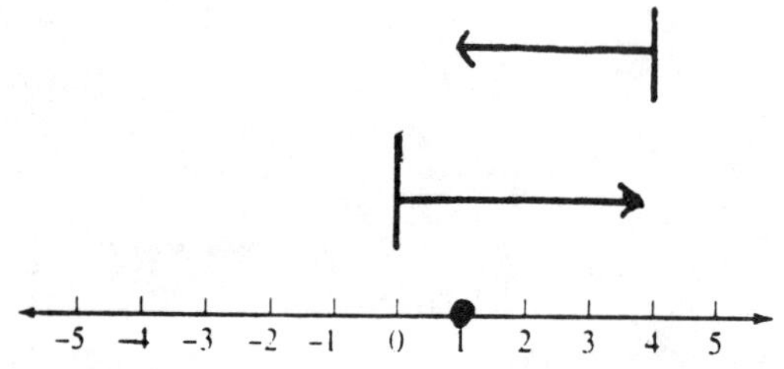

11. same; absolute

13. opposite

15. $m + n = n + m$

NOTATION

17. $(-13 + 6) + 4 = -13 + (6 + 4)$
$= -13 + 10$
$= -3$

19. $x + 23 = -12$
$x + 23 - 23 = -12 - 23$
$x = -12 + (-23)$
$x = -35$

PRACTICE

21. $6 + 3 = 9$

23. $-6 + 8 = 2$

25. $-65 + (-12) = -77$

27. $-10.5 + 2.3 = -8.2$

29. $-\frac{9}{16} + \frac{7}{16} = \frac{-2}{16}$
$= \frac{-1}{8}$

31. $-\frac{1}{4} + \frac{2}{3} = \frac{-1 \cdot 3}{4 \cdot 3} + \frac{2 \cdot 4}{3 \cdot 4}$
$= \frac{-3}{12} + \frac{8}{12}$
$= \frac{5}{12}$

33. $8 + (-5) + 13 = 3 + 13$
$= 16$

35. $21 + (-27) + (-9)$
$= (-6) + (-9)$
$= -15$

37. $(-27) + (-3) + (-13) + 22$
$= (-30) + (-13) + 22$
$= (-43) + 22$
$= -21$

39. $3{,}718 + (-5237) = -1{,}519$

41. $-237.37 + (-315.07) + (-27.4)$
$= (-552.44) + (-27.4)$
$= -579.84$

43. $8 - (-3) = 8 + 3 = 11$

45. $-12 - 9 = -12 + (-9)$
$= -21$

47. $-19 - (-17) = -19 + 17$
$= -2$

49. $-1.5 - 0.8 = -1.5 + (-0.8)$
$= -2.3$

51. $-25 - (-25) = -25 + 25$
$= 0$

53. $49 - (-12) = 49 + 12$
$= 61$

55. $-\frac{1}{8}-\frac{3}{8}=-\frac{1}{8}+(-\frac{3}{8})$
$=-\frac{4}{8}$
$=-\frac{1}{2}$

57. $-\frac{9}{16}-(-\frac{1}{4})=-\frac{9}{16}+\frac{1}{4}$
$=-\frac{9}{16}+\frac{1\cdot 4}{4\cdot 4}$
$=-\frac{9}{16}+\frac{4}{16}$
$=-\frac{5}{16}$

59. 12,466

61. −44571.251

63. Evaluate $x=-8$:
$x+4=(-8)+4=-4$
Evaluate $x=-4$:
$x+4=(-4)+4=0$
Evaluate $x=20$:
$x+4=(20)+4=24$

65. Evaluate $x=-6$:
$8+x=8+(-6)=2$
Evaluate $x=2$:
$8+x=8+(2)=10$
Evaluate $x=-13$:
$8+x=8+(-13)=-5$

67. $57+x=12$
$57+(-45)=12$
$12=12$
Yes, -45 is a solution.

69. $-23+t=-12$
$-23+(-11)=-12$
$-34=-12$
No, -11 is not a solution.

71. $x+7=-12$
$x+7-7=-12-7$
$x=-12+(-7)$
$x=-19$

73. $20=-31+x$
$20+31=-31+31+x$
$51=x$

75. $x-9=-23$
$x-9+9=-23+9$
$x=-14$

77. $a-7=-3$
$a-7+7=-3+7$
$a=4$

APPLICATIONS

79. Key Word: retreated
Translation: subtraction
Key Word: advanced
Translation: addition
Net gain is the sum of the advances less the amount of the retreat.
Net gain = (2,400 + 1,250) – 1,500
Net gain = (3,650) – 1,500
Net gain = 2,150 m

81. $E1+E2+E3=(2)+(-1)+(-2)$
$=1+(-2)$
$=-1$

83. 776 BC can be written as – 776 AD
The number of years = 1996 – (– 776)
= 1996 + 776
= 2,772 years

85. 108 – (– 52) = 108 + 52
= 160°F

87. Below sea level can be considered as – 282 and – 1,312.
Difference = – 282 – (– 1,312)
= – 282 + 1,312
= 1030 ft.

89. The effect would be the difference between revenue and cost.
effect = 6.7 – 4.5 = 2.2
Because the result is positive, the effect is a gain of $2.2 million.

91. The net movement would be the sum of the movement for all of the years.

net movement =

$(-240) + 110 + 30 + (-55) + 100 + (-77)$
$= (-130) + 30 + (-55) + 100 + (-77)$
$= -100 + (-55) + 100 + (-77)$
$= -155 + 100 + (-77)$
$= -55 + (-77)$
$= -132$

The net movement of the border of the Sahara desert is 132 km to the south.

93. Balance = beginning balance increased by the sum of deposits and decreased by the sum of checks.

sum of deposits
$= 713.87 + 1{,}245.57$
$= 1{,}959.44$

sum of checks
$= 813.45 + 937.49 + 1{,}532.79$
$= 3{,}283.73$

Balance
$= 3{,}432.17 + 1{,}959.44 - 3{,}283.73$
$= 5{,}391.61 - 3{,}283.73$
$= 2{,}107.88$

The ending balance is \$2,107.88.

REVIEW

99. $c \cdot c \cdot c \cdot c = c^4$

101. $a \cdot a \cdot b \cdot b \cdot b = a^2 b^3$

STUDY SET Section 2.2

VOCABULARY

1. product

3. like

5. commutative

7. undefined

CONCEPTS

9. $-5 \cdot (?) = 25$

11. unlike

13. 0

15. The difference between the points on the number line is 3.

17. a) NEG
b) Not possible to tell. Depends o how big the positive number is compared to the negative number.
c) Not possible to tell. Depends o how big the positive number is compared to the negative number.
d) NEG

19. $\frac{x}{2} = -3$
$\frac{(-6)}{2} = -3$
$-3 = -3$
Yes, -6 is a solution.

NOTATION

21. $(-37 \cdot 5)(2) = -37(5 \cdot 2)$
$= -37(10)$
$= -370$

23. $-3x = 36$
$\frac{-3x}{-3} = \frac{36}{-3}$
$x = -12$

PRACTICE

25. $(-6)(-9) = +(6 \cdot 9)$
$= 54$

27. $12(-5) = -(12 \cdot 5)$
$= -60$

29. $3(-4)(-5) = -(3 \cdot 4)(-5)$
$= -12(-5)$
$= +(12 \cdot 5)$
$= 60$

31. $(-4)(3)(-7) = -(4 \cdot 3)(-7)$
$= -12(-7)$
$= +(12 \cdot 7)$
$= 84$

33. $-0.6(-4) = +(0.6 \cdot 4)$
$= 2.4$

35. $1.2(-0.4) = -(1.2 \cdot 0.4)$
$= -0.48$

37. $\frac{1}{2}(-\frac{3}{4}) = -(\frac{1}{2} \cdot \frac{3}{4})$
$= -\frac{3}{8}$

39. $-1\frac{1}{4}(-\frac{3}{4}) = -\frac{5}{4}(-\frac{3}{4})$
$= +(\frac{5}{4} \cdot \frac{3}{4})$
$= \frac{15}{16}$

41. $(-2)(-3)(-4)(-5)$
$= +(2 \cdot 3)(-4)(-5)$
$= 6(-4)(-5)$
$= -(6 \cdot 4)(-5)$
$= -24(-5)$
$= +(24 \cdot 5)$
$= 120$

43. $\frac{80}{-20} = -4$

45. $\frac{-110}{-110} = +\frac{110}{110} = 1$

47. $\frac{-160}{40} = -40$

49. $\frac{320}{-16} = -20$

51. $\frac{0}{150} = 0$

53. $\frac{-17}{0}$ = undefined

55. $-\frac{1}{3} \div \frac{4}{5} = -\frac{1}{3} \bullet \frac{5}{4}$
$= -(\frac{1}{3} \bullet \frac{5}{4})$
$= -\frac{5}{12}$

57. $-\frac{3}{16} \div (-\frac{2}{3}) = +(\frac{3}{16} \bullet \frac{2}{3})$
$= \frac{6}{48}$
$= \frac{1}{8}$

59. $(-23.5)(47.2) = -(23.5 \bullet 47.2)$
$= -1{,}109.2$

61. $(-6.37)(-7.2)(-9.1)$
$= +(6.37 \bullet 7.2)(-9.1)$
$= 45.864(-9.1)$
$= -(45.864 \bullet 9.1)$
$= -417.3624$

63. $\frac{204.6}{-37.2} = -5.5$

65. Evaluate $x = -1$:
$3x = 3(-1) = -3$
Evaluate $x = -5$:
$3x = 3(-5) = -15$
Evaluate $x = -10$:
$3x = 3(-10) = -30$

67. Evaluate $x = -2$:
$\frac{x}{2} = \frac{(-2)}{2} = -1$
Evaluate $x = -6$:
$\frac{x}{2} = \frac{(-6)}{2} = -3$
Evaluate $x = -8$:
$\frac{x}{2} = \frac{(-8)}{2} = -4$

69. $3x = -3$
$\frac{3x}{3} = \frac{-3}{3}$
$x = -1$

71. $-54 = -18z$
$\frac{-54}{-18} = \frac{-18z}{-18}$
$3 = z$

73. $\frac{b}{3} = -5$
$(3)\frac{b}{3} = (3)(-5)$
$b = -15$

75. $-6 = \frac{t}{-7}$
$(-6)(-7) = \frac{t}{-7}(-7)$
$+(6 \bullet 7) = t$
$42 = t$

APPLICATIONS

77. Because the temperature is decreasing, use $-6°$ as the change in temperature per hour.
Key Phrase: change per hour
Translation: multiply
Total change $= (-6)(12)$
$= -(6 \bullet 12)$
$= -72°$

79. Let lost \$40 be -40 per hour.
Key Phrase: lost per hour
Translation: multiplication
Financial Condition $= (-40)(12)$
$= -(40 \bullet 12)$
$= -480$
His financial condition is that he has lost \$480.

81. Amount of time $= \frac{\text{number of gallons}}{\text{rate}}$
$= \frac{396}{3}$
$= 132$ minutes

83. To find average, we need the total rating points.
Total $= 0.6 + (-0.3) + 1.7 + 1.5 + (-0.2) + 1.1 + (-0.2)$
$= 4.2$

Daily average $= \frac{\text{total rating points}}{\text{number of days}}$
$= \frac{4.2}{7}$
$= 0.6$
Because the average is positive, the network had an average gain of 0.6 rating points.

85. Depreciation =

$$\begin{aligned} &\text{beginning value} - \text{ending value} \\ &= 615{,}500 - 544{,}915 \\ &= 70{,}585 \end{aligned}$$

REVIEW

89.

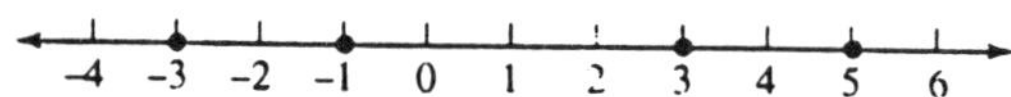

91. Evaluate $s = 4$:

$$\begin{aligned} V &= s^2 = (4)^2 \\ &= 16 \text{ in}^3 \end{aligned}$$

93. A real number is any number that can be written as a decimal.

95. 34% of 612 = 34%•612

$$\begin{aligned} &= 0.34 \cdot 612 \\ &= 208 \end{aligned}$$

STUDY SET Section 2.3

VOCABULARY

1. order

3. evaluate

5. base; 3

7. parentheses; brackets; absolute values

CONCEPTS

9. exponential

11. a) Evaluate $x = -3$:
$$x^2 = (-3)^2 = (-3)(-3) = 9$$
Evaluate $x = -4$:
$$x^2 = (-4)^2 = (-4)(-4) = 16$$
Evaluate $x = -5$:
$$x^2 = (-5)^2 = (-5)(-5) = 25$$
A negative number raised to an even power will give a positive result.

b) Evaluate $x = -3$:
$$x^3 = (-3)^3 = (-3)(-3)(-3) = (9)(-3) = -27$$
Evaluate $x = -4$:
$$x^3 = (-4)^3 = (-4)(-4)(-4) = (16)(-4) = -64$$
Evaluate $x = -5$:
$$x^3 = (-5)^3 = (-5)(-5)(-5) = (25)(-5) = -125$$
A negative raised to an odd power will give a negative result.

13. $-3[5^2 - 6(3-2)]$
1. subtraction in parentheses
2. multiply by 6
3. power 2
4. subtraction in brackets
5. multiply by -3

There are 5 operations in the expression.

NOTATION

15. $(-6)^2 - 2(5 - 4 \cdot 2)$
$$= (-6)^2 - 2(5-8) = (-6)^2 - 2(-3) = 36 - 2(-3) = 36 - (-6) = 36 + 6 = 42$$

PRACTICE

17. $(-6)^2 = (-6)(-6) = (6 \cdot 6) = 36$

19. $-4^4 = -(4)(4)(4)(4) = -(16)(4)(4) = -(64)(4) = -256$

21. $(-5)^3 = (-5)(-5)(-5) = (25)(-5) = -125$

23. $-(-6)^4 = -(-6)(-6)(-6)(-6) = -(36)(-6)(-6) = -(-216)(-6) = -(1296) = -1296$

25. $(-0.4)^2 = (-0.4)(-0.4) = 0.16$

27. $\left(-\frac{2}{5}\right)^3 = \left(-\frac{2}{5}\right)\left(-\frac{2}{5}\right)\left(-\frac{2}{5}\right) = \left(\frac{4}{25}\right)\left(-\frac{2}{5}\right) = -\frac{8}{125}$

29. $3 - 5 \cdot 4 = 3 - 20 = -17$

31. $-3(5-4) = -3(1)$
$= -3$

33. $3+(-5)^2 = 3+25$
$= 28$

35. $(-3-5)^2 = (-8)^2$
$= (-8)(-8)$
$= 64$

37. $2+3(-\frac{25}{5})-(-4)$
$= 2+3(-5)-(-4)$
$= 2+(-15)-(-4)$
$= (-13)-(-4)$
$= (-13)+4$
$= -9$

39. $(-2)^3\left(\frac{-6}{-2}\right)(-1)$
$= (-2)(-2)(-2)\left(\frac{-6}{-2}\right)(-1)$
$= (4)(-2)\left(\frac{-6}{-2}\right)(-1)$
$= (-8)\left(\frac{-6}{-2}\right)(-1)$
$= (-8)(3)(-1)$
$= (-24)(-1)$
$= 24$

41. $\frac{-7-3^2}{2\cdot 4} = \frac{-7-9}{8}$
$= \frac{-16}{8}$
$= -2$

43. $\left(\frac{1}{2}\right)\left(\frac{1}{8}\right)+\left(-\frac{1}{4}\right)^2$
$= \left(\frac{1}{2}\right)\left(\frac{1}{8}\right)+\left(-\frac{1}{4}\right)\left(-\frac{1}{4}\right)$
$= \frac{1}{16}+\frac{1}{16}$
$= \frac{2}{16}$
$= \frac{1}{8}$

45. $-2|4-8| = -2|-4|$
$= -2(4)$
$= -8$

47. $|7-8(4-7)| = |7-8(-3)|$
$= |7-(-24)|$
$= |7+24|$
$= |31|$
$= 31$

49. $3+2[-1-4(5)] = 3+2[-1-20]$
$= 3+2[-21]$
$= 3+(-42)$
$= -39$

51. $-3[5^2-(7-3)^2] = -3[5^2-(4)^2]$
$= -3[25-16]$
$= -3[9]$
$= -27$

53. $-2(2\cdot 3-4)^3 = -2(6-4)^3$
$= -2(2)^3$
$= -2(2)(2)(2)$
$= -4(2)(2)$
$= -8(2)$
$= -16$

55. $\frac{(3+5)^2+|-2|}{-2(5-8)} = \frac{(8)^2+|-2|}{-2(-3)}$
$= \frac{64+|-2|}{-2(-3)}$
$= \frac{64+2}{6}$
$= \frac{66}{6}$
$= 11$

57. $\frac{2[-4-2(3-1)]}{3[(3)(2)]} = \frac{2[-4-2(2)]}{3[6]}$
$= \frac{2[-4-4]}{18}$
$= \frac{2[-8]}{18}$
$= \frac{-16}{18} = -\frac{8}{9}$

59. $2x-y = 2(3)-(-2)$
$= 6-(-2)$
$= 6+2$
$= 8$

61. $-y+yz = -(-2)+(-2)(-4)$
$= -(-2)+8$
$= 2+8$
$= 10$

63. $(3+x)y = (3+[3])(-2)$
$= (6)(-2)$
$= -12$

65. $(x+y)(x-y)$
$= (3+[-2])(3-[-2])$
$= (3+[-2])(3+2)$
$= (1)(5)$
$= 5$

67. $(4x)^2+3y^2 = (4[3])^2+3(-2)^2$
$= (12)^2+3(-2)^2$
$= 144+3(4)$
$= 144+12$
$= 156$

69. $\frac{2x+y^3}{y+2z} = \frac{2(3)+(-2)^3}{(-2)+2(-4)}$
$= \frac{2(3)+(-8)}{(-2)+(-8)}$
$= \frac{6+(-8)}{-10}$
$= \frac{-2}{-10}$
$= \frac{1}{5}$

71. $b^2-4ac = (5)^2-4(-1)(-2)$
$= 25-4(-1)(-2)$
$= 25-(-4)(-2)$
$= 25-8$
$= 17$

73. $a+ab+b = (-5)^2+2(-5)(-1)+(-1)^2$
$= 25+2(-5)(-1)+1$
$= 25+(-10)(-1)+1$
$= 25+10+1$
$= 36$

75. $\frac{n}{2}[2a+(n-)d]$
$= \frac{10}{(2)}[2(-4)+(10-1)(6)]$
$= \frac{10}{2}[2(-4)+(9)(6)]$
$= 5[2(-4)+(10-1)(6)]$
$= 5[(-8)+54]$
$= 5[46]$
$= 230$

77. 983.4496

81. $(23.1)^2-(14.7)(-61.9)$
$= 533.61-(14.7)(-61.9)$
$= 533.61-(-909.93)$
$= 533.61+909.93$
$= 1443.54$

APPLICATIONS

83. Evaluate $-34°$F:
$\frac{5(F-32)}{9} = \frac{5([-34]-32)}{9}$
$= \frac{5(-66)}{9}$
$= \frac{-330}{9}$
$\approx -37°$C

Evaluate $-84°$F:
$\frac{5(F-32)}{9} = \frac{5([-84]-32)}{9}$
$= \frac{5(-116)}{9}$
$= \frac{-580}{9}$
$\approx -64°$C

85. Low tide reading is high tide reading less amount of decrease.
Amount of decrease $= 5(21)$.
Low tide reading $= 80-5(21)$
$= 80-105$
$= -25$
The low tide reading is -25 cm.

87. $\pi[b^2+d^2+(b+d)s]$
$= \pi[(1)^2+(2.5)^2+(1+2.5)5]$
$= \pi[(1)^2+(2.5)^2+(3.5)5]$
$= \pi[1+6.25+(3.5)5]$
$= \pi[1+6.25+17.5]$
$= \pi[7.25+17.5]$
$= \pi[24.75]$
≈ 77.8 in^2

REVIEW

91.

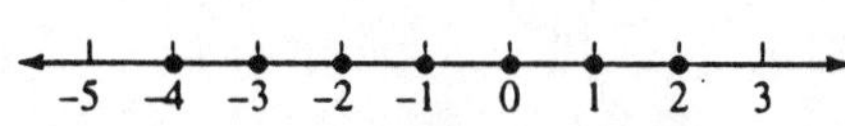

93. $aabbb = a^2b^3$

95. $12+x=17$
$12-12+x = 17-12$
$x = 5$

STUDY SET Section 2.4

VOCABULARY

1. simplify

3. coefficient; variable

5. remove

CONCEPTS

7. the distributive property

9. $2(3+4) = 2 \cdot 3 + 2 \cdot 4$

11. $x + (20 - x) = x + 20 - x$
$= x - x + 20$
$= 20$ ft.

13. Second angle is 90° less the first angle.
Second angle $= 90° - a°$
$= (90 - a)°$

NOTATION

15. $-7(a^2 + 2a - 5)$
$= (-7) \cdot a^2 + (-7) \cdot 2a - (-7) \cdot 5$
$= -7a^2 + (-7 \cdot 2)a - (-35)$
$= -7a^2 + (-14)a - (-35)$
$= -7a^2 - 14a + 35$

17. a) No, K is not the same as k.
b) Yes, both have a d.

PRACTICE

19. $9(7m) = (9 \cdot 7)m$
$= (63)m$
$= (63)m$

21. $5(-7q) = (5 \cdot -7)q$
$= (-35)q$
$= -35q$

23. $(-5p)(-4b) = (-5)(-4)(p)(b)$
$= (-20)(pb)$
$= -20pb$

25. $-5(4r)(-2r)$
$= (-5)(4)(-2)(r)(r)$
$= (-20)(-2)r^2$
$= (40)(r^2)$
$= 40r^2$

27. $5(x+3) = 5 \cdot x + 5 \cdot 3$
$= 5x + 15$

29. $-2(b-1) = (-2) \cdot b - (-2)(1)$
$= -2b - (-2)$
$= -2b + 2$

31. $(3t - 2)8 = (3t) \cdot 8 - 2 \cdot 8$
$= (3 \cdot 8)t - 2 \cdot 8$
$= (24)t - 16$
$= 24t - 16$

33. $(2y - 1)6 = 2y \cdot 6 - 1 \cdot 6$
$= (2 \cdot 6)y - 1 \cdot 6$
$= (12)y - 6$
$= 12y - 6$

35. $0.4(x - 4) = 0.4 \cdot x - (0.4)(4)$
$= 0.4x - 1.6$

37. $-\frac{2}{3}(3w - 6)$
$= \left(-\frac{2}{3}\right) \cdot 3w - \left(-\frac{2}{3}\right)6$
$= \left(-\frac{2}{3} \cdot 3\right)w - \left(-\frac{2}{3} \cdot 6\right)$
$= (-2)w - (-4)$
$= -2w + 4$

39. $r(r - 10) = r \cdot r - r \cdot 10$
$= r^2 - r \cdot 10$
$= r^2 - 10r$

41. $-(x - 7) = -1(x - 7)$
$= (-1)x - (-1)(7)$
$= -1x - (-7)$
$= -x + 7$

43. $17(2x - y + 2)$
$= 17 \cdot 2x - 17 \cdot y + 17 \cdot 2$
$= (17 \cdot 2)x - 17 \cdot y + 17 \cdot 2$
$= 34x - 17y + 34$

45. $-(-14+3p-t)$
$= -1(-14+3p-t)$
$= (-1)(-14)+(-1)\cdot 3p-(-1)t$
$= 14+(-3)p-(-t)$
$= 14+(-3p)-(-t)$
$= 14-3p+t$

47. $-5r$ and $4s$

49. $-15r^2s$

51. -5 and 1

53. -15

55. -1

57. $\frac{1}{4}$

59. term, because it is by itself within a subtraction

61. factor, because the 3 is multiplying it.

63. $3x+17x=(3+17)x$
$=20x$

65. $8x^2-5x^2=(8-5)x^2$
$=3x^2$

67. $-4x+4x=(-4+4)x$
$=0x$
$=0$

69. $-7b^2+7b^2=(-7+7)b^2$
$=(0)b^2$
$=0$

71. $3x+5x-7x=(3+5-7)x$
$=(8-7)x$
$=(1)x$
$=x$

73. $13x^2+2x^2-5x^2=(13+2-5)x^2$
$=(15-5)x^2$
$=10x^2$

75. $1.8h-0.7h=(1.8-0.7)h$
$=1.1h$

77. $\frac{3}{5}t+\frac{1}{5}t=(\frac{3}{5}+\frac{1}{5})t$
$=\frac{4}{5}t$

79. $4(y+9)-6y=4\cdot y+(4)(9)-6y$
$=4y+36-6y$
$=4y-6y+36$
$=(4-6)y+36$
$=(-2)y+36$
$=-2y+36$

81. $2z+5(z-3)=2z+5\cdot z-5(3)$
$=2z+5z-15$
$=(2+5)z-15$
$=7z-15$

83. $8(c+7)-2(c-3)$
$=8(c+7)+(-2)(c-3)$
$=8\cdot c+8(7)+(-2)c-(-2)(3)$
$=8c+56+(-2)c-(-6)$
$=8c+56-2c+6$
$=8c-2c+56+6$
$=(8-2)c+56+6$
$=6c+62$

85. $108-(-52)=108+52=160$°F

87. $(a+2)-(a-b)$
$=(a+2)+(-1)(a-b)$
$=a+2+(-1)a-(-1)b$
$=a+2-a+b$
$=a-a+b+2$
$=0+b+2$
$=b+2$

89. $x(x+3)-3x^2=x\cdot x+x\cdot 3-3x^2$
$=x^2+3x-3x^2$
$=x^2-3x^2+3x$
$=1x^2-3x^2+3x$
$=(1-3)x^2+3x$
$=(-2)x^2+3x$
$=-2x^2+3x$

APPLICATIONS

91. There are 12 sides of the cross and each side has length x. The perimeter is $12x$.

93. There are two sides that have length $(x + 4)$ and two sides that have length x. The perimeter is the sum of all of the sides.

$$\begin{aligned} \text{perimeter} &= 2(x+4) + 2x \\ &= 2 \bullet x + 2(4) + 2x \\ &= 2x + 8 + 2x \\ &= 2x + 2x + 8 \\ &= (2+2)x + 8 \\ &= 4x + 8 \end{aligned}$$

The perimeter of the ping-pong table is $(4x + 8)$ ft.

REVIEW

99. $x^2 z(y^3 - z)$

$$\begin{aligned} &= (-3)^2(0)([-5]^3 - 0) \\ &= (9)(0)(-125 - 0) \\ &= (9)(0)(-125) \\ &= (0)(-125) \\ &= 0 \end{aligned}$$

101. $\frac{x-y^2}{2y-1+x} = \frac{(-3)-(-5)^3}{2(-5)-1+(-3)}$

$$\begin{aligned} &= \frac{(-3)-(-125)}{(-10)-1+(-3)} \\ &= \frac{(-3)+125}{(-10)-1-3} \\ &= \frac{122}{-11-3} \\ &= \frac{122}{-14} \\ &= -\frac{61}{7} \end{aligned}$$

STUDY SET Section 2.5

VOCABULARY

1. equation

3. number

5. distributive

CONCEPTS

7. subtraction; multiplication

9. subtracting

11. a) $3x + 5 - x = 3x - x + 5$
$= 2x + 5$

b) $3x + 5 - x = 9$
$3x - x + 5 = 9$
$2x + 5 = 9$
$2x + 5 - 5 = 9 - 5$
$2x = 4$
$\frac{2x}{2} = \frac{4}{2}$
$x = 2$

c) Evaluate $x = 9$:
$3x + 5 - x = 3(9) + 5 - (9)$
$= 27 + 5 - 9$
$= 32 - 9$
$= 23$

13. a) Because each expression includes an x and x is the number of seats in the balcony, the seating is based on the number of balcony seats.

b) 200

c) Total number of seats is the sum of the seats in all the areas.
Total = main + box + balcony
$= (x + 200) + (x - 50) + x$
$= x + 200 + x - 50 + x$
$= x + x + x + 200 - 50$
$= 3x + 150$

15. 30, because 3•5•2 = 30

NOTATION

17. $2x - 7 = 21$
$2x - 7 + 7 = 21 + 7$
$2x = 28$
$\frac{2x}{2} = \frac{28}{2}$
$x = 14$

19. a) $-x = -1x$
b) $\frac{3x}{5} = \frac{3}{5}x$

PRACTICE

21. $2x + 5 = 17$
$2x + 5 - 5 = 17 - 5$
$2x = 12$
$\frac{2x}{2} = \frac{12}{2}$
$x = 6$
check: $2(6) + 5 = 17$
$12 + 5 = 7$
$7 = 7$

23. $-5q - 2 = 1$
$-5q - 2 + 2 = 1 + 2$
$-5q = 3$
$\frac{-5q}{-5} = \frac{3}{-5}$
$q = -\frac{3}{5}$
check: $-5\left(-\frac{3}{5}\right) - 2 = 1$
$3 - 2 = 1$
$1 = 1$

25. $0.6 = 4.1 - x$
$0.6 - 4.1 = 4.1 - 4.1 - x$
$-3.5 = -x$
$-3.5 = -1x$
$\frac{-3.5}{-1} = \frac{-1x}{-1}$
$3.5 = x$
check: $0.6 = 4.1 - (3.5)$
$0.6 = 0.6$

27. $-g = -4$
$-1g = -4$
$\frac{-1g}{-1} = \frac{-4}{-1}$
$g = 4$
check: $-(4) = -4$
$-4 = -4$

29. $-8-3c=0$
$-8+8-3c=0+8$
$-3c=8$
$\frac{-3c}{-3}=\frac{8}{-3}$
$c=-\frac{8}{3}$
check: $-8-3\left(-\frac{8}{3}\right)=0$
$-8-(-8)=0$
$-8+8=0$
$0=0$

31. $-\frac{5}{6}k=10$
$6(-\frac{5}{6})k=6(10)$
$-5k=60$
$\frac{-5k}{-5}=\frac{60}{-5}$
$k=-12$
check: $-\frac{5}{6}(-12)=10$
$\frac{60}{6}=10$
$10=10$

33. $-\frac{t}{3}+2=6$
$-\frac{t}{3}+2-2=6-2$
$-\frac{t}{3}=4$
$-3(-\frac{t}{3})=-3(4)$
$t=-12$
check: $-\frac{(-12)}{3}+2=6$
$-(-4)+2=6$
$4+2=6$
$6=6$

35. $\frac{2x}{3}-2=4$
$\frac{2x}{3}-2+2=4+2$
$\frac{2x}{3}=6$
$3(\frac{2x}{3})=3(6)$
$2x=18$
$\frac{2x}{2}=\frac{18}{2}$
$x=9$
check: $\frac{2(9)}{3}-2=4$
$\frac{18}{3}-2=4$
$6-2=4$
$4=4$

37. $\frac{x+5}{3}=11$
$3(\frac{x+5}{3})=3(11)$
$x+5=33$
$x+5-5=33-5$
$x=28$
check: $\frac{(28)+5}{3}=11$
$\frac{33}{3}=11$
$11=11$

39. $\frac{y-2}{7}=-3$
$7(\frac{y-2}{7})=7(-3)$
$y-2=-21$
$y-2+2=-21+2$
$y=-19$
check: $\frac{(-19)-2}{7}=-3$
$\frac{-21}{7}=-3$
$-3=-3$

41. $3(x+2)-x=12$
$3x+3•2-x=12$
$3x+6-x=12$
$3x-x+6=12$
$2x+6=12$
$2x+6-6=12-6$
$2x=6$
$\frac{2x}{2}=\frac{6}{2}$
$x=3$
check: $3([3]+2)-(3)=12$
$3(5)-3=12$
$15-3=12$
$12=12$

43. $-3(2y-2)-y=5$
$(-3•2)y-(-3)(2)-y=5$
$-6y-(-6)-y=5$
$-6y+6-y=5$
$-6y-y+6=5$
$-7y+6=5$
$-7y+6-6=5-6$
$-7y=-1$
$\frac{-7y}{-7}=\frac{-1}{-7}$
$y=\frac{1}{7}$

check: $-3(2[\frac{1}{7}]-2)-(\frac{1}{7})=5$
$-3(\frac{2}{7}-2)-\frac{1}{7}=5$
$-3(\frac{2}{7}-\frac{14}{7})-\frac{1}{7}=5$
$-3(-\frac{12}{7})-\frac{1}{7}=5$
$\frac{36}{7}-\frac{1}{7}=5$
$\frac{35}{7}=5$
$5=5$

45. $3x+2.5=2x$
$3x-2x+2.5=2x-2x$
$x+2.5=0$
$x+2.5-2.5=0-2.5$
$x=-2.5$
check: $3(-2.5)+2.5=2(-2.5)$
$-7.5+2.5=-5$
$-5=-5$

47. $9y-3=6y$
$9y-9y-3=6y-9y$
$-3=-3y$
$\frac{-3}{-3}=\frac{-3y}{-3}$
$1=y$
check: $9(1)-3=6(1)$
$9-3=6$
$6=6$

49. $\frac{1}{2}+\frac{x}{5}=\frac{3}{4}$
LCD $=20$
$20\left(\frac{1}{2}+\frac{x}{5}\right)=20\left(\frac{3}{4}\right)$
$20\left(\frac{1}{2}\right)+20(\frac{x}{5})=20(\frac{3}{4})$
$10+4x=15$
$10-10+4x=15-10$
$4x=5$
$\frac{4x}{4}=\frac{5}{4}$
$x=\frac{5}{4}$
check: $\frac{1}{2}+\frac{x}{5}=\frac{3}{4}$
$\frac{1}{2}+\frac{1}{5}x=\frac{3}{4}$
$\frac{1}{2}+\frac{1}{5}\left(\frac{5}{4}\right)=\frac{3}{4}$
$\frac{1}{2}+\frac{1}{4}=\frac{3}{4}$
$\frac{2}{4}+\frac{1}{4}=\frac{3}{4}$
$\frac{3}{4}=\frac{3}{4}$

51. $\frac{2}{3}=-\frac{2x}{3}+\frac{3}{4}$
LCD $=12$
$12\left(\frac{2}{3}\right)=12\left(-\frac{2x}{3}+\frac{3}{4}\right)$
$12\left(\frac{2}{3}\right)=12\left(-\frac{2x}{3}\right)+12\left(\frac{3}{4}\right)$
$8=-8x+9$
$8-9=-8x+9-9$
$-1=-8x$
$\frac{-1}{-8}=\frac{-8x}{-8}$
$\frac{1}{8}=x$
check: $\frac{2}{3}=-\frac{2x}{3}+\frac{3}{4}$
$\frac{2}{3}=-\frac{2}{3}x+\frac{3}{4}$
$\frac{2}{3}=-\frac{2}{3}\left(\frac{1}{8}\right)+\frac{3}{4}$
$\frac{2}{3}=-\frac{1}{12}+\frac{3}{4}$
$\frac{2}{3}=-\frac{1}{12}+\frac{9}{12}$
$\frac{2}{3}=\frac{8}{12}$
$\frac{2}{3}=\frac{2}{3}$

53. $3(a+2)=2(z-7)$
$3a+3(2)=2a-2(7)$
$3a+6=2a-14$
$3a-2a+6=2a-2a-14$
$a+6=-14$
$a+6-6=-14-6$
$a=-20$
check: $3([-20]+2=2([-20]-7)$
$3(-18)=2(-27)$
$-54=-54$

55. $9(x+11)+5(13-x)=0$
$9x+9(11)+5(13)-5x=0$
$9x+99+65-5x=0$
$9x-5x+99+65=0$
$4x+164=0$
$4x+164-164=0-164$
$4x=-164$
$\frac{4x}{4}=\frac{-164}{4}$
$x=-41$
check:
$9([-41]+11)+5(13-[-41])=0$
$9(-41+11)+5(13+41)=0$
$9(-30)+5(54)=0$
$-270+270=0$
$0=0$

57. $\frac{3t-21}{2} = t - 6$

$2(\frac{3t-21}{2}) = 2(t - 6)$

$3t - 21 = 2t - 2(6)$

$3t - 21 = 2t - 12$

$3t - 2t - 21 = 2t - 2t - 12$

$t - 21 = -12$

$t - 21 + 21 = -12 + 21$

$t = 9$

check: $\frac{3(9)-21}{2} = (9) - 6$

$\frac{27-21}{2} = 3$

$\frac{6}{2} = 3$

$3 = 3$

59. $\frac{10-5s}{3} = x + 6$

$3\left(\frac{10-5s}{3}\right) = 3(s + 6)$

$10 - 5s = 3s + 3(6)$

$10 - 5s = 3s + 18$

$10 - 5s - 3s = 3s - 3s + 18$

$10 - 8s = 18$

$10 - 10 - 8s = 18 - 10$

$-8s = 8$

$\frac{-8s}{-8} = \frac{8}{-8}$

$s = -1$

check: $\frac{10-5(-1)}{3} = (-1) + 6$

$\frac{10-(-5)}{3} = -1 + 6$

$\frac{10+5}{3} = -1 + 6$

$\frac{15}{3} = 5$

$5 = 5$

61. $2 - 3(x - 5) = 4(x - 1)$

$2 + (-3)(x - 5) = 4(x - 1)$

$2 + (-3)x - (-3)(5) = 4x - 4(1)$

$2 + (-3x) - (-15) = 4x - 4$

$2 - 3x + 15 = 4x - 4$

$2 + 15 - 3x = 4x - 4$

$17 - 3x = 4x - 4$

$17 - 3x + 3x = 4x + 3x - 4$

$17 = 7x - 4$

$17 + 4 = 7x - 4 + 4$

$21 = 7x$

$\frac{21}{7} = \frac{7x}{7}$

$3 = x$

check: $2 - 3([3] - 5) = 4([3] - 1)$

$2 - 3(-2) = 4(2)$

$2 - (-6) = 8$

$2 + 6 = 8$

$8 = 8$

63. $8x + 3(2 - x) = 5(x + 2) - 4$

$8x + 3(2) - 3x = 5x + 5(2) - 4$

$8x + 6 - 3x = 5x + 10 - 4$

$8x - 3x + 6 = 5x + 10 - 4$

$5x + 6 = 5x + 6$

$5x - 5s + 6 = 5x - 5x + 6$

$6 = 6$

Identity

65. $s(s + 2) = s^2 + 2s + 1$

$s \bullet s + s(2) = s^2 + 2s + 1$

$s^2 + 2s = s^2 + 2s + 1$

$s^2 - s^2 + 2s = s^2 - s^2 + 2s + 1$

$2s = 2s + 1$

$2s - 2s = 2s - 2s + 1$

$0 = 1$

Impossible equation

67. $2(3z + 4) = 2(3z - 2) + 13$

$2 \bullet 3z + 2(4) = 2 \bullet 3z - 2(2) + 13$

$6z + 8 = 6z - 4 + 13$

$6z + 8 = 6z + 9$

$6z - 6z + 8 = 6z - 6z + 9$

$9 = 9$

Impossible equation

69. $4(y - 3) - y = 3(y - 4)$

$4y - 4(3) - y = 3y - 3(4)$

$4y - 12 - y = 3y - 12$

$4y - y - 12 = 3y - 12$

$3y - 12 = 3y - 12$

$3y - 3y - 12 = 3y - 3y - 12$

$-12 = -12$

Identity

APPLICATIONS

71. Analyze: The length of the two sections of the board must be equal to the original length of the board. The second section is twice the length of the first section.
Form: Let x be the length of the first section and $2x$ represent the length of the second section.
The sum of the sections is equal to the original board.
1st section + 2nd section = original
$x + 2x = 12$
Solve: $x + 2x = 12$
$3x = 12$
$\frac{3x}{3} = \frac{12}{3}$
$x = 4$
State: The length of the first section is 4 ft. and the second section is 8 ft.
Check: $4 + 8 = 12$

73. Analyze: The crane has 3 sections. The second section is 4 ft. longer than the first. The third section is 1 ft. shorter than the first.
Form: Let x = length of first section.
$x + 4$ = length of second section.
$x - 1$ = length of third section.
The sum of the sections is equal to the length of the extended crane.
1st + 2nd + 3rd = extended
$x + (x + 4) + (x - 1) = 18$
Solve: $x + (x + 4) + (x - 1) = 18$
$x + x + x + 4 - 1 = 18$
$3x + 3 = 18$
$3x + 3 - 3 = 18 - 3$
$3x = 15$
$\frac{3x}{3} = \frac{15}{3}$
$x = 5$
State: The length of the first section is 5 ft., the second section is 9 ft., and the third section is 4 ft.
Check: $5 + 9 + 4 = 18$
$14 + 4 = 18$
$18 = 18$

75. Analyze: The total tour will be separated into 3 country tours. The Japan tour last 4 weeks more than the Australian tour. The Sweden tour lasts 2 weeks less than the Australian tour.
Form: The information is given in terms of the Australian tour. Let the Australian tour be x. $x + 4$ represent the Japan tour. $x - 2$ represent the Sweden tour. The sum of the 3 country tours is the total tour.
Australian + Japan + Sweden = total
$x + (x + 4) + (x - 2) = 38$
Solve: $x + (x + 4) + (x - 2) = 38$
$x + x + x + 4 - 2 = 38$
$3x + 2 = 38$
$3x + 2 - 2 = 38 - 2$
$3x = 36$
$\frac{3x}{3} = \frac{36}{3}$
$x = 12$
State: The Australian tour lasts 12 weeks, the Japan tour lasts 16 weeks, and the Sweden tour lasts 10 weeks.
Check: $12 + 16 + 10 = 38$
$28 + 10 = 38$
$38 = 38$

77. **Analyze:** The dessert is made up of two ingredients: pie and ice cream. The pie has 100 calories more than twice the calories of the ice cream.
Form: The calories of the pie is in terms of the ice cream. Let x represent the number of calories in the ice cream.
$2x + 100$ represent the calories in the pie.
The calories of the dessert is the sum of the calories in the pie and ice cream.
dessert = pie + ice cream.
$850 = (2x + 100) + x$
Solve: $850 = (2x + 100) + x$
$850 = 2x + x + 100$
$850 = 3x + 100$
$850 - 100 = 3x + 100 - 100$
$750 = 3x$
$\frac{750}{3} = \frac{3x}{3}$
$250 = x$
State: The ice cream has 250 calories and the pie has 600 calories.
Check: $850 = 600 + 250$
$850 = 850$

79. **Analyze:** The settlement was split among three people. The client got half of the settlement. The attorney got \$12,000 and the assistant got \$1,000.
Form: Need to find the settlement. Let x = the amount of the settlement.
Half of the settlement was given to the attorney and assistant.
half = attorney + assistant
$\frac{x}{2} = 12,000 + 1,000$
Solve: $\frac{x}{2} = 12,000 + 1,000$
$\frac{x}{2} = 13,000$
$2\left(\frac{x}{2}\right) = 2(13,000)$
$x = 26,000$
State: The out-of-court settlement was \$26,000.
Check: $\frac{(26{,}000)}{2} = 13{,}000$
$13{,}000 = 13{,}000$

81. **Analyze:** Two solar panels are combined. The second panel is 3.4 ft. wider than the first panel.
Form: The second panel is given in terms of the first panel.
Let x = the width of the first panel.
$x + 3.4$ = the width of the second panel.
The total width of the panels is the sum of the two panels.
total = first panel + second panel
$18 = x + (x + 3.4)$
Solve: $18 = x + (x + 3.4)$
$18 = 2x + 3.4$
$18 - 3.4 = 2x + 3.4 - 3.4$
$14.6 = 2x$
$\frac{14.6}{2} = \frac{2x}{2}$
$7.3 = x$
State: The first panel has a width of 7.3 ft. and the second panel has a width of 10.7 ft.
Check: $7.3 + 10.7 = 18$
$18 = 18$

83. **Analyze:** Two bottles of a vitamin are purchased. The second bottle is half the first bottle.
Form: The second bottle is given in terms of the first bottle.
Let x = the price of the first bottle.
$\frac{x}{2}$ = the width of the second panel.
The total price of the bottles is the sum of the prices of the two bottles.
total = first bottle + second bottle
$2.25 = x + \frac{x}{2}$
Solve: $2.25 = x + \frac{x}{2}$
$2(2.25) = 2\left(x + \frac{x}{2}\right)$
$4.50 = 2x + 2\left(\frac{x}{2}\right)$
$4.50 = 2x + x$
$4.50 = 3x$
$\frac{4.50}{3} = \frac{3x}{3}$
$1.50 = x$
State: The price of the first bottle of vitamins is \$1.50 and the second is \$0.75.
Check: $1.50 + 0.75 = 2.25$
$2.25 = 2.25$

85. **Analyze:** Five employees will be laid off each month until the total number of employees becomes 465. There are currently 510 employees.
Form: The number of months is the undecided number.
Let $x =$ the number of months.
$5x =$ the number of employees laid off.
The current number of employees is equal to the sum of the employees laid off and the future number of employees
current = laid off + future
$510 = 5x + 465$
Solve: $510 = 5x + 465$

$$510 - 465 = 5x + 465 - 465$$
$$45 = 5x$$
$$\frac{45}{5} = \frac{5x}{5}$$
$$9 = x$$

State: It will take 9 months to reach the future level of 465 employees.
Check: $510 = 5(9) + 465$

$$510 = 45 + 465$$
$$510 = 510$$

REVIEW

89. $29.05 = x \cdot 415$

$$29.05 = 415x$$
$$\frac{29.05}{415} = \frac{415x}{415}$$
$$0.07 = x$$
$$0.07 = 7\%$$

91. Cost is the sum of the price of the purchases and the tax.
Tax is 6% of the purchase.
Purchases $= 98.95 + 7.95$
$= 106.90$
Tax $= 6\% \cdot 106.90$
$= 0.06 \cdot 106.90$
$= 6.414$
≈ 6.41 (round to nearest cent)
Cost $= 106.90 + 6.41$
$= 113.31$

STUDY SET Section 2.6

VOCABULARY

1. formula

3. perimeter

5. radius

7. circumference

CONCEPTS

9. a) volume
b) area
c) perimeter

11. $A = \frac{1}{2}bh$
$= \frac{1}{2}(6x - 4)(4)$
$= \frac{1}{2}(4)(6x - 4)$
$= 2(6x - 4)$
$= (2 \cdot 6)x - 2(4)$
$= 12x - 8$
Area of the triangle is $(12x - 8)$ mm^2.

13. a) Isolate w on one side of the equation.
b) There is an a on both sides of the equation. a isolated on one side.

15. a) Since diameter is twice the radius, multiply r by 2 to get D.
b) Since the radius is half the diameter, divide D by 2 to get r.

NOTATION

17. $V = \frac{1}{3}Bh$
$3(V) = 3(\frac{1}{3}Bh)$
$3V = Bh$
$\frac{3V}{h} = \frac{Bh}{h}$
$\frac{3V}{h} = B$

PRACTICE

19. $s = p - d$
$1,624.95 = p - 350$
$1,624.95 + 350 = p - 350 + 350$
$1,974.95 = p$
The original price of the television set was \$1,974.95.

21. $p = r - c$
$8.7 = 11 - c$
$8.7 - 11 = 11 - 11 - c$
$-2.3 = -c$
$-2.3 = -1c$
$\frac{-2.3}{-1} = \frac{-1c}{-1}$
$2.3 = c$
The cost of the film was \$2.3 million.

23. $r = c + m$
$11,750 = c + 3,700$
$11,750 - 3,700 = c + 3,700 - 3,700$
$8,050 = c$
The wholesale cost of the boat was \$8,050.

25. $d = rt$
$1,826 = r(742)$
$1,826 = 742r$
$\frac{1,826}{742} = \frac{742r}{742}$
$2.4609 \approx r$
$2.5 \approx r$
The average rate of swimming was 2.5 mph.

27. $C = \frac{5(F-32)}{9}$
$2,212 = \frac{5(F-32)}{9}$
$9(2,212) = 9\left(\frac{5(F-32)}{9}\right)$
$19,908 = 5(F - 32)$
$\frac{19,908}{5} = \frac{5(F-32)}{5}$
$3981.6 = F - 32$
$3981.6 + 32 = F - 32 + 32$
$4013.6 = F$
$4014 \approx F$
Silver boils at 4,014°F.

29. perimeter: $P = a + b + c$
$= 10 + 16 + 10$
$= 36$
Perimeter is 36 ft.
area: $A = \frac{1}{2}bh$
$= \frac{1}{2}(16)(6)$
$= 8(6)$
$= 48$
Area is 48 ft^2.

31. circumference: $C = 2\pi r$
$= 2\pi(8)$
$= 2(8)\pi$
$= 16\pi$
≈ 50.3
Circumference is 50.3 in.
area: $A = \pi r^2$
$= \pi(8)^2$
$= \pi(64)$
≈ 201.1
Area is 201.1 in^2.

33. perimeter: $P = a + b + c + d$
$= 10 + 24 + 10 + 12$
$= 34 + 10 + 12$
$= 44 + 12$
$= 56$
The perimeter of the trellis is 56 in.
area: $A = \frac{1}{2}h(b + d)$
$= \frac{1}{2}(8)(24 + 12)$
$= \frac{1}{2}(8)(36)$
$= 4(36)$
$= 144$
The area of the trellis is 144 in^3.

35. $E = IR$
$\frac{E}{I} = \frac{IR}{I}$
$\frac{E}{I} = R$
$R = \frac{E}{I}$

37. $V = lwh$
$V = lhw$
$V = (lh)w$
$\frac{V}{lh} = \frac{(lh)w}{lh}$
$\frac{V}{lh} = w$
$w = \frac{V}{lh}$

39. $y = mx + b$
$y - b = mx + b - b$
$y - b = mx$
$\frac{y-b}{m} = \frac{mx}{m}$
$\frac{y-b}{m} = x$
$x = \frac{y-b}{m}$

41. $A = P + Prt$
$A - P = P - P + Prt$
$A - P = Prt$
$A - P = (Pr)t$
$\frac{A-P}{Pr} = \frac{(Pr)t}{Pr}$
$\frac{A-P}{Pr} = t$
$t = \frac{A-P}{Pr}$

43. $V = \frac{1}{3}\pi r^2 h$
$3(V) = 3(\frac{1}{3}\pi r^2 h)$
$3V = \pi r^2 h$
$3V = (\pi r^2)h$
$\frac{3V}{\pi r^2} = \frac{(\pi r^2)h}{\pi r^2}$
$\frac{3V}{\pi r^2} = h$
$h = \frac{3V}{\pi r^2}$

45. $F = \frac{GMm}{d^2}$
$d^2(F) = d^2\left(\frac{GMm}{d^2}\right)$
$d^2F = GMm$
$Fd^2 = M(Gm)$
$\frac{Fd^2}{Gm} = \frac{M(Gm)}{Gm}$
$\frac{Fd^2}{Gm} = M$
$M = \frac{Fd^2}{Gm}$

APPLICATIONS

47. Since triangles are the same,
total area is twice the area of one triangle.
Area of triangle $= \frac{1}{2}bh$
$= \frac{1}{2}(245)(10)$
$= \frac{1}{2}(10)(245)$
$= 5(245)$
$= 1{,}225$
Total area $= 2(1{,}225)$
$= 2,450 \text{ ft}^2$

49. perimeter of square $= 4s$
$= 4(2.75)$
$= 11.00$ ft
area of square $= s^2$
$= (2.75)^2$
$= 7.5625 \text{ ft}^2$

51. area of rectangular base $= lw$

$= (8)(4)$

$= 32 \text{ ft}^2$

volume $= lwh$

$= (8)(4)(4)$

$= (32)(4)$

$= 128 \text{ ft}^3$

53. The igloo is similar to half of a sphere.
The inside height is the radius.

Volume of igloo $= \frac{1}{2}\left(\frac{4}{3}\pi r^3\right)$

$= \frac{1}{2}\left(\frac{4}{3}\pi(5.5)^3\right)$

$= \left(\frac{1}{2}\bullet\frac{4}{3}\right)\pi(5.5)^3$

$= \left(\frac{1}{2}\bullet\frac{4}{3}\right)\pi(166.375)$

$= \left(\frac{2}{3}\right)\pi(166.375)$

$\approx 348 \text{ ft}^3$

55. radius of barbecue $= \frac{18}{2} = 9$ in

area of circular barbecue $= \pi r^2$

$= \pi(9)^2$

$= \pi(81)$

$\approx 254\, in^2$

57. Solve for I:

$E = IR$

$\frac{E}{R} = \frac{IR}{R}$

$\frac{E}{R} = I$

$I = \frac{E}{R}$

Evaluate for $E = 48$ and $R = 12$:

$I = \frac{(48)}{(12)}$

$I = 4$ amperes

59. Solve for R:

$P = I^2R$

$\frac{P}{I^2} = \frac{I^2R}{I^2}$

$\frac{P}{I^2} = R$

$R = \frac{P}{I^2}$

Evaluate for $P = 2700$ and $I = 14$:

$R = \frac{(2700)}{(14)^2}$

$R = \frac{2700}{196}$

$R \approx 13.78$ ohms

61. Solve for p:

$G = U - TS + pV$

$G - U = U - U - TS + pV$

$G - U = -TS + pV$

$G - U + TS = -TS + TS + pV$

$G - U + TS = pV$

$\frac{G-U+TS}{V} = \frac{pV}{V}$

$\frac{G-U+TS}{V} = p$

$p = \frac{G-U+TS}{V}$

REVIEW

65. Evaluate $x = -2$:

$x^2 - 3 = (-2)^2 - 3$

$= 4 - 3$

$= 1$

Evaluate $x = 0$:

$x^2 - 3 = (0)^2 - 3$

$= 0 - 3$

$= -3$

Evaluate $x = 3$:

$x^2 - 3 = (3)^2 - 3$

$= 9 - 3$

$= 6$

67. equation

69. expression

STUDY SET Section 2.7

VOCABULARY

1. perimeter

3. vertex

CONCEPTS

5. 180°

7. 1-inch paint brush:
$\frac{x}{2}\bullet\$4 = \$\frac{x}{2}\bullet 4$
$= \$2x$
2-inch paint brush:
$x\bullet\$5 = \$5x$
3-inch paint brush:
$(x+10)\bullet\$7 = \$7(x+10)$
$= \$(7x + 7\bullet 10)$
$= \$(7x + 70)$

9. principal = \$30,000
rate = 14%
time = 2 years

11. a) Key phrase: poured into empty
Translation: addition
3rd barrel = 1st + 2nd
$= (x + 42)$ gallons

b) Some number in between 20% and 40%. So, 32% is a good estimate.

13. a) Blueberries:
$\$0.38\bullet x = \$0.38x$
Bran Flakes:
$\$0.08\bullet 14 = \$1.12$
Blueberries & Bran Flakes:
$\$0.21(14 + x)$
b) cost per ounce = \$0.21

PRACTICE

15. $0.08x + 0.07(15,000 - x) = 1,110$
$100[0.08x + 0.07(15,000 - x)] = 100(1,110)$
$100\bullet 0.08x + (100\bullet 0.07)(15,000 - x) = 100(1,110)$
$8x + 7(15,000 - x) = 111,000$
$8x + 7\bullet 15,000 - 7x = 111,000$
$8x + 105,000 - 7x = 111,000$
$8x - 7x + 105,000 = 111,000$
$x + 105,000 = 111,000$
$x + 105,000 - 105,000 = 111,000 - 105,000$
$x = 6,000$

17. 1st angle + 2nd angle = 90°
$(6x + 2) + 2x = 90$
$6x + 2x + 2 = 90$
$8x + 2 = 90$
$8x + 2 - 2 = 90 - 2$
$8x = 88$
$\frac{8x}{8} = \frac{88}{8}$
$x = 11°$

19. 1st angle = 2nd angle
$2x + 5 = 3x - 10$
$2x - 2x + 5 = 3x - 2x - 10$
$5 = x - 10$
$5 + 10 = x - 10 + 10$
$15° = x$

APPLICATIONS

21. perimeter: $P = a + b + c$
$57 = x + x + x$
$57 = 3x$
$\frac{57}{3} = \frac{3x}{3}$
$19 = x$

23. Two sides are in terms of the 3rd side.
Let x = 3rd side.
$x - 4$ = other two sides
perimeter: $P = a + b + c$
$25 = x + (x - 4) + (x - 4)$
$25 = x + x + x - 4 - 4$
$25 = 3x - 8$
$25 + 8 = 3x - 8 + 8$
$33 = 3x$
$\frac{33}{3} = \frac{3x}{3}$
$11 = x$

25. **Analyze:** Length is in terms of width. The perimeter of the pool is 1,110 m.
Form: Let w = width of pool
$6w + 30$ = length of pool
perimeter: $P = 2l + 2w$
$1,110 = 2(6w + 30) + 2(w)$
Solve: $1,110 = 2(6w + 30) + 2(w)$
$1,110 = 2(6w) + 2(30) + 2(w)$
$1,110 = 12w + 60 + 2w$
$1,110 = 12w + 2w + 60$
$1,110 = 14w + 60$
$1,110 - 60 = 14w + 60 - 60$
$1,050 = 14w$
$\frac{1,050}{14} = \frac{14w}{14}$
$75 = w$
State: The width of the pool is 75 m and the length is 480m.
Check: $1,110 = 2(6[75] + 30) + 2(75)$
$1,110 = 2(450 + 30) + 150$
$1,110 = 2(480) + 150$
$1,110 = 960 + 150$
$1,110 = 1,110$

27. **Analyze:** Two of the angles are in term of the third angle. Two angles have equal measure and both are four times the third angle. The sum of the angles of a triangle equal 180°.
Form: Let a = measure of third angle
$4a$ = measure of other two angles
sum of angles = 180°
$4a + 4a + a = 180$
Solve: $4a + 4a + a = 180$
$9a = 180$
$\frac{9a}{9} = \frac{180}{9}$
$a = 20$
State: The measure of the third angle is 20°.
Check: $4(20) + 4(20) + (20) = 180$
$80 + 80 + 20 = 180$
$180 = 180$

29. **Analyze:** Console and big-screen TV's are in terms of portable TV's. Cost of TV's are given in chart. Storage cost is equal to price times number of TV's: $c = pn$. Total storage cost is $276.
Form: Let n = number of portable TV's.
$n - 25$ = number of console TV's.
$n - 40$ = number of big-screen TV's.
cost of portable = $\$1.50 \cdot n$
cost of console = $\$4.00 \cdot (n - 25)$
cost of big-screen = $\$7.50 \cdot (n - 40)$
Total cost
= portable + console + big-screen
$276 = 1.5n + 4(n - 25) + 7.5(n - 40)$
$276 = 1.5n + 4n - 4(25) + 7.5n - 7.5(40)$
$276 = 1.5n + 4n - 100 + 7.5n - 300$
$276 = 5.5n - 100 + 7.5n - 300$
$276 = 5.5n + 7.5n - 100 - 300$
$276 = 13n - 400$
$276 + 400 = 13n - 400 + 400$
$676 = 13n$
$\frac{676}{13} = \frac{13n}{13}$
$52 = n$
State: There are 52 portable, 27 console, and 12 big-screen TV's.
Check:
$276 = 1.5(52) + 4([52] - 25) + 7.5([52] - 40)$
$276 = 1.5(52) + 4(27) + 7.5(12)$
$276 = 78 + 108 + 90$
$276 = 276$

31. **Analyze:** Spreadsheets and database were sold in equal numbers. There were 15 more word processing sold than the other two combined. Total sales was $72,000. We want to find the number of spreadsheets sold.
Form: Let s = the number of spreadsheets sold.
s = the number of databases sold.
$2s + 15$ = number of word processing sold.
sales = price • number
sales of spreadsheet = $\$150s$
sales of database = $\$195s$
sales of wordprocessing = $\$210(2s + 15)$
total sales = sum of sales
$72,000 = 150s + 195s + 210(2s + 15)$

Solve: $72,000 = 345s + 210(2s + 15)$
$72,000 = 345s + 210•2s + 210•15$
$72,000 = 345s + 420s + 3,150$
$72,000 = 765s + 3,150$
$72,000 - 3,150 = 765s + 3,150 - 3,150$
$68,850 = 765s$
$\frac{68,850}{765} = \frac{765s}{765}$
$90 = s$
State: There were 90 spreadsheets were sold.
Check:
$72,000 = 150(90) + 195(90) + 210(2(90) + 15)$
$72,000 = 13,500 + 17,550 + 210(180 + 15)$
$72,000 = 13,500 + 17,550 + 210(195)$
$72,000 = 13,500 + 17,550 + 40,950$
$72,000 = 72,000$

33. **Analyze:** We know that \$1249.50 of interest is made from money in 3 accounts at rates: 7%, 8%, 10.5%. The amount invested in each account is the same. We need to find how much was invested.
Form: Let x = amount invested in each account.
interest in 7% account $= 0.07x$
interest in 8% account $= 0.8x$
interest in 10.5% account $= 0.105x$
Total interest = sum of interest
$1249.50 = 0.7x + 0.08x + 0.105x$
Solve: $1249.50 = 0.255x$
$\frac{1249.50}{0.255} = \frac{0.255x}{0.255}$
$4900 = x$
State: The amount invested in each account was \$4900 and the total invested was \$14,700
Check:
$1249.50 =$
$0.7(4900) + 0.08(4900) + 0.105(4900)$
$1249.50 = 343 + 392 + 514.5$
$1249.50 = 1249.50$

35. **Analyze:** We know the same amount invested in tow accounts with rates: 11% and 13%. The interest in the 13% account is 150 more than the interest in the 11% account. We need to find the amount invested.
Form: Let x = amount invested in each account.
interest in 11% account $= 0.11x$
interest in 13% account $= 0.13x$
interest in 13% = interest in 11% + 150
$0.13x = 0.11x + 150$
Solve: $0.13x = 0.11x + 150$
$0.13x - 0.11x = 0.11x - 0.11x + 150$
$0.02x = 150$
$\frac{0.02x}{0.02} = \frac{150}{0.02}$
$x = 7500$
State: The amount invested in each account is \$7,500.
Check: $0.13(7500) = 0.11(7500) + 150$
$975 = 825 + 150$
$975 = 975$

37. **Analyze:** We know the rate for each van (25 mph and 20 mph). The distance apart is 90 miles. We need to find the time.
Form: Let t = time.
distance = rate + time
distance of east going van $= 25•t$
distance of west going van $= 20•t$
total distance = sum of distances
$90 = 25t + 20t$
Solve: $90 = 25t + 20t$
$90 = 45t$
$\frac{90}{45} = \frac{45t}{45}$
$2 = t$
State: The vans will lose radio contact after 2 hours.
Check: $90 = 25(2) + 20(2)$
$90 = 50 + 40$
$90 = 90$

39. **Analyze:** We know the distance apart is 330 m and the time traveled is 3 hours. The second train is 20 mph faster than the first train. We want to find the speeds of the trains.
Form: Let r = speed of first train.
$r + 20$ = speed of second train.
Distance of first train $= 3r$
Distance of second train $= 3(r + 20)$
Total distance = first train + second train

$330 = 3r + 3(r + 20)$

Solve: $330 = 3r + 3(r + 20)$

$330 = 3r + 3r + 3(20)$

$330 = 3r + 3r + 60$

$330 = 6r + 60$

$330 - 60 = 6r + 60 - 60$

$270 = 6r$

$\frac{270}{6} = \frac{6r}{6}$

$45 = r$

State: The speed of the first train is 45 mph and the speed of the second train is 65 mph.

Check: $330 = 3(45) + 3((45) + 20)$

$330 = 3(45) + 3(65)$

$330 = 135 + 195$

$330 = 330$

41. Analyze: We know the speeds of both planes (450 mph and 500 mph) and the total distance is 3800 miles. We need to find the time they will meet.

Form: Let t = time planes meet.

Dist. for Montreal bound plane $= 450t$.

Dist. for Berlin bound plane $= 500t$.

Total distance = sum of distances.

$3800 = 450t + 500t$

Solve: $3800 = 450t + 500t$

$3800 = 950t$

$\frac{3800}{950} = \frac{950t}{950}$

$4 = t$

State: The planes will meet 4 hours into the flight.

Check: $3800 = 450(4) + 500(4)$

$3800 = 1800 + 2000$

$3800 = 3800$

43. Analyze: We want to find how much of a 3% solution is needed to turn 59 gallons of a 7% solution into a 5% solution.

Form: Let x = amount of 3% solution.

$50 + x$ = amount of 5% solution.

amt. of salt in 3% sol. $= 0.03x$

amt. of salt in 5% sol. $= 0.05(50 + x)$

amt. of salt in 7% sol. $= 0.07 \cdot 50 = 3.5$

amt. of salt in 5% sol. = amt. of salt in 3% sol. + amt. of salt in 7% sol.

$0.05(50+x) = 0.03x + 3.5$

Solve: $0.05(50 + x) = 0.03x + 3.5$

$100 \cdot 0.05(50 + x) = 100(0.03x + 3.5)$

$5(50 + x) = 100(0.03x + 3.5)$

$5 \cdot 50 + 5x = 100 \cdot 0.03x + 100(3.5)$

$250 + 5x = 3x + 350$

$250 + 5x - 3x = 3x - 3x + 350$

$250 + 2x = 350$

$250 - 250 + 2x = 350 - 250$

$2x = 100$

$\frac{2x}{2} = \frac{100}{2}$

$x = 50$

State: Fifty gallons of 3% solution will be needed to make the 5% solution.

Check: $0.05(50+50) = 0.03(50) + 3.5$

$0.05(100) = 0.03(50) + 3.5$

$5 = 1.5 + 3.5$

$5 = 5$

45. Analyze: We need to find out how much water to add to 30 ounces of 10% solution of benzalkonium to make a 8% solution.

Form: Let x = amount of water.

$30 + x$ = amount of 8% solution.

amt. of benz. in water $= 0$

amt. of benz. in 10% sol. $= 0.10 \cdot 30 = 3$

amt. of benz. in 8% sol. $= 0.08(30 + x)$

amt. in 8% = amt. in 10% + amt. in water

$0.08(30 + x) = 3 + 0$

Solve: $0.08(30 + x) = 3$

$0.08 \cdot 30 + 0.08x = 3$

$2.4 + 0.08x = 3$

$2.4 - 2.4 + 0.08x = 3 - 2.4$

$0.08x = 0.6$

$\frac{0.08x}{0.08} = \frac{0.6}{0.08}$

$x = 7.5$

State: It takes 7.5 ounces of water to make an 8% solution.

Check: $0.08((30) + 7.5) = 3$

$0.08(37.5) = 3$

$3 = 3$

47. Analyze: We need to find how many gallons of \$1.15 gas is needed to be added to 20 gallons of \$0.85 gas in order to make gas worth \$1.00.

Form: Let x = gallons of \$1.15 gas
$20 + x$ = gallons of \$1.00 gas
value of \$1.15 gas = \$1.15x
value of \$0.85 gas = \$0.85(20) = \$17
value of \$1.00 gas = \$1.00($20 + x$)
value of \$1.00 gas =
val of \$0.85 gas + val of \$1.15 gas
$1.00(20 + x) = 17 + 1.15x$
Solve: $1.00(20 + x) = 17 + 1.15x$
$1.00 \cdot 20 + 1.00x = 17 + 1.15x$
$20 + 1.00x = 17 + 1.15x$
$20 - 20 + 1.00x = 17 - 20 + 1.15x$
$1.00x = -3 + 1.15x$
$1.00x - 1.15x = -3 + 1.15x - 1.15x$
$-0.15x = -3$
$\frac{-0.15x}{-0.15} = \frac{-3}{-0.15}$
$x = 20$
State: We will need 20 gallons of \$1.15 gas to make the gas worth \$1.00
Check:
$1.00(20 + (20)) = 17 + 1.15(20)$
$1.00(40) = 17 + 1.15(20)$
$40 = 17 + 23$
$40 = 40$

49. **Analyze:** We know that lemon drops are \$1.90 and jelly beans are \$1.20 per pound. A 100 pounds of mixture will be made. The mixture will be worth \$1.48.
Form: Let n = amount of lemon drops.
$100 - n$ = amount of jelly beans.
Value of lemon drops = $1.90n$
Value of jelly beans = $1.20(100 - n)$
Value of mixture = $1.48(100) = 148$
Value of mixture = value of lemon drops + value of jelly beans
$148 = 1.90n + 1.20(100 - n)$
Solve: $148 = 1.90n + 1.20(100 - n)$
$148 = 1.90n + 1.20 \cdot 100 - 1.20n$
$148 = 1.90n + 120 - 1.20n$
$148 = 1.90n - 120n + 120$
$148 = 0.70n + 120$
$148 - 120 = 0.70n + 120 - 120$
$28 = 0.70n$
$\frac{28}{0.70} = \frac{0.70n}{0.70}$
$40 = n$
State: The mixture is made up of 40 pounds of lemon drops and 60 pounds of jelly beans.
Check: $148 = 1.90(40) + 1.20(100 - (40))$
$148 = 1.90(40) + 1.20(60)$
$148 = 76 + 72$
$148 = 148$

51. **Analyze:** We know regular coffee costs \$4 and gourmet costs \$7. We will use 40 pounds of gourmet cooked to make a \$5 blend. We need to find how many pounds of regular.
Form: Let n = pounds of regular coffee.
$n + 40$ = pounds of blend coffee.
Value of regular = $4n$
Value of gourmet = $7(40) = 280$
Value of blend = $5(n + 40)$
Val of blend = val of reg. + val of gour.
$5(n + 40) = 4n + 280$
Solve: $5(n + 40) = 4n + 280$
$5n + 5(40) = 4n + 280$
$5n + 200 = 4n + 280$
$5n - 4n + 200 = 4n - 4n + 280$
$n + 200 = 280$
$n + 200 - 200 = 280 - 200$
$n = 80$
State: To make the mixture, 80 pounds of regular coffee is needed.
Check: $4(80) + 280 = 5((80) + 40)$
$4(80) + 280 = 5(120)$
$320 + 280 = 600$
$600 = 600$

REVIEW

57. $-2.5(2x - 5)$
$= -2.5(2x) - (-25)(5)$
$= -50x - (-125)$
$= -50x + 125$

59. $8p - 9q + 11p + 20q$
$= 8p + 11p - 9q + 20q$
$= 8p + 11p + 20q - 9q$
$= (8 + 11)p + (20 - 9)q$
$= 19p + 11q$

STUDY SET Section 2.8

VOCABULARY

1. inequality

3. solution

CONCEPTS

5. same

7. opposite

9. a) true statement.
b) false statement

11. $2x - 4 > 12$
$2x - 4 + 4 > 12 + 4$
$2x > 16$
$\frac{2x}{2} > \frac{16}{2}$
$x > 8$
a) all numbers greater than 8
b) ------(------>
8
c) $(8, \infty)$

NOTATION

13. is less than

15. is greater than

17. is not equal to

19. $x > -2$

21. $-2 \leq 17$

23. $4x - 5 \geq 7$
$4x - 5 + 5 \geq 7 + 5$
$4x \geq 12$
$\frac{4x}{4} \geq \frac{12}{4}$
$x \geq 3$

PRACTICE

25. <-----)----- ; $(-\infty, 5)$
5

27. ---(-------]--- ; $(-3, 1]$
-3 1

29. $x < -1$; $(-\infty, -1)$

31. $-7 < x \leq 2$; $(-7, 2]$

33. $x + 2 > 5$
$x + 2 - 2 > 5 - 2$
$x > 3$
------(----->
3

35. $3 + x < 2$
$3 - 3 + x < 2 - 3$
$x < 2$
<-----)-----
2

37. $3 + x < 2$
$3 - 3 + x < 2 - 3$
$x < -1$
<-----)-----
-1

39. $2x - 0.3 \leq 0.5$
$2x - 0.3 + 0.3 \leq 0.5 + 0.3$
$2x \leq 0.8$
$\frac{2x}{2} \leq \frac{0.8}{2}$
$x \leq 0.4$
<-----)-----
0.4

41. $-3x - 7 > -1$
$-3x - 7 + 7 > -1 + 7$
$-3x > 6$
$\frac{-3x}{-3} > \frac{6}{-3}$
$x > -2$
------(----->
-2

43. $-4x + 6 > 17$
$-4x + 6 - 6 > 17 - 6$
$-4x > 11$
$\frac{-4x}{-4} > \frac{11}{-4}$
$x > -\frac{11}{4}$
------(----->
$-\frac{11}{4}$

45. $\frac{2}{3}x \geq 2$
$3\left(\frac{2}{3}x\right) \geq 3(2)$
$2x \geq 6$
$\frac{2x}{2} \geq \frac{6}{2}$
$x \geq 3$
------[----->
3

47. $-\frac{7}{8}x \leq 21$
$8(-\frac{7}{8}x) \leq 8(21)$
$-7x \leq 168$
$\frac{-7x}{-7} \leq \frac{168}{-7}$
$x \geq -24$
-------[----->
-24

49. $2x + 9 \leq x + 8$
$2x + 9 - 9 \leq x + 8 - 9$
$2x \leq x - 1$
$2x - x \leq x - x - 1$
$x \leq -1$
<-----]-----
-1

51. $9x + 13 \geq 8x$
$9x - 8x + 13 \geq 8x - 8x$
$x + 13 \geq 0$
$x + 13 - 13 \geq 0 - 13$
$x \geq -13$
-------[----->
-13

57. $7 - x \leq 3x - 2$
$7 - x + x \leq 3x + x - 2$
$7 \leq 4x - 2$
$7 + 2 \leq 4x - 2 + 2$
$9 \leq 4x$
$\frac{9}{4} \leq \frac{4x}{4}$
$\frac{9}{4} \leq x$
------[----->
$\frac{9}{4}$

59. $3(x - 8) < 5x + 6$
$3x - 3(8) < 5x + 6$
$3x - 24 < 5x + 6$
$3x - 3x - 24 < 5x - 3x + 6$
$-24 < 2x + 6$
$-24 - 6 < 2x + 6 - 6$
$-30 < 2x$
$\frac{-30}{2} < \frac{2x}{2}$
$-15 < x$
-------[----->
-15

61. $8(5 - x) \leq 10(8 - x)$
$8(5) - 8x \leq 10(8) - 10x$
$40 - 8x \leq 80 - 10x$
$40 - 8x + 8x \leq 80 - 10x + 8x$
$40 \leq 80 - 2x$
$40 - 80 \leq 80 - 80 - 2x$
$-40 \leq -2x$
$\frac{-40}{-2} \geq \frac{-2x}{-2}$
$20 \geq x$
<-----]-----
20

63. $2 < x - 5 < 5$
$2 + 5 < x - 5 + 5 < 5 + 5$
$7 < x < 10$
---(---------)---
7 10

65. $-5 < x + 4 \leq 7$
$-5 - 4 < x + 4 - 4 \leq 7 - 4$
$-9 < x \leq 3$
---(---------]---
-9 3

67. $0 \leq x + 10 \leq 10$
$0 - 10 \leq x + 10 - 10 \leq 10 - 10$
$-10 \leq x \leq 0$
---[---------]---
-10 0

69. $4 < -2x < 10$
$\frac{4}{-2} < \frac{-2x}{-2} < \frac{10}{-2}$
$-2 > x > -5$
$-5 < x < -2$
---(---------)---
-5 -2

71. $-3 \le \frac{x}{2} \le 5$

$2(-3) \le 2(\frac{x}{2}) \le 2(5)$

$-6 \le x \le 10$

---[---------]--

-6 10

73. $3 \le 2x - 1 < 5$

$3 + 1 \le 2x - 1 + 1 < 5 + 1$

$4 \le 2x < 6$

$\frac{4}{2} \le \frac{2x}{2} \le \frac{6}{2}$

$2 \le x \le 3$

---[---------]--

2 3

75. $0 < 10 - 5x \le 15$

$0 - 10 < 10 - 10 - 5x \le 15 - 10$

$-10 < -5x \le 5$

$\frac{-10}{-5} < \frac{-5x}{-5} \le \frac{5}{-5}$

$2 > x \ge -1$ or $-1 \le x < 2$

---[---------)--

-1 2

APPLICATIONS

77. Analyze: We know 3 scores: 68, 75, and 79. We need to find the last score to earn at least an 80 average.

Form: Let x = last score.

average ≥ 80

$\frac{68+75+79+x}{4} \ge 80$

Solve:

$\frac{222+x}{4} \ge 80$

$4\left(\frac{222+x}{4}\right) \ge 4(80)$

$222 + x \ge 320$

$222 - 222 + x \ge 320 - 222$

$x \ge 98$

State: To earn at least a B, the student needs at least a 98% on the last test.

Check: Pick 99.

$\frac{222+(99)}{4} \ge 80$

$\frac{321}{4} \ge 80$

$80.25 \ge 80$

79. Analyze: We know 2 ratings: 17 mpg and 19 mpg. We need to find the rating of the third car to have an average rating at least 21 mpg.

Form: Let x = rating of third car.

average ≥ 21

$\frac{17+19+x}{3} \ge 21$

Solve:

$\frac{36+x}{3} \ge 21$

$3\left(\frac{36+x}{3}\right) \ge 3(21)$

$36 + x \ge 63$

$36 - 36 + x \ge 63 - 36$

$x \ge 27$

State: The third car needs to have a rating of at least 27 mpg.

Check: Pick 36.

$\frac{36+36}{3} \ge 21$

$\frac{72}{3} \ge 21$

$24 \ge 21$

81. 2 hours per day = 2(60) minutes per day

= 120 minutes per day

7(120) = 840 minutes per week.

Let w = homework hours.

$w \ge 840$ minutes.

83. a) $0° \le a \le 18°$

b) $18° \le a \le 50°$

c) $30° \le a \le 37°$

d) $75° \le a \le 90°$

85. a) 470 ft $\le x \le$ 13,143 ft

b) $\frac{470}{5{,}280}$ mi $\le x \le \frac{13{,}143}{5{,}280}$ mi

0.1 mi $\le x \le$ 2.5 mi

87. For plug:

$1.497 - 0.001 \le w \le 1.497 + 0.001$

1.496 in $\le w \le$ 1.498 in

For opening:

$1.5005 - 0.0005 \le w \le 1.497 + 0.0005$

1.5000 in $\le w \le$ 1.5010

REVIEW

91. $-5^3 = -(5)(5)(5)$

$= -125$

93. $-4^4 = -(4)(4)(4)(4)$

$= -256$

CHAPTER 2 REVIEW

1. a) $12 + 33 = 45$

b) $-45 + (-37) = -82$

c) $-15 + 37 = 22$

d) $25 + (-13) = 25 - 13 = 12$

e) $12 + (-8) + (-15)$
$= 12 - 8 - 15$
$= 4 - 15$
$= -11$

f) $-25 + (-14) + 35 = -39 + 35$
$= -4$

g) $-9.9 + (-2.4) = -12.3$

h) $\frac{5}{16} + (-\frac{1}{2}) = \frac{5}{16} + (-\frac{8}{16})$
$= \frac{5}{16} - \frac{8}{16}$
$= -\frac{3}{16}$

i) $35 + (-13) + (-17) + 6$
$= 22 + (-17) + 6$
$= 5 + 6$
$= 11$

j) $-21 + (-11) + 32 + (-45)$
$= -32 + 32 + (-45)$
$= 0 + (-45)$
$= -45$

k) $0 + (-7) = -7$

l) $-7 + 7 = 0$

3. a) $45 - 64 = -19$

b) $-17 - 32$
$= -17 + (-32)$
$= -49$

c) $-27 - (-12)$
$= -27 + 12$
$= -15$

d) $3.6 - (-2.1)$
$= 3.6 + 2.1$
$= 5.7$

5. a) $x + 12 = -17$
$x + 12 - 12 = -17 - 12$
$x = -17 + (-12)$
$x = -29$

b) $-1.7 = y - 1.3$
$-1.7 + 1.3 = y - 1.3 + 1.3$
$-0.4 = y$

c) $17 + p = -8$
$17 - 17 + p = -8 - 17$
$p = -8 + (-17)$
$p = -25$

d) $-8 + q = -5$
$-8 + 8 + q = -5 + 8$
$q = 3$

7. difference in
elevation = highest − lowest
$= 29,028 - (-36,205)$
$= 29,028 + 36,205$
$= 65,233$ ft

9. a) associative property of multiplication
b) commutative property of multiplication

11. a) $-12x = 24$
$\frac{-12x}{-12} = \frac{24}{-12}$
$x = -2$

b) $36a = -108$
$\frac{36a}{36} = \frac{-108}{36}$
$a = -3$

c) $\frac{b}{5} = -4$
$5\left(\frac{b}{5}\right) = 5(-4)$
$b = -20$

d) $\frac{f}{17} = -3$

$17\left(\frac{f}{17}\right) = 17(-4)$

$f = -51$

13. a) $4^3 + 2(-6 - 2{\bullet}2)$
$= 4^3 + 2(-6 - 4)$
$= 4^3 + 2(-10)$
$= 64 + 2(-10)$
$= 64 + (-20)$
$= 44$

b) $-5[-3 - 2(5 - 7^2)] - 5$
$= -5[-3 - 2(5 - 49)] - 5$
$= -5[-3 - 2(-44)] - 5$
$= -5[-3 - (-88)] - 5$
$= -5[-3 + 88] - 5$
$= -5[85] - 5$
$= -425 - 5$
$= -425 + (5)$
$= -430$

c) $\frac{-4(4+2)-4}{18-4(-5)} = \frac{-4(6)-4}{18-(-20)}$
$= \frac{-24-4}{18+20}$
$= \frac{-28}{38}$
$= -\frac{14}{19}$

d) $(-3)^3(-\frac{8}{2}) + 5$
$= (-27)(-\frac{8}{2}) + 5$
$= (-27)(-4) + 5$
$= 108 + 5$
$= 113$

e) $\frac{|-35|-2(-7)}{2^4-23} = \frac{35-2(-7)}{2^4-23}$
$= \frac{35-(-14)}{16-23}$
$= \frac{35+14}{-7}$
$= \frac{49}{-7}$
$= -7$

f) $-9^2 + (-9)^2 = -81 + 81$
$= 0$

15. $r = 2:$

$\frac{4\pi r^3}{3} = \frac{4\pi(2)^3}{3}$
$= \frac{4\pi{\bullet}8}{3} \approx 33.5 \text{ in}^3$

17. a) $-4(7w) = -4{\bullet}7w = -28w$

b) $-3r(-5r) = (-3)(-5)r{\bullet}r$
$= 15r^2$

c) $3(-2x)(-4y) = 3(-2)(-4)x{\bullet}y$
$= 24xy$

d) $0.4(5.2f) = (0.4{\bullet}5.2)f$
$= 2.08f$

19. a) 3 terms
b) 1 term

21. a) $8p + 5p - 4p$
$= 13p - 4p$
$= 9p$

b) $-5m + 2n - 2m - 2n$
$= -5m - 2m + 2n - 2n$
$= -7m + 0n$
$= -7m$

c) $6a + 2b - 8a - 12b - 2a - 10b$
$= 6a - 8a - 2a + 2b - 12b - 10b$
$= -2a - 2a - 10b + 10b$
$= -4a - 20b$

d) $5(p - 2q) - 2(3p + 4q) - p - 18q$
$= 5p - 5(2)q + (\text{-}3)(3)p + (2)(4)q - p - 18q$
$= 5\text{p} - 10q + (\text{-}6)p + (\text{-}8)q - p - 18q$
$= 5p + (\text{-}6) - p - 10q + (\text{-}8)q - 18q$
$= (-1)p - p - 18q - 18q$
$= -2p - 36q$

e) $x^2 - x(x - 2)$
$= x^2 + (\text{-}x){\bullet}x - (\text{-}x){\bullet}1$
$= x^2 + (-x^2) - (-x)$
$= 0 - (-x)$
$= 0 + x$
$= x$

f) $8a^3 + 4a^3 - 20a^3$
$= 12a^3 - 20a^3$
$= -8a^3$

23. a) $5x + 4 = 14$
$5x + 4 - 4 = 14 - 4$
$5x = 10$
$\frac{5x}{5} = \frac{10}{5}$
$x = 2$

b) $-12y + 8 = 20$
$-12y + 8 - 8 = 20 - 8$
$-12y = 12$
$\frac{-12y}{-12} = \frac{12}{-12}$
$y = -1$

c) $\frac{n}{5} - 2 = 4$
$\frac{n}{5} - 2 + 2 = 4 + 2$
$\frac{n}{5} = 6$
$5\left(\frac{n}{5}\right) = 5(6)$
$n = 30$

d) $\frac{b-5}{4} = -6$
$4\left(\frac{b-5}{4}\right) = 4(-6)$
$b - 5 = -24$
$b - 5 + 5 = -24 + 5$
$b = -19$

e) $5(2x - 4) - 5x = 0$
$5 \bullet 2x - 5(4) - 5x = 0$
$10x - 20 - 5x = 0$
$10 - 5x - 20 = 0$
$5x - 20 = 0$
$5x - 20 + 20 = 0 + 20$
$5x = 20$
$\frac{5x}{5} = \frac{20}{5}$
$x = 4$

f) $-2(x - 5) = 5(-3x + 4) + 3$
$-2x - (-2)5 = 5(-3)x + 5(4) + 3$
$-2x - (-10) = -15x + 20 + 3$
$-2x + 10 = -15x + 23$
$-2x + 15x + 10 = -15x + 15x + 23$
$13x + 10 = 23$
$13x + 10 - 10 = 23 - 10$
$13x = 13$
$\frac{13x}{13} = \frac{13}{13}$
$x = 1$

g) $\frac{3}{4} = \frac{1}{2} + \frac{d}{5}$
$LCD = 20$
$20\left(\frac{3}{4}\right) = 20\left(\frac{1}{2} + \frac{d}{5}\right)$
$20\left(\frac{3}{4}\right) = 20\left(\frac{1}{2}\right) + 20\left(\frac{d}{5}\right)$
$15 = 10 + 4d$
$15 - 10 = 10 - 10 + 4d$
$5 = 4d$
$\frac{5}{4} = \frac{4d}{4}$
$\frac{5}{4} = d$

h) $-\frac{2}{3}f = 4$
$3\left(-\frac{2}{3}f\right) = 3(4)$
$-2f = 12$
$\frac{-2f}{-2} = \frac{12}{-2}$
$f = -6$

i) $3(z + 8) = 6(a + 4) - 3a$
$3a + 3(8) = 6a + 6(4) - 3a$
$3a + 24 = 6a + 24 - 3a$
$3a + 24 = 6a - 3a + 24$
$3a + 24 = 3a + 24$
$3a - 3a + 24 = 3a - 3a + 24$
$24 = 24$
Identity

j) $2(y + 10) + y = 3(y + 8)$
$2y + 2(10) + y = 3y + 3(8)$
$2y + 20 + y = 3y + 24$
$2y + y + 20 = 3y + 24$
$3y + 20 = 3y + 24$
$3y - 3y + 20 = 3y - 3y + 24$
$20 = 24$

25. $s = p - d$
$13,998 = p - 2,100$
$13,998 + 2,100 = p - 2,100 + 2,100$
$16,098 = p$

27. $p = r - c$
$1,700 = 13,500 - c$
$1,700 - 13,500 = 13,500 - 13,500 - c$
$-11,800 = -c$
$\frac{-11,800}{-1} = \frac{-c}{-1}$
$\$11,800 = c$

29. $C = \frac{5(F-32)}{9}$
$40 = \frac{5(F-32)}{9}$
$9(40) = 9\left(\frac{5(F-32)}{9}\right)$
$360 = 5(F-32)$
$\frac{360}{5} = \frac{5(F-32)}{5}$
$72 = F - 32$
$72 + 32 = F - 32 + 32$
$104° = F$

31. $A = lw$
$= 60•24$
$= 1440 \text{ in}^2$

33. $A = \frac{1}{2}h(b+d)$
$= \frac{1}{2}(12)(11+13)$
$= \frac{1}{2}(12)(24)$
$= 6(24)$
$= 144 \text{ in}^2$

35. $A = \pi r^2$
$= \pi(8)^2$
$= \pi(64)$
$\approx 201.06 \text{ cm}^2$

37. $V = Bh$
$= (\pi r^2)h$
$= \pi(0.5)^2(12)$
$= \pi(0.25)(12)$
$= \pi(3)$
$\approx 9.4 \text{ ft}$

39. $r = \frac{30}{2} = 15$
$V = \frac{4}{3}\pi r^3$
$= \frac{4}{3}\pi(15)^3$
$= \frac{4}{3}\pi(3375)$
$\approx 14,137.17 \text{ in}^3$

41. **Analyze:** The electric bill consists of a charge of \$17.50 and a rate of \$0.18 per kilowatt hour. The total bill was \$43.96. We want to find the number of kilowatt hours used.
Form: Let h = kilowatt hours.
bill = charge + rate•kilowatt hours
$43.96 = 17.50 + 0.18•h$
Solve: $43.96 = 17.50 + 0.18•h$
$43.96 - 17.50 = 17.50 - 17.50 + 0.18h$
$26.46 = 0.18h$
$\frac{26.46}{0.18} = \frac{0.18h}{0.18}$
$147 = h$
State: Last month, 147 kilowatt hours were used.
Check: 43.96=17.50+0.18(147)
43.96=17.50+26.46
43.96=43.96

43. **Analyze:** The measure of one angle is 27°. Since two sides of are the same, the opposite angles are the same. The sum of the measures of the angles in a triangle is 180°. We want to find the measure of the other two angles.
Form: Let x = measure of 1st angle.
total = 1st + 2nd + 3rd angles
$180 = 27 + x + x$
Solve: $180 = 27 + x + x$
$180 = 27 + 2x$
$180 - 27 = 27 - 27 + 2x$
$153 = 2x$
$\frac{153}{2} = \frac{2x}{2}$
$76.5 = x$
State: The measure of the remaining two angles is 76.5°.
Check: $180 = 27 + 76.5 + 76.5$
180=180

45. **Analyz:** The amount invested was \$27,000. The accounts had rates of 7% and 9%. After 1 year, interest earned was \$2,110. We want to know how much was invested in each account.

Form: Let $x =$ amount in 7% account.

$27{,}000 - x =$ amt. in 9% account.

amt. of interest in 7% acct. $= x \cdot 0.07 \cdot 1$

amt. of interest in 7% acct.

$= (27,000 - x) \cdot 0.09 \cdot 1$

total interest

$=$ interest(7%) $+$ interest(9%)

$2,110$

$= x \cdot 0.07 \cdot 1 + (27,000 - x) \cdot 0.09 \cdot 1$

Solve:

$2,110 = x \cdot 0.07 \cdot 1 + (27,000 - x) \cdot 0.09 \cdot 1$

$2,110 = 0.07x + 0.09(27,000 - x)$

$2,110 = 0.07x + 0.09(27,000) - 0.09x$

$2,110 = 0.07x + 2,430 - 0.09x$

$2,110 = 0.07x - 0.09x + 2,430$

$2,110 = -0.02x + 2,430$

$2,110 - 2,430 = -0.02x + 2,430 - 2,430$

$-320 = -0.02x$

$\frac{-320}{-0.02} = \frac{-0.02x}{-0.02}$

$16,000 = x$

State: The amount invested into the 7% account was \$16,000 and into the 9% account was \$11,000.

Check:

$2,110 = 0.07(16{,}000)$
$+ 0.09(27,000 - 16{,}000)$

$2,110 = 0.07(16{,}000)$
$+ 0.09(11,000)$

$2,110 = 1,120 + 990$

$2,110 = 2,110$

47. **Solve:** We know that candy is \$0.90 and gumdrops are \$1.20 per pound. Twenty pounds of mixture will be made. The mixture will be worth \$1.50. We want to find how much of each ingredient does the manager need.

Form: Let $n =$ amount of candy.

$100 - n =$ amount of gumdrops.

Value of candy $= 0.90n$

Value of gumdrops $= 1.50(20 - n)$

Value of mixture $= 1.28(20) = 24$

Value of mixture $=$ value of candy
$+$ value of gumdrops

$24 = 0.90n + 1.50(20 - n)$

Solve: $24 = 0.90n + 1.50(20 - n)$

$24 = 0.90n + 1.50 \cdot 20 - 1.50n$

$24 = 0.90n + 30 - 1.50n$

$24 = 0.90n - 1.50n + 30$

$24 = -0.60n + 30$

$24 - 30 = -0.60n + 30 - 30$

$-6 = -0.60n$

$\frac{-6}{-0.60} = \frac{-0.60n}{-0.60}$

$10 = n$

State: The mixture is made up of 10 pounds of candy and 10 pounds of gumdrops.

Check: $24 = 0.90(10) + 1.50(20 - (10))$

$24 = 0.90(10) + 1.50(10)$

$24 = 9 + 15$

$24 = 24$

49. a) $3x + 2 < 5$

$3x + 2 - 2 < 5 - 2$

$3x < 3$

$\frac{3x}{3} < \frac{3}{3}$

$x < 1$

```
<====)-----
     3
```

b) $-5x - 8 > 7$

$-5x - 8 + 8 > 7 + 8$

$-5x > 15$

$\frac{-5x}{-5} < \frac{15}{-5}$

$x < -3$

```
<====)-----
     -3
```

c) $5x - 3 \geq 2x + 9$

$5x - 2x - 3 \geq 2x - 2x + 9$

$3x - 3 \geq 9$

$3x - 3 + 3 \geq 9 + 3$

$3x \geq 6$

$\frac{3x}{3} \geq \frac{6}{3}$

$x \geq 2$

```
------[====>
      2
```

d) $7x + 1 \leq 8x - 5$

$7x - 8x + 1 \leq 8x - 8x - 5$

$-x + 1 \leq -5$

$-x + 1 - 1 \leq -5 - 1$

$-x \leq -6$

$\frac{-x}{-1} \leq \frac{-6}{-1}$

$x \leq 6$

<====]-----

6

e) $5(3 - x) \leq 3(x - 3)$

$5(3) - 5x \leq 3x - 3(3)$

$15 - 5x \leq 3x - 9$

$15 - 5x - 3x \leq 3x - 3x - 9$

$15 - 8x \leq -9$

$15 - 15 - 8x \leq -9 - 15$

$-8x \leq -24$

$\frac{-8x}{-8} \geq \frac{-24}{-8}$

$x \geq 3$

------[====>

3

f) $-\frac{3}{4}x \geq -9$

$4(-\frac{3}{4}x) \geq 4(-9)$

$-3x \geq -36$

$\frac{-3x}{-3} \leq \frac{-36}{-3}$

$x \leq 12$

<====]-----

12

g) $8 < x + 2 < 13$

$8 - 2 < x + 2 - 2 < 13 - 2$

$6 < x < 11$

---(=====)---

6 11

h) $0 \leq 2 - 2x \leq 4$

$0 - 2 \leq 2 - 2 - 2x \leq 4 - 2$

$-2 \leq -2x \leq 2$

$\frac{-2}{-2} \geq \frac{-2x}{-2} \geq \frac{2}{-2}$

$1 \geq x \geq -1$

$-1 \leq x \leq 1$

---[=====]---

-1 1

51. Let x = weight of Ping-Pong balls.
2.40 g $\leq w \leq$ 2.53 g

STUDY SET Section 3.1

VOCABULARY

1. ordered

3. origin

5. rectangular

CONCEPTS

7. origin; left; up

9. no

11. Left and up is quadrant II.

13.

x	y
4	3
0	5
-3	4
-4	-3
0	-5
3	-4

15. Ten minutes before the workout, her heart rate was 60 beats/min.

17. At 30 minutes during the workout, her heart rate was 150 beats/min.

19. She had a heart rate of 100 beats/min at two times: 5 minutes and 50 minutes during the workout.

21. Before the workout, she had a heart rate of 60 beats/min. After the cool down, her heart rate was 70 beats/min. The difference $= 70 - 60 = 10$ minutes.

NOTATION

23. (3,5) is an ordered pair or an interval depending of the context. 3(5) is multiplication. $5(3+5)$ is an expression containing grouping symbols. These indicate the various ways that parentheses can be used.

25. By changing each pair into decimal form, we can determine whether they are the same point.
$(2.5,\ -\frac{7}{2}) = (2.5,\ -3.5)$
$(2\frac{1}{2},\ -3.5) = (2.5,\ -3.5)$
$(2.5,\ -3\frac{1}{2}) = (2.5,\ -3.5)$
Yes, they are the same point.

PRACTICE

27.

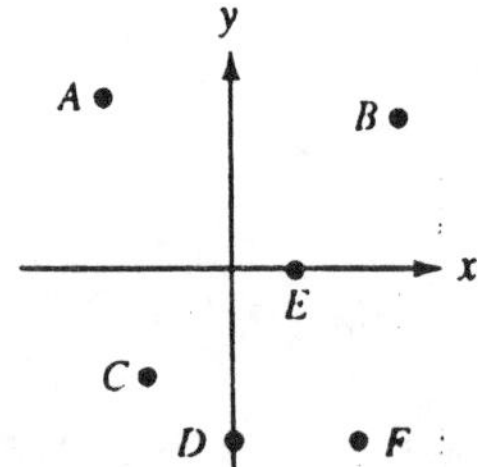

APPLICATIONS

29.

rivets	welds	anchors
(-6, 0)	(-4, 3)	(-6, -3)
(-2, 0)	(0, 3)	(6, -3)
(2, 0)	(4, 3)	
(6, 0)		

31. Working left/right which is also top/down for this graph, we can identify the points as A: (-3, 10) B: (-2, 7) C: (-1, 5) D: (0, 3) E: (1, 2) F: (2.5, 0.5) G: (4, 0)

33. a) \$2
b) \$4
c) \$7
d) \$2
For each of the values, we used the black dot rather than the open dot to determine the value.

35.

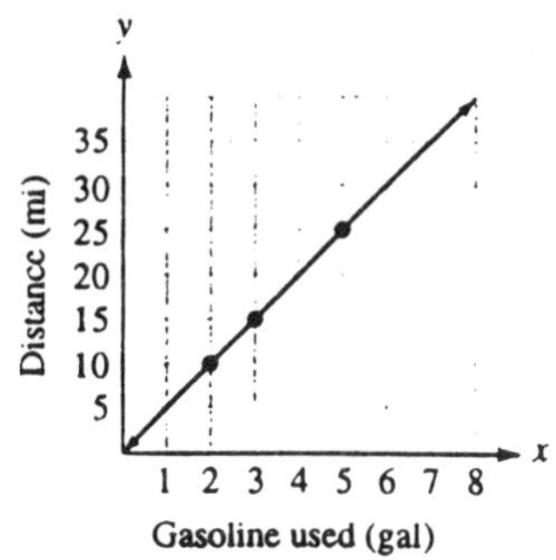

37. Carbondale: (3, J)
Champaign: (4, D)
Chicago: (5, B)
Peoria: (3, C)
Rockford: (3, A)
Springfield: (2, E)
St. Louis (2, H)

REVIEW

43. $-3-3(-5) = -3-(-15)$
$= -3+15$
$= 12$

45. $-(-8) = 8$

47. $-4x+7 = -21$
$-4x+7-7 = -21-7$
$-4x = -28$
$\frac{-4x}{-4} = \frac{-28}{-4}$
$x = 7$

49. Evaluate $x = -2,\ y = -5:$
$(x+1)(x+y)^2$
$= ((-2)+1)((-2)+(-5))^2$
$= (-1)(-7)^2$
$= (-1)(49)$
$= -49$

STUDY SET Section 3.2

VOCABULARY

1. two

3. independent; dependent

CONCEPTS

5. a) x and y are the two variables.

b) Substitute $x = 4 \,\&\, y = -2$:

$y = -2x + 6$
$(-2) = -2(4) + 6$
$-2 = -8 + 6$
$-2 = -2$

Yes, $(4, -2)$ satisfies the equation.

c) Substitute $x = -3 \,\&\, y = 12$:

$y = -2x + 6$
$(12) = -2(-3) + 6$
$12 = 6 + 6$
$12 = 12$

Yes, $x = -3$ and $y = 12$ is a solution.

d) There are infinitely many solutions.

7. a) The result is a true statement because the coordinates of M would satisfy the equation because it is on the line.
b) The result is a false statement because the coordinates of N would not satisfy the equation because N is not on the line.

9. Because the equation has an exponent of 2, the graph would not be a straight line. Not enough ordered pairs were found to determine that the graph is a parabola.

11. x is the only variable.

$x + 2 = 6$
$x + 2 - 2 = 6 - 2$
$x = 4$

4 is the only solution.

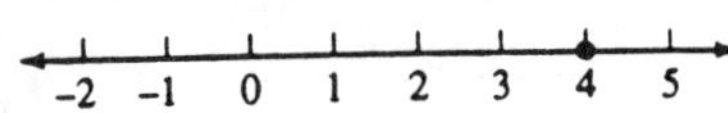

13. x and y are the two variables.
There are infinitely many solutions.

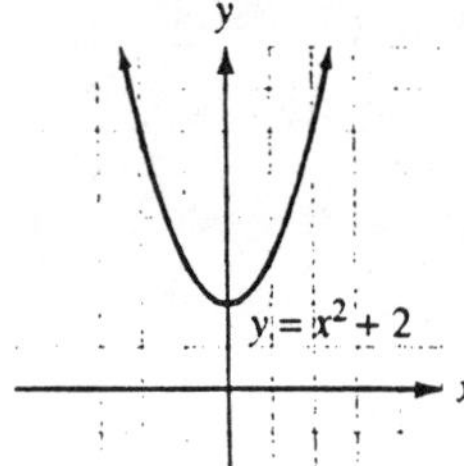

NOTATION

15. $y = -x + 4$
$(6) = -(-2) + 4$
$6 = 2 + 4$
$6 = 6$

17. $x - 2y = -4$
$(4) - 2(4) = -4$
$4 - 8 = -4$
$-4 = -4$

Yes, (4, 4) satisfies the equation.

19. $y = |5 - 2x|$
$(-3) = |5 - 2(4)|$
$-3 = |5 - 8|$
$-3 = |-3|$
$-3 = 3$

No, (4, - 3) does not satisfy the equation.

21. $y = x - 3$

Evaluate $x = 0$:
$y = (0) - 3$
$y = -3$

Evaluate $x = 1$:
$y = (1) - 3$
$y = -2$

Evaluate $x = -2$:
$y = (-2) - 3$
$y = (-2) + (-3)$
$y = -5$

23. $y = x^2 - 3$

Evaluate $x = 0$:

$y = (0)^2 - 3$

$y = 0 - 3 = -3$

Evaluate $x = 1$:

$y = (1)^2 - 3$

$y = 1 - 3 = -2$

Evaluate $x = 2$:

$y = (2)^2 - 3$

$y = 4 - 3 = 1$

Evaluate $x = 3$:

$y = (3)^2 - 3$

$y = 9 - 3 = 6$

Evaluate $x = -1$:

$y = (-1)^2 - 3$

$y = 1 - 3 = -2$

Evaluate $x = -2$:

$y = (-2)^2 - 3$

$y = 4 - 3 = 1$

Evaluate $x = -3$:

$y = (-3)^2 - 3$

$y = 9 - 3 = 6$

25.

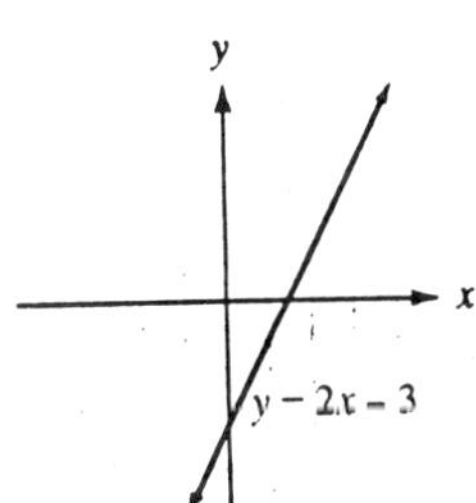

27.

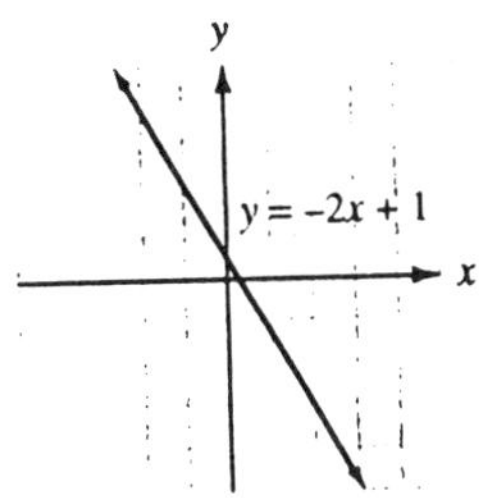

29. The graph is the same shape, but one unit higher.

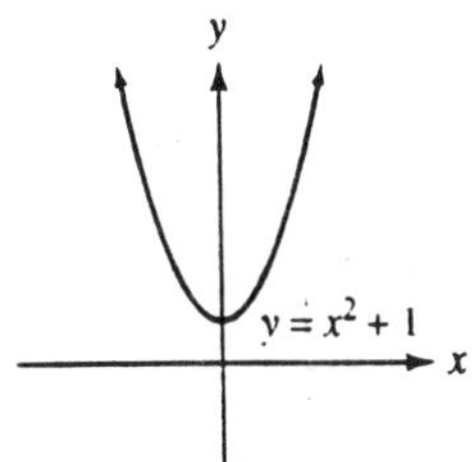

31. The graph is the same shape, but two units to the right.

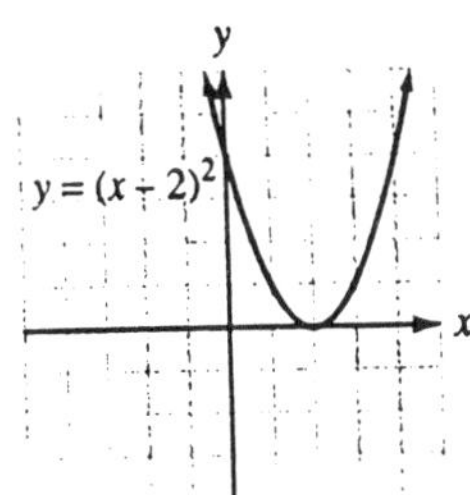

33. The graph is the same shape, but flipped upside down.

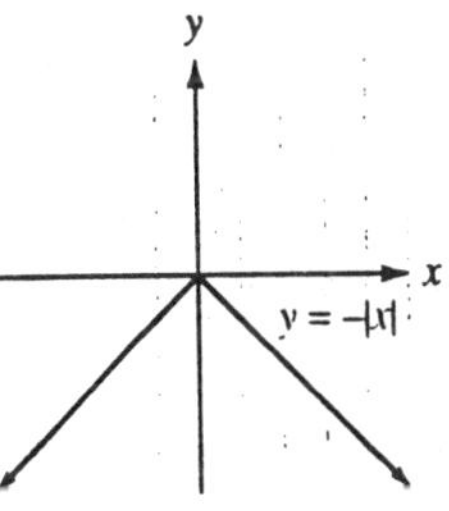

35. The graph is the same shape, but two units to the left.

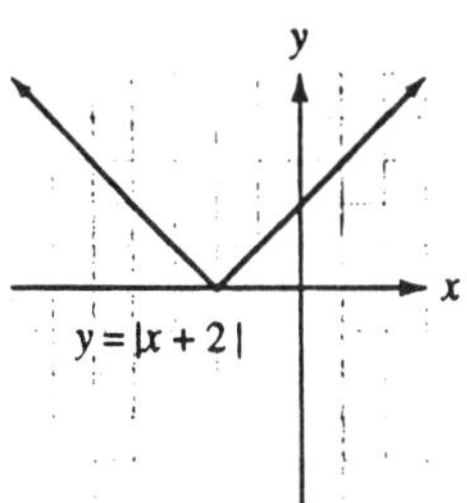

37. The graph is the same shape, but flipped upside down.

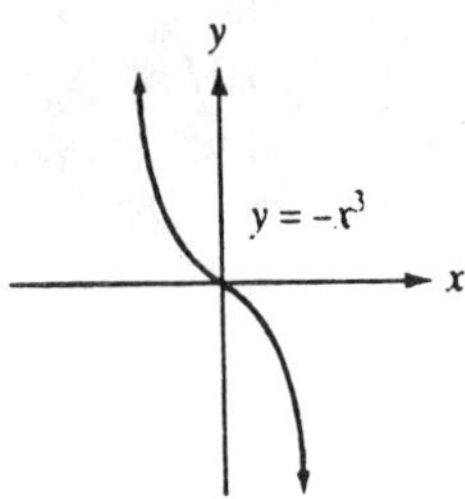

39. The graph is the same shape, but two units lower.

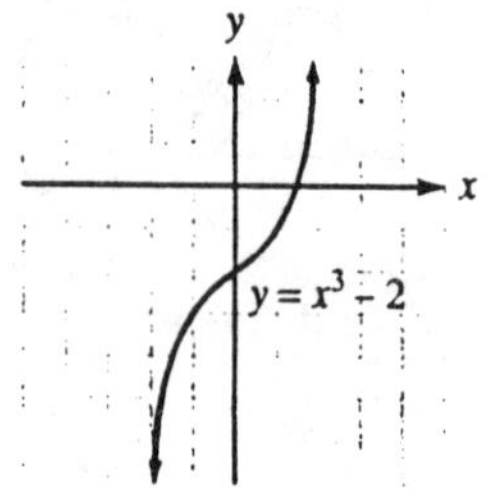

APPLICATION

41.

x	0	2	4	-2	-4
y	0	1	4	1	4

43. a) It costs 8¢ to make a 2-in bolt.
b) It costs 12¢ to make a 7-in bolt.
c) At a cost of 4¢, the 4-in bolt is the least expensive bolt to make.
d) The cost decreases as the length approaches 4-in., then increases as the length increases to 7-in.

45. a) To determine the purchase price, we look on the graph at $t = 0$. The purchase price is $90,000.
b) The lowest value occurred in the 3rd year.
c) After the 6th year, the value was greater than the purchase price.
d) The value decreased for the first three years, then increased in value since.

REVIEW

51. $\frac{x}{8} = -12$
$8\left(\frac{x}{8}\right) = 8(-12)$
$x = -96$

53. $\frac{x+5}{6}$ is an expression.

55. What number is 0.5% of 250?
$x = 0.5\% \cdot 250$
$x = 0.005 \cdot 250$
$x = 1.25$

STUDY SET Section 3.3

VOCABULARY

1. linear

3. y-intercept

5. parallel

CONCEPTS

7. a) Since the exponents on x in $y = x^3$ is 3, the equation is not linear.
b) Since it is in the form $Ax + By = C$, $2x + 3y = 6$ is a linear equation.
c) Since the absolute values contain the variable x, the equation is not linear.
d) Since the equation $x = -2$, can be written as $1x + 0y = -2$, it is linear.
e) Since the exponent on one of the x's in $y = 2x - x^2$ is 2, the equation is not linear.

9. $5y = 2x + 10$
$\frac{5y}{5} = \frac{2x+10}{5}$
$y = \frac{2x+10}{5}$
Evaluate $x = 10$:
$y = \frac{2(10)+10}{5}$
$y = \frac{20+10}{5}$
$y = \frac{30}{5}$
$y = 6$
Evaluate $x = -5$:
$y = \frac{2(-5)+10}{5}$
$y = \frac{-10+10}{5}$
$y = \frac{0}{5}$
$y = 0$
Evaluate $x = 5$:
$y = \frac{2(5)+10}{5}$
$y = \frac{10+10}{5}$
$y = \frac{20}{5}$
$y = 4$

11. Because A is on the line, the coordinates will satisfy the equation that is related to the line.

13. Both x and y are to the first power

15. The x-intercept is (-3, 0) and the y-intercept is (0, -1).

17. A horizontal line has no x-term, so $y = b$.

19. The student has made a mistake because the points should lie on a straight line.

21. To find the y-intercept of the graph of a linear equation, we let $y = 0$ and solve for x.

23. a) Let $y = 0$.
$(0) = 6x$
$\frac{0}{6} = \frac{6x}{6}$
$0 = x$
x-intercept: (0, 0)
b) Let $x = 0$.
$y = 6(0)$
$y = 0$
y-intercept: (0, 0)
c) It takes two distinct points to determine a line.

NOTATION

25. a) $-4x = -2y - 6$
$-4x + 2y = -2y + 2y - 6$
$-4x + 2y = -6$
b) $y = \frac{1}{2}x$
$2(y) = 2(\frac{1}{2}x)$
$2y = x$
$-x + 2y = x - x$
$-x + 2y = 0$
or
$-1(-x + 2y) = -1(0)$
$-1(-x) + (-1)(2y) = -1(0)$
$x - 2y = 0$

PRACTICE

27.

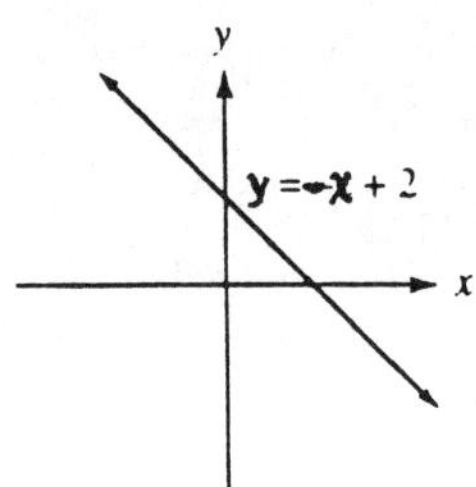

29.

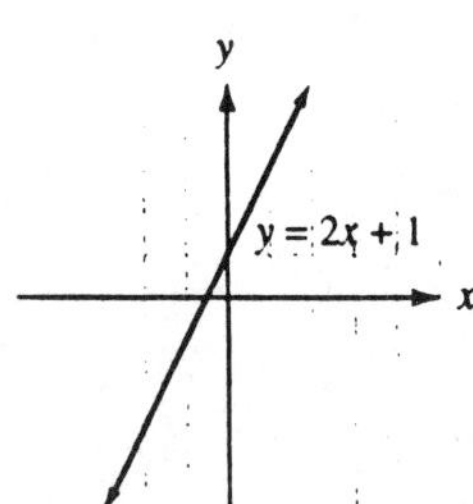

31. $2y = 4x - 6$

$\frac{2y}{2} = \frac{4x-6}{2}$

$y = \frac{4x}{2} - \frac{6}{2}$

$y = 2x - 3$

If $x = -1$,

$y = 2(-1) - 3 = -2 - 3$

$= -2 + (-3) = -5$

$(-1, -5)$

If $x = 0$,

$y = 2(0) - 3 = 0 - 3 = -3$

$(0, -3)$

If $x = 1$,

$y = 2(1) - 3 = 2 - 3 = -1$

$(1, -1)$

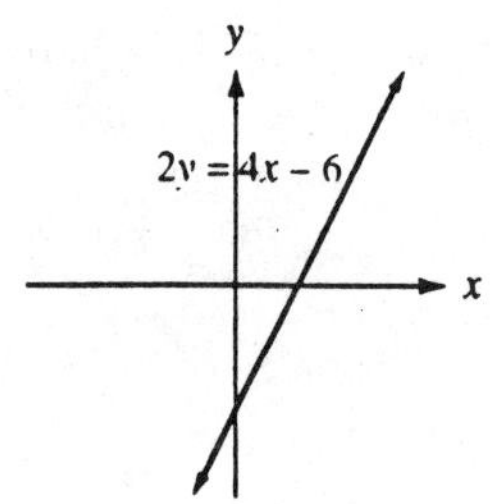

33. $2y = x - 4$

$\frac{2y}{2} = \frac{x-4}{2}$

$y = \frac{x}{2} - \frac{4}{2}$

$y = \frac{1}{2}x - 2$

If $x = -2$,

$y = \frac{1}{2}(-2) - 2 = (-1) - 2$

$= (-1) + (-2) = -3$

$(-2, -3)$

If $x = 0$,

$y = \frac{1}{2}(0) - 2 = 0 - 2 = -2$

$(0, -2)$

If $x = 2$,

$y = \frac{1}{2}(2) - 2 = 1 - 2 = -1$

$(2, -1)$

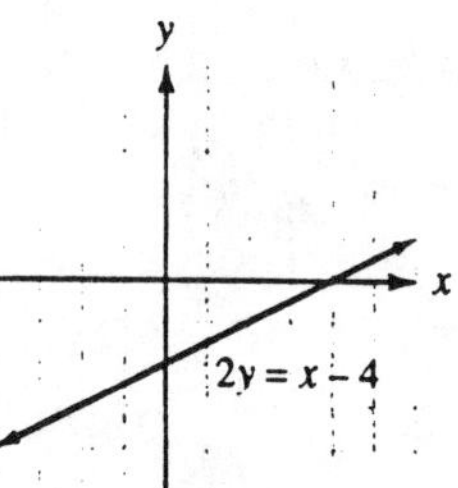

35.

Let $y = 0$:	Let $x = 0$:
$2(0) - 2x = 6$	$2y - 2(0) = 6$
$0 - 2x = 6$	$2y - 0 = 6$
$-2x = 6$	$2y = 6$
$\frac{-2x}{-2} = \frac{6}{-2}$	$\frac{2y}{2} = \frac{6}{2}$
$x = -3$	$y = 3$
x-intercept: (-3, 0)	y-intercept: (0, 3)

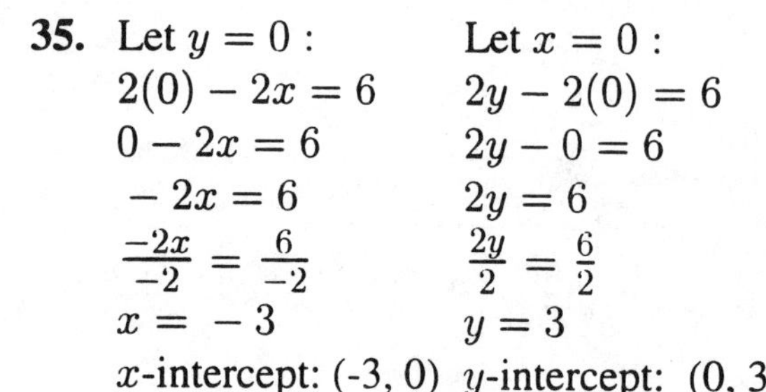

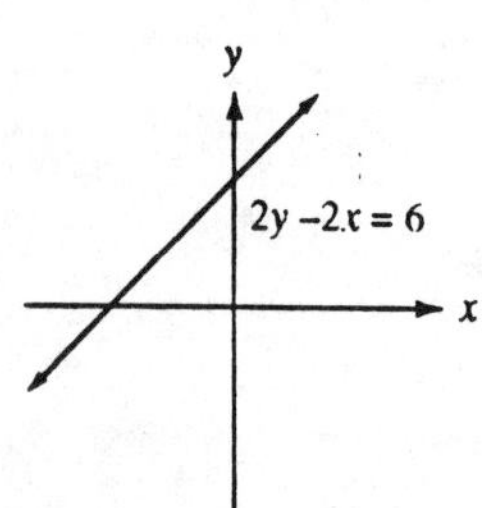

37. Let $y = 0$: Let $x = 0$:

$15(0) + 5x = -15$ $\quad$ $15y + 5(0) = -15$

$0 + 5x = -15$ $\quad$ $15y + 0 = -15$

$5x = -15$ $\quad$ $15y = -15$

$\frac{5x}{5} = \frac{-15}{5}$ $\quad$ $\frac{15y}{15} = \frac{-15}{15}$

$x = -3$ $\quad$ $y = -1$

x-intercept: (-3, 0) y-intercept: (0, -1)

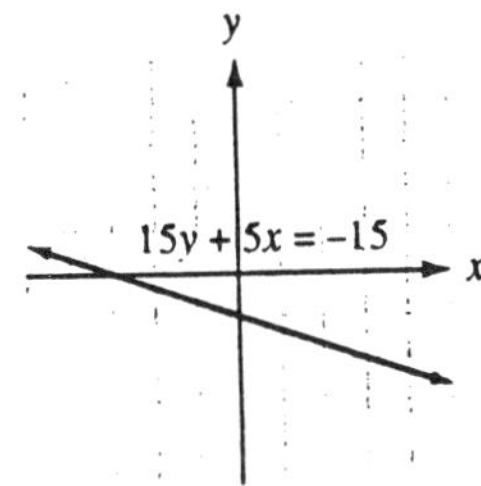

39. Let $y = 0$: Let $x = 0$:

$3x + 4(0) = 8$ $\quad$ $3(0) + 4y = 8$

$3x + 0 = 8$ $\quad$ $0 + 4y = 8$

$3x = 8$ $\quad$ $4y = 8$

$\frac{3x}{3} = \frac{8}{3}$ $\quad$ $\frac{4y}{4} = \frac{8}{4}$

$x = \frac{8}{3}$ $\quad$ $y = 2$

x-intercept: $(\frac{8}{3}, 0)$ y-intercept: (0, 2)

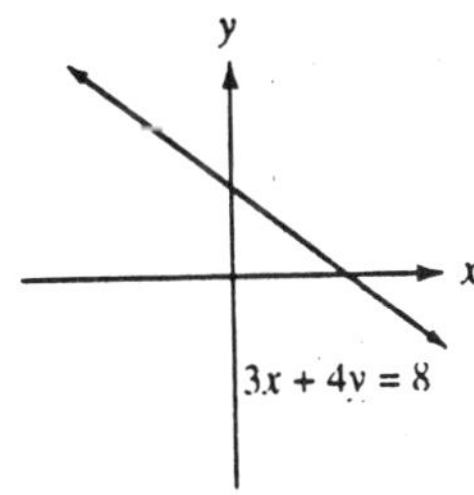

41. Let $y = 0$: Let $x = 0$:

$-4(0) + 9x = -9$ $\quad$ $-4y + 9(0) = -9$

$0 + 9x = -9$ $\quad$ $-4y + 0 = -9$

$9x = -9$ $\quad$ $-4y = -9$

$\frac{9x}{9} = \frac{-9}{9}$ $\quad$ $\frac{-4y}{-4} = \frac{-9}{-4}$

$x = -1$ $\quad$ $y = \frac{9}{4}$

x-intercept: (-1, 0) y-intercept: $(0, \frac{9}{4})$

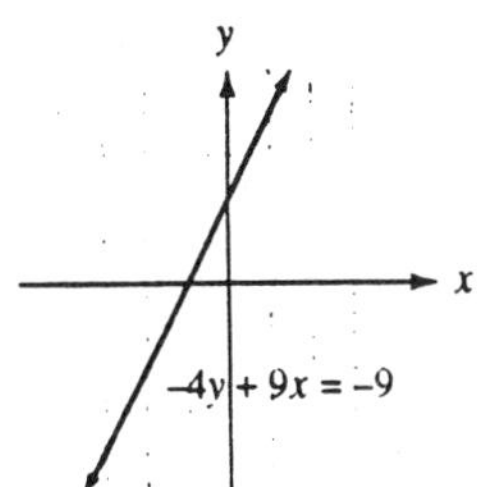

43. The graph is a horizontal line with an y-intercept of (0, 4).

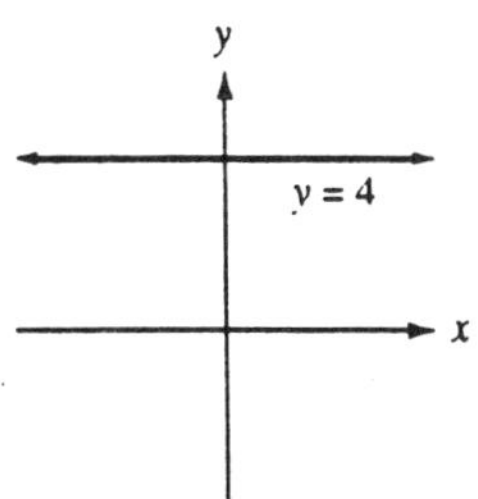

45. The graph is a vertical line with an x-intercept of $(-2, 0)$

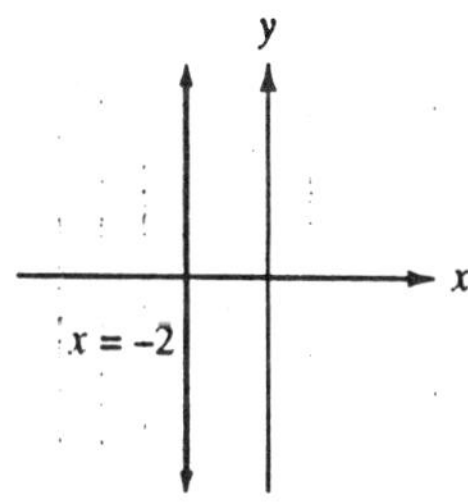

47. The graph is a horizontal line with an y-intercept of $(0, -\frac{1}{2})$.

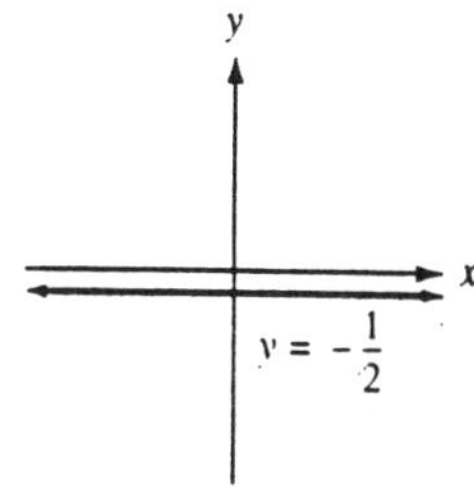

49. The graph is a vertical line with an x-intercept of $(\frac{4}{3}, 0)$

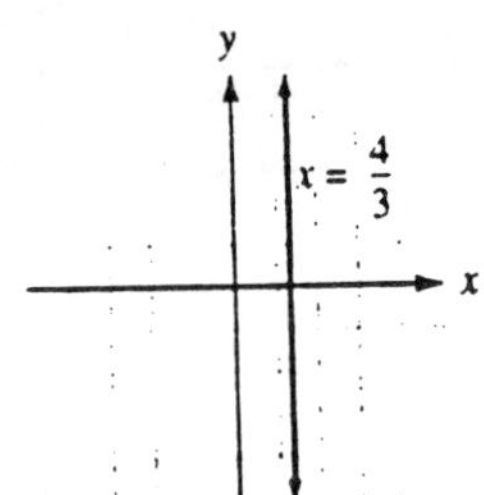

51. If $x = -2$:
$y = 2(-2) = -4$ $(-2, -4)$
If $x = 0$:
$y = 2(0) = 0$ $(0, 0)$
If $x = 2$:
$y = 2(2) = 4$ $(2, 4)$

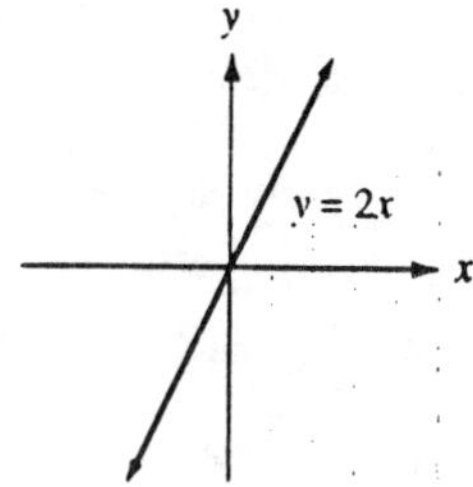

53. If $x = -2$:
$y = -2(-2) = 4$ $(-2, 4)$
If $x = 0$:
$y = -2(0) = 0$ $(0, 0)$
If $x = 2$:
$y = -2(2) = -4$ $(2, -4)$

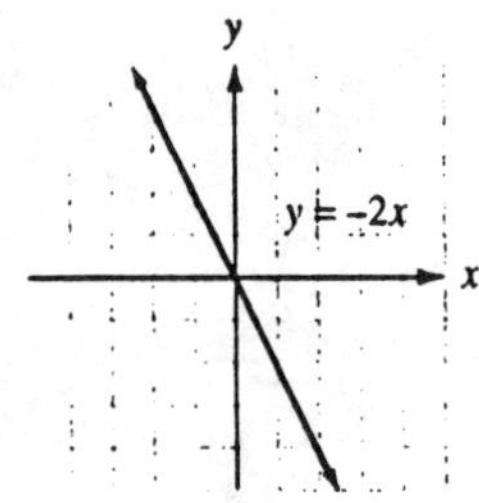

APPLICATION

55. a). Analyze: Service fee is \$50 and \$25 per unit.
Form: Since enrollment cost is in term of the number of units, let u = number of units and c = cost for each term.
Unit fee $= 25 \bullet u$.
Total cost = service fee plus unit fee.
$c = 50 + 25u$

b) If $u = 4$:
$c = 50 + 25(4)$
$c = 50 + 100$
$c = \$150$
If $u = 8$:
$c = 50 + 25(8)$
$c = 50 + 200$
$c = \$250$
If $u = 14$:
$c = 50 + 25(14)$
$c = 50 + 350$
$c = \$400$

c) The number of units is 0, so the y-intercept is just the service fee. The service fee is \$50.
d) The total cost is the sum of the cost of the two terms.
If $u = 18$:
$c = 50 + 25(18)$
$c = 50 + 450$
$c = 500$
If $u = 12$:
$c = 50 + 25(12)$
$c = 50 + 300$
$c = 350$
Total for the two terms is $500 + 350 = \$850$.

57. a) If $r = 7$:

$$h = 3.9(7) + 28.9$$
$$h = 27.3 + 28.9$$
$$h = 56.2$$

If $r = 8.5$:

$$h = 3.9(8.5) + 28.9$$
$$h \approx 33.2 + 28.9$$
$$h \approx 62.1$$

If $r = 9$:

$$h = 3.9(9) + 28.9$$
$$h = 35.1 + 28.9$$
$$h = 64$$

b) Because the graph is increasing, the longer the radius bone, the taller the woman is.

c) Reading the graph, the woman's height would be 58 in. Checking the answer with the equation:

$$58 = 3.9(7.5) + 28.9$$
$$58 \approx 29.3 + 28.9$$
$$58 \approx 58.2$$

58 is a good estimate.

REVIEW

63. $-(5 - 4c) = -1(5 - 4c)$
$$= (-1)(5) - (-1)(4c)$$
$$= -5 - (-4c)$$
$$= -5 + 4c$$

65. $\frac{x+6}{2} = 1$

$$2\left(\frac{x+6}{2}\right) = 2(1)$$
$$x + 6 = 2$$
$$x + 6 - 6 = 2 - 6$$
$$x = -4$$

67. profit = revenue − cost

$$p = r - c$$

69. $1 + 2[-3 - 4(2 - 8^2)]$
$$= 1 + 2[-3 - 4(2 - 64)]$$
$$= 1 + 2[-3 - 4(-62)]$$
$$= 1 + 2[-3 - (-248)]$$
$$= 1 + 2[-3 + 248]$$
$$= 1 + 2[245]$$
$$= 1 + 490$$
$$= 491$$

STUDY SET Section 3.4

VOCABULARY

1. ratio

3. slope

5. change

CONCEPTS

7. a) l_2, because it is increasing.
b) l_1, because it is decreasing.
c) l_4, because it is horizontal.
d) l_3, because it is vertical.

9. Let $x_1 = -4$ and $y_1 = 2$
$x_2 = 5$ and $y_2 = -7$.
$$m = \frac{y_2 - y_1}{x_2 - x_1} = \frac{(-7)-2}{5-(-4)} = \frac{(-7)+(-2)}{5+4} = \frac{-9}{9} = -1$$

11. Two points obtained from the graph: (2, 31) and (5, 40).
Let $x_1 = 2$ and $y_1 = 31$
$x_2 = 5$ and $y_2 = 40$.
$$m = \frac{y_2 - y_1}{x_2 - x_1} = \frac{40-31}{5-2} = \frac{9}{3} = 3$$
The rate of change is 3 inches per year.

13. a) The *Lost World* because it reached $100 million in the fewest days, 6.
b) $LW: \frac{100}{6} \approx$ \$16.7 million per day
$ID: \frac{100}{7} \approx$ \$14.3 million per day
$JP: \frac{100}{9} \approx$ \$11.1 million per day

NOTATION

15. $m = \frac{y_2 - y_1}{x_2 - x_1} = \frac{y_1 - y_2}{x_1 - x_2}$

PRACTICE

17. Let $x_1 = 2$ and $y_1 = 4$
$x_2 = 1$ and $y_2 = 3$.
$$m = \frac{y_2 - y_1}{x_2 - x_1} = \frac{3-4}{1-2} = \frac{-1}{-1} = 1$$

19. Let $x_1 = 3$ and $y_1 = 4$
$x_2 = 2$ and $y_2 = 7$.
$$m = \frac{y_2 - y_1}{x_2 - x_1} = \frac{7-4}{2-3} = \frac{3}{-1} = -3$$

21. Let $x_1 = 0$ and $y_1 = 0$
$x_2 = 4$ and $y_2 = 5$.
$$m = \frac{y_2 - y_1}{x_2 - x_1} = \frac{5-0}{4-0} = \frac{5}{4}$$

23. Let $x_1 = -3$ and $y_1 = 5$
$x_2 = -5$ and $y_2 = 6$.
$$m = \frac{y_2 - y_1}{x_2 - x_1} = \frac{6-5}{-5-(-3)} = \frac{1}{-5+3} = \frac{1}{-2} = -\frac{1}{2}$$

25. Let $x_1 = -2$ and $y_1 = -2$
$x_2 = -12$ and $y_2 = -8$.
$$m = \frac{y_2 - y_1}{x_2 - x_1} = \frac{-8-(-2)}{-12-(-2)} = \frac{-8+2}{-12+2} = \frac{-6}{-10} = \frac{3}{5}$$

27. Let $x_1 = 5$ and $y_1 = 7$
$x_2 = -4$ and $y_2 = 7$.
$$m = \frac{y_2 - y_1}{x_2 - x_1} = \frac{7-7}{-4-5} = \frac{0}{-4+(-5)} = \frac{0}{-9} = 0$$

29. Let $x_1 = 8$ and $y_1 = -4$
$x_2 = 8$ and $y_2 = -3$.
$$m = \frac{y_2 - y_1}{x_2 - x_1} = \frac{-3-(-4)}{8-8} = \frac{-3+4}{0} = \frac{1}{0}$$
Undefined.

31. Let $x_1 = -6$ and $y_1 = 0$
$x_2 = 0$ and $y_2 = -4$.
$$m = \frac{y_2 - y_1}{x_2 - x_1} = \frac{-4-0}{0-(-6)} = \frac{-4}{0+6} = \frac{-4}{6} = -\frac{2}{3}$$

33. Let $x_1 = -2.5$ and $y_1 = 1.75$
$x_2 = -0.5$ and $y_2 = -7.75$.
$$m = \frac{y_2 - y_1}{x_2 - x_1} = \frac{-7.75-1.75}{-0.5-(-2.5)} = \frac{-7.75+(-1.75)}{-0.5+2.5} = \frac{-9.50}{2} = -4.75$$

35. Two points obtained from the graph: (2, 1) and (−1, −1).
Let $x_1 = 2$ and $y_1 = 1$
$x_2 = -1$ and $y_2 = -1$.
$$m = \frac{y_2 - y_1}{x_2 - x_1} = \frac{-1-1}{-1-2} = \frac{-1+(-1)}{-1+(-2)} = \frac{-2}{-3} = \frac{2}{3}$$

37. Two points obtained from the graph: $(-4, 4)$ and $(4, -3)$.
Let $x_1 = -4$ and $y_1 = 4$
$x_2 = 4$ and $y_2 = -3$.

$$m = \frac{y_2 - y_1}{x_2 - x_1} = \frac{-3-4}{4-(-4)} = \frac{-3+(-4)}{4+4} = \frac{-7}{8}$$

39. $m = 2 = \frac{2}{1}$. So, rise is 2 and run is 1.

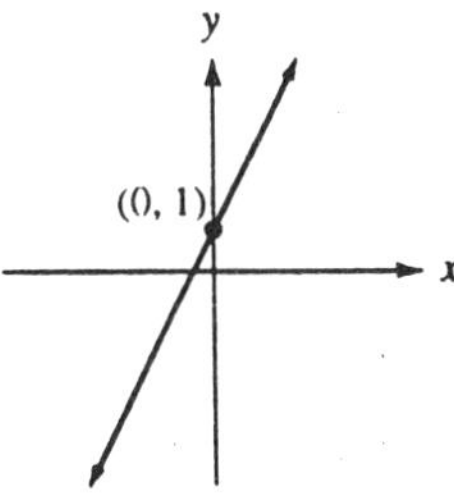

41. Rise is -3 and run is 2.

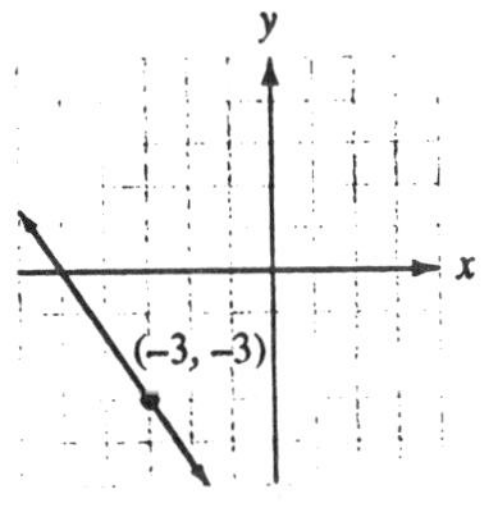

43. Rise is 3 and run is 4.

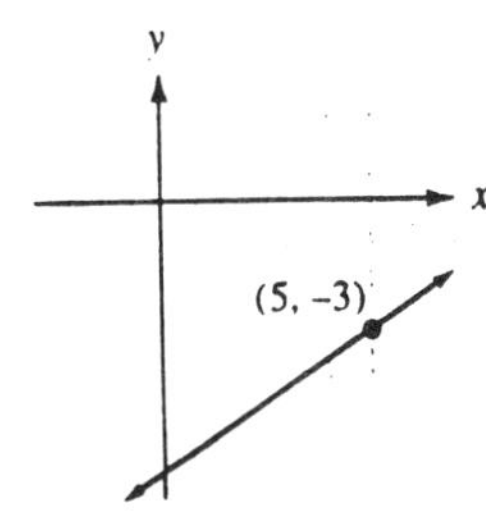

45. $m = -4 = \frac{-4}{1}$. So, rise is -4 and run is 1.

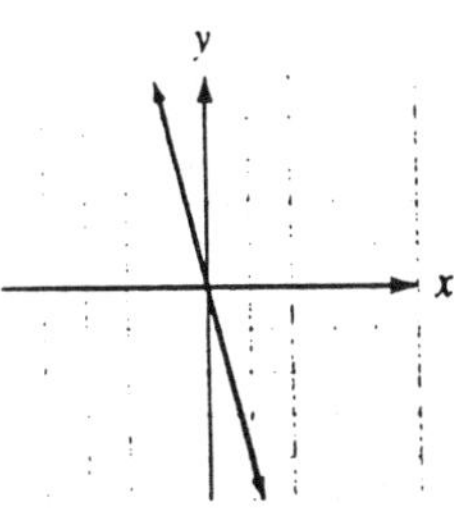

47. With slope of $m = 0$, the graph will be a horizontal line with y-intercept at (0, 1).

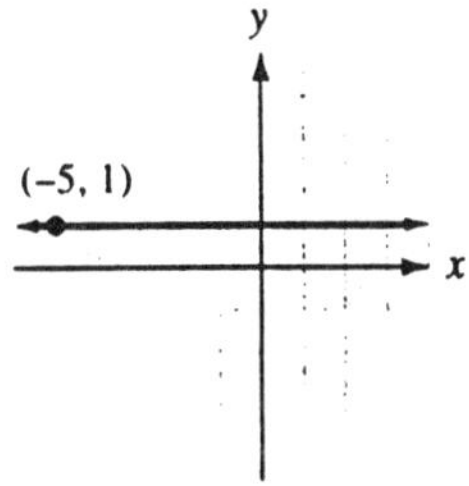

49. With undefined slope, the graph will be a vertical line with x-intercept at (1, 0).

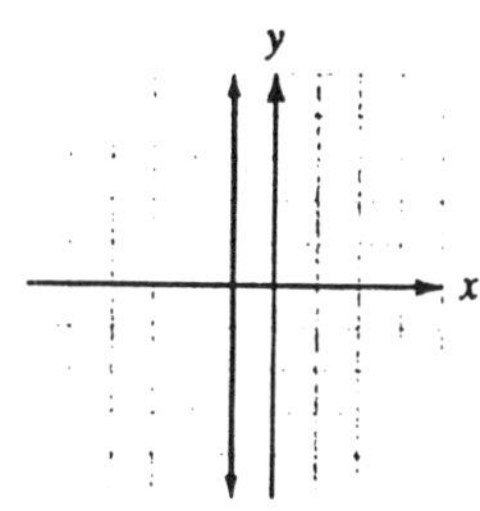

APPLICATION

51. The edge of the shallow end can be represented by the coordinates (5, -3). The edge of the deep end can be represented by (20, -9).
Let $x_1 = 5$ and $y_1 = -3$
$x_2 = 20$ and $y_2 = -9$.
$m = \frac{y_2 - y_1}{x_2 - x_1} = \frac{-9-(-3)}{20-5} = \frac{-9+3}{15}$
$= \frac{-6}{15} = -\frac{2}{5}$
The slope of the drop off is $-\frac{2}{5}$.

53. slope $= \frac{\text{rise}}{\text{run}} = \frac{364 \text{ ft}}{1 \text{ mile}} = \frac{364 \text{ ft}}{5280 \text{ ft}} = \frac{1}{20}$

$= 0.05 = 5\%$

55. a) slope $= \frac{\text{rise}}{\text{run}} = \frac{2 \text{ ft}}{16 \text{ ft}} = \frac{1}{8}$

b) Each ramp is 4 ft less than the ramp in design #1: 16 – 4 = 12 ft.
slope $= \frac{\text{rise}}{\text{run}} = \frac{1 \text{ ft}}{12 \text{ ft}} = \frac{1}{12}$

c) Design #1 requires less material so is cheaper. But, it is steeper.
Design #2 is more expensive, but is less steep.

57. Two points obtained from the graph: (2400, 150) and (4800, 330).
Let $x_1 = 2{,}400$ and $y_1 = 150$
$x_2 = 4{,}800$ and $y_2 = 330$.
$m = \frac{y_2 - y_1}{x_2 - x_1} = \frac{330-150}{4800-2400}$
$= \frac{180}{2400} = \frac{3}{40}$
The rate of change is 3 hp for each 40 rpm.

REVIEW

63. The point $(-3, 6)$ lies to the left and up from the origin. So, the point lies in quadrant II.

65. $y = x^2 + 17$
$(-2) = (-1)^2 + 17$
$-2 = 1 + 17$
$-2 = 18$
No, (-1, -2) is not a solution.

67. $y = 2x + 2$
$y - 2x = 2x - 2x + 2$
$-2x + y = 2$
The equation is linear because it can be written as $Ax + By = C$ where $A = -2$, $B = 1$, and $C = 2$.

STUDY SET Section 3.5

VOCABULARY

1. slope-intercept

3. Parallel

5. recipricals

CONCEPTS

7. No, because all of the points do not lie on one line.

9. a) The initial number of stores was 10.

b) Two points obtained from the graph: (0, 10) and (3, 25).

Let $x_1 = 0$ and $y_1 = 10$

$x_2 = 3$ and $y_2 = 25$.

$m = \frac{y_2 - y_1}{x_2 - x_1} = \frac{25-10}{3-0} = \frac{15}{3} = 5$

The growth rate was 5 stores per year.

c) Since $m = 5$ and the y-intercept is (0, 10), use $y = mx + b$.

$y = 5x + 10$.

11. a) (0, 0)

b) l_1 and l_2 have the same slope but have different y-intercept.

13. a) $y = \frac{-2x}{3} - 2 - -\frac{2}{3}x - 2$

$m = -\frac{2}{3}$

b) $y = 2 - 8x = -8x + 2$

$m = -8$

c) $y = x = 1x + 0$

$m = 1$

NOTATION

15. $6x - 2y = 10$

$6x - 6x - 2y = -6x + 10$

$-2y = -6x + 10$

$\frac{-2y}{-2} = \frac{-6x+10}{-2}$

$y = \frac{-6x}{-2} + \frac{10}{-2}$

$y = 3x + (-5)$

$y = 3x - 5$

The slope is 3 and the y-intercept is (0,-5)

PRACTICE

17. $m = 5$ and $b = -3$

$y = mx + b$

$y = 5x - 3$

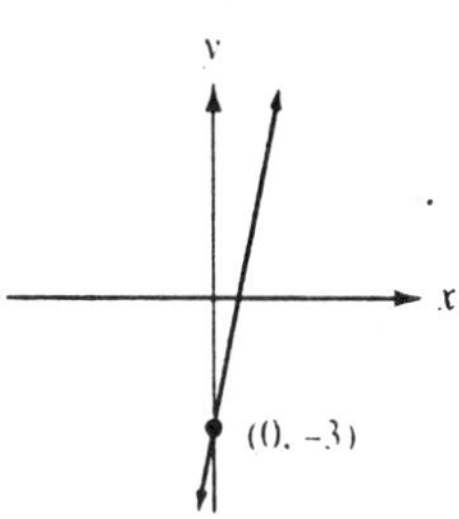

19. $m = -3$ and $b = 6$

$y = mx + b$

$y = -3x + 6$

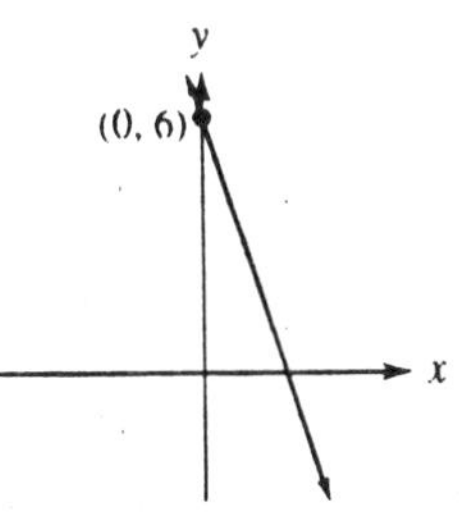

21. $m = \frac{1}{4}$ and $b = -2$

$y = mx + b$

$y = \frac{1}{4}x - 2$

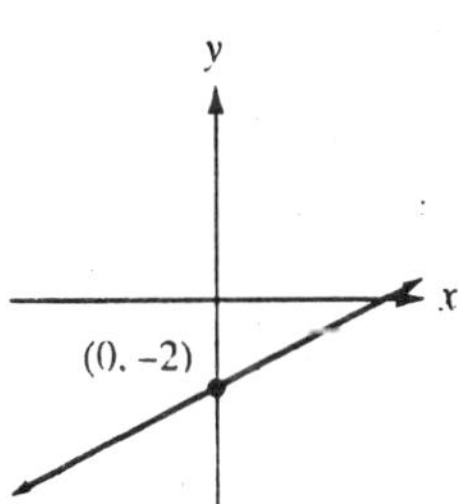

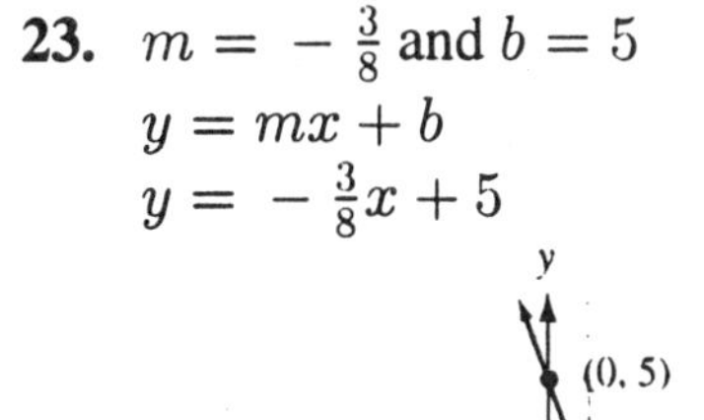

23. $m = -\frac{3}{8}$ and $b = 5$

$y = mx + b$

$y = -\frac{3}{8}x + 5$

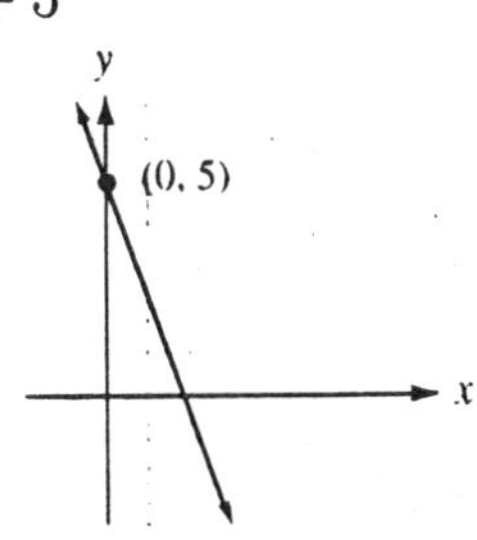

25. $3x + 4y = 16$
$3x - 3x + 4y = -3x + 16$
$4y = -3x + 16$
$\frac{4y}{4} = \frac{-3x+16}{4}$
$y = -\frac{3x}{4} + \frac{16}{4}$
$y = -\frac{3}{4}x + 4$
$m = -\frac{3}{4}$ and $b = 4$
y-intercept: (0, 4)

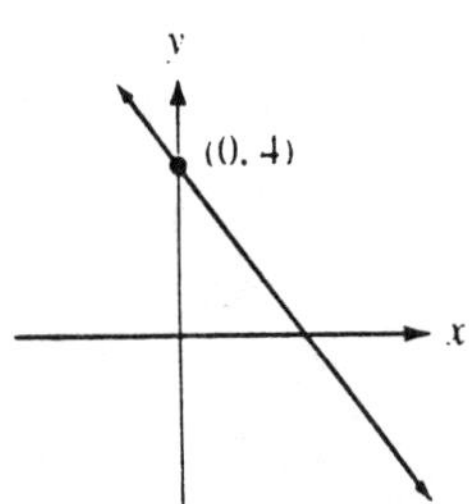

27. $10x - 5y = 5$
$10x - 10x - 5y = -10x + 5$
$-5y = -10x + 5$
$\frac{-5y}{-5} = \frac{-10x+5}{-5}$
$y = \frac{-10x}{-5} + \frac{5}{-5}$
$y = \frac{-10}{-5}x + (-1)$
$y = 2x - 1$
$m = 2$ and $b = -1$
y-intercept: (0, − 1)

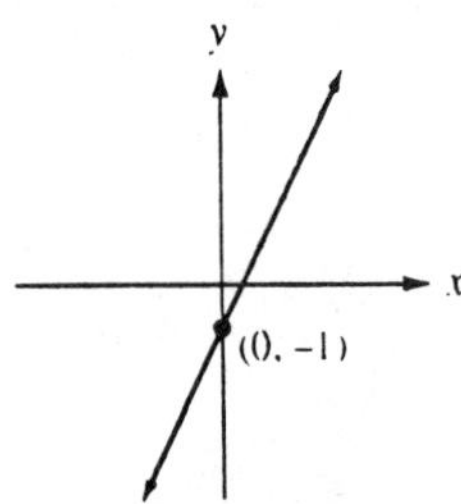

APPLICATION

29. a) Let x = number of hours filming and y = production costs.
$m = 2,000$ and y-intercept: (0, 5,000).
$y = mx + b$
$y = 2,000x + 5,000$
b) If $x = 8$,
$y = 2,000(8) + 5,000$
$= 16,000 + 5,000$
$= \$21,000.$

31. Let x = minutes and y = temperature.
$m = 5$ and y-intercept: (0, − 10).
$y = mx + b$
$y = 5x - 10$

33. a) Let x = ounces and y = cost.
$m = 20¢ = \$0.20$
and y-intercept: (0, 1.00).
$y = mx + b$
$y = 0.2x + 1.00$
b)

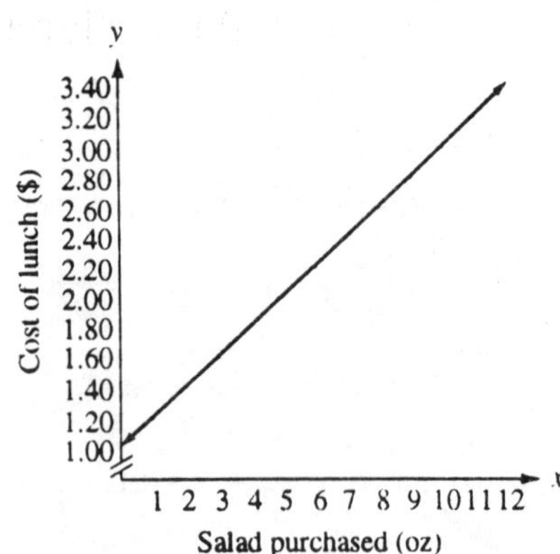

c) The new graph would have the same slope but different y-intercept. The graph would be raised by $1.00.

d) The new graph would have the same y-intercept, but have a steeper slope.

35. Let x = number of months the secretary works at LIZCO. and y = cost to the secretary of getting the job.
$m = -20$ (because her cost is decreased by the money that LIZCO pays her per month) and y-intercept: (0, 500) (the original cost to the secretary for getting the job).
$y = mx + b$
$y = -20x + 500$

37. $m_1 = -7.615$ and $m_2 = 0.128$
To check if perpendicular, multiply slopes.
$(-7.615)(0.128) = -0.9747 \neq 1$
Not perpendicular because the product of the slopes was not − 1.

REVIEW

43. Let $x_1 = 6\ y_1 - 2$

$x_2 = -6\ y_2 = 1$

$$m = \frac{y_2 - y_1}{x_2 - x_1} = \frac{1-(-2)}{-6-6} = \frac{1+2}{-6+(-6)}$$

$$= \frac{3}{-12} = -\frac{1}{4}$$

45. $-4-(-4) = -4+4 = 0$

47. $[-2(4-8)+4^2]$

Subtraction should be done first, because it is in the innermost parentheses.

49. What percent of 6 is 1.5?

1.5 is what percent of 6.

$1.5 = x \cdot 6$

$\frac{1.5}{6} = \frac{x \cdot 6}{6}$

$0.25 = x$

$0.25 = 25\%$

STUDY SET Section 3.6

VOCABULARY

1. point-slope

3. x-coordinate; y-coordinate

CONCEPTS

5. a) The slope is 2 and the y-intercept is $(0, -3)$.
b) The graph passes through (5, 4) and the slope is 6.

7. a) In a group of 8,000 women, there were 124 births in 1993.
b) Because we can assume 0 births for 0 women, the slope is
$m = \frac{124-0}{8{,}000-0} = \frac{124}{8{,}000} = \frac{31}{2{,}000}$
The birthrate is 31 births per 2,000 women or 15.5 births per 1,000 women.

9. a) No, because we need another point or the slope to determine the line.
b) No, we need a point.
c) Yes, two points is enough information to determine a line.

NOTATION

11. we read x_1 as x sub one

13.
$y-2=-3(x-4)$
$y-2=-3x-(-3)(4)$
$y-2=-3x-(-12)$
$y-2=-3x+12$
$y-2+2=-3x+12+2$
$y=-3x+14$

15. Let $m=-2,\ x_1=-1$, and $y_1=5$
$y-y_1=m(x-x_1)$
$y-(5)=(-2)(x-(-1))$
$y-5=-2x-(-2)(-1)$
$y-5=-2x-2$
$y-5+5=-2x-2+5$
$y=-2x+3$

PRACTICE

17. Let $m=3,\ x_1=2$, and $y_1=1$
$y-y_1=m(x-x_1)$
$y-(1)=(3)(x-(2))$
$y-1=3(x-2)$

19. Let $m=-\frac{4}{5},\ x_1=-5$, and $y_1=-1$
$y-y_1=m(x-x_1)$
$y-(-1)=(-\frac{4}{5})(x-(-5))$
$y+1=-\frac{4}{5}(x+5)$

21. Let $m=\frac{1}{5},\ x_1=10$, and $y_1=1$
$y-y_1=m(x-x_1)$
$y-(1)=(\frac{1}{5})(x-(10))$
$y-1=(\frac{1}{5})(x-10)$
$y-1=\frac{1}{5}x-(\frac{1}{5})(10)$
$y-1=\frac{1}{5}x-2$
$y-1+1=\frac{1}{5}x-2+1$
$y=\frac{1}{5}x-1$

23. Let $m=-5,\ x_1=-9$, and $y_1=8$
$y-y_1=m(x-x_1)$
$y-(8)=(-5)(x-(-9))$
$y-8=-5x-(-5)(-9)$
$y-8=-5x-45$
$y-8+8=-5x-45+8$
$y=-5x-37$

25. Let $m=-\frac{4}{3},\ x_1=6$, and $y_1=-4$
$y-y_1=m(x-x_1)$
$y-(-4)=(-\frac{4}{3})(x-(6))$
$y+4=-\frac{4}{3}x-(-\frac{4}{3})(6)$
$y+4=-\frac{4}{3}x-(-8)$
$y+4=-\frac{4}{3}x+8$
$y+4-4=-\frac{4}{3}x+8-4$
$y=-\frac{4}{3}x+4$

27. Let $m=-\frac{2}{3},\ x_1=3$, and $y_1=0$
$y-y_1=m(x-x_1)$
$y-(0)=(-\frac{2}{3})(x-(3))$
$y=-\frac{2}{3}x-(-\frac{2}{3})(3)$
$y=-\frac{2}{3}x-(-2)$
$y=-\frac{2}{3}x+2$

29. Let $m = 8,\ x_1 = 0$, and $y_1 = 4$
$y - y_1 = m(x - x_1)$
$y - (4) = (8)(x - (0))$
$y - 4 = 8x$
$y - 4 = 8x$
$y - 4 + 4 = 8x + 4$
$y = 8x + 4$

31. Let $m = -3,\ x_1 = 0$, and $y_1 = 0$
$y - y_1 = m(x - x_1)$
$y - (0) = (-3)(x - (0))$
$y = -3x$

33. Let $x_1 = 1\ y_1 = 7$
$x_2 = -2\ y_2 = 1$
$m = \frac{y_2 - y_1}{x_2 - x_1} = \frac{1-7}{-2-1} = \frac{-6}{-3} = 2$
$y - y_1 = m(x - x_1)$
$y - (7) = (2)(x - (1))$
$y - 7 = 2x - (2)(1)$
$y - 7 = 2x - 2$
$y - 7 + 7 = 2x - 2 + 7$
$y = 2x + 5$

35. Let $x_1 = -4\ y_1 = 3$
$x_2 = 2\ y_2 = 0$
$m = \frac{y_2 - y_1}{x_2 - x_1} = \frac{0-3}{2-(-4)} = \frac{-3}{2+4}$
$= \frac{-3}{6} = -\frac{1}{2}$
$y - y_1 = m(x - x_1)$
$y - (3) = \left(-\frac{1}{2}\right)(x - (-4))$
$y - 3 = -\frac{1}{2}x - \left(-\frac{1}{2}\right)(-4)$
$y - 3 = -\frac{1}{2}x - 2$
$y - 3 + 3 = -\frac{1}{2}x - 2 + 3$
$y = -\frac{1}{2}x + 1$

37. Let $x_1 = 5\ y_1 = 5$
$x_2 = 7\ y_2 = 5$
$m = \frac{y_2 - y_1}{x_2 - x_1} = \frac{5-5}{7-5} = \frac{0}{2} = 0$
$y - y_1 = m(x - x_1)$
$y - (5) = (0)(x - (5))$
$y - 5 = 0$
$y - 5 + 5 = 0 + 5$
$y = 5$

39. A vertical line has undefined slope. So, the equation is $x = 4$.

41. A horizontal line has slope equal to 0.
Let $m = 0,\ x_1 = 4$, and $y_1 = 5$.
$y - y_1 = m(x - x_1)$
$y - (5) = (0)(x - (4))$
$y - 5 = 0$
$y - 5 + 5 = 0 + 5$
$y = 5$

APPLICATIONS

43. a) part 1: (0, 0) (-5, 2)
part 2: (0, 0) (-3, 6)
part 3: (0, 0) (-1, 7)
part 4: (0, 0) (0, 10)

b) part 1: $x_1 = 0\ y_1 = 0$
$x_2 = -5\ y_2 = 2$
$m = \frac{y_2 - y_1}{x_2 - x_1} = \frac{2-0}{-5-0} = \frac{2}{-5} = -\frac{2}{5}$
$y - y_1 = m(x - x_1)$
$y - (0) = \left(-\frac{2}{5}\right)(x - (0))$
$y = -\frac{2}{5}x$

part 3: $x_1 = 0\ y_1 = 0$
$x_2 = \text{-}1\ \ y_2 = 7$
$m = \frac{y_2 - y_1}{x_2 - x_1} = \frac{7-0}{-1-0} = \frac{7}{-1} = -7$
$y - y_1 = m(x - x_1)$
$y - (0) = (-7)(x - (0))$
$y = -7x$

part 4: $x_1 = 0\ y_1 = 0$
$x_2 = 0\ \ y_2 = 10$
$m = \frac{y_2 - y_1}{x_2 - x_1} = \frac{10-0}{0-0} = \frac{10}{0}$ undefined
The equation of a vertical line with x-intercept of (0, 0) is $x = 0$.

c) The pole is not in the shape of a straight line.

45. a) The coordinates are (months, waste). Thus, we have (3, 800) and (5, 720).
Let $x_1 = 3$ $y_1 = 800$
$x_2 = 5$ $y_2 = 720$
$m = \frac{y_2 - y_1}{x_2 - x_1} = \frac{720-800}{5-3}$
$= \frac{-80}{2} = -40$
$y - y_1 = m(x - x_1)$
$y - (800) = (-40)(x - (3))$
$y - 800 = -40x - (-40)(3)$
$y - 800 = -40x - (-120)$
$y - 800 = -40x + 120$
$y - 800 + 800 = -40x + 120 + 800$
$y = -40x + 920$

b) Let $x = 12$ (1 year = 12 months)
$y = -40(12) + 920$
$y = -480 + 920$
$y = 440 \text{ yd}^3$

47. The coordinates are (years, clients). Thus, we have (1, 75) and (2, 105).
Let $t_1 = 1$ $c_1 = 75$
$t_2 = 2$ $c_2 = 105$
$m = \frac{c_2 - c_1}{t_2 - t_1} = \frac{105-75}{2-1} = \frac{30}{1} = 30$
$c - c_1 = m(t - t_1)$
$c - (75) = (30)(t - (1))$
$c - 75 = 30t - (30)(1)$
$c - 75 = 30t - 30$
$c - 75 + 75 = 30t - 30 + 75$
$c = 30t + 45$

49. We have (0, 32) and (100, 212).
Let $C_1 = 0$ $F_1 = 32$
$C_2 = 100$ $F_2 = 212$
$m = \frac{F_2 - F_1}{C_2 - C_1} = \frac{212-32}{100-0} = \frac{180}{100} = \frac{9}{5}$
$F - F_1 = m(C - C_1)$
$F - (32) = (\frac{9}{5})(C - (0))$
$F - 32 = \frac{9}{5}C$
$F - 32 + 32 = \frac{9}{5}C + 32$
$F = \frac{9}{5}C + 32$

51. Let x = minutes and y = temperature.
slope $= \frac{\text{change in temperature}}{\text{change in time}} = \frac{-4}{15}$
Because half an hour is 30, have the coordinate: (30, 75)
Let $m = -\frac{4}{15}$, $x_1 = 30$, and $y_1 = 75$.
$y - y_1 = m(x - x_1)$
$y - (75) = (-\frac{4}{15})(x - (30)$
$y - 75 = -\frac{4}{15}x - (-\frac{4}{15})(30)$
$y - 75 = -\frac{4}{15}x - (-8)$
$y - 75 = -\frac{4}{15}x + 8$
$y - 75 + 75 = -\frac{4}{15}x + 8 + 75$
$y = -\frac{4}{15}x + 83$

REVIEW

57. Let $x_1 = 2$ $y_1 = 4$
$x_2 = -6$ $y_2 = 8$
$m = \frac{y_2 - y_1}{x_2 - x_1} = \frac{8-4}{-6-2} = \frac{4}{-8} = -\frac{1}{2}$

59. Area of circle $= \pi r^2$
$r = \frac{diameter}{2} = \frac{12}{2} = 6 \text{ ft}$
$A = \pi(6)^2$
$A = \pi(36)$
$A \approx 113.1 \text{ ft}$

61. $(-1)^5$
$= (-1)(-1)(-1)(-1)(-1)$
$= (1)(-1)(-1)(-1)(-1)$
$= (-1)(-1)(-1)(-1)$
$= (1)(-1)(-1)$
$= (-1)(-1)$
$= 1$

63. The second term of $-4x^2 + 6x - 13$ is $6x$. The coefficient of $6x$ is 6.

STUDY SET Section 3.7

VOCABULARY

1. function

3. independent; dependent

5. constant

CONCEPTS

7. a) positive number. For example, $f(5) = (5)^2 = (5)(5) = 25$.

b) positive number. For example, $f(-5) = (-5)^2 = (-5)(-5) = 25$.

c) 0. For example, $f(0) = (0)^2 = (0)(0) = 0$.

d) Domain: all real numbers.
Range: real numbers greater than or equal to zero, $y \geq 0$.

9. a) $(2, 5)$ and $(2, -5)$
b) No, because more than one y-value is assigned to the x-value 2.

11. a) $f(2) = 0$
b) $f(0) = 1$
c) $f(-4) = 3$

13. a) yes.
b) no, because of the exponent 2.
c) no, because of the exponent 3.
d) yes.

15. $g = kf$

NOTATION

17. The independent variable is x

19. The function notation $f(4) = -5$ states that when 4 is substituted for x in function f, the result is -5. This fact canbe graphed by plotting the point $(4, -5)$.

21. $x > 0,\ y \leq 0$.

PRACTICE

23. yes

25. yes

27. no. (4, 2) and (4, $-$ 2)

29. yes

31. yes

33. no, (3, 4) and (3, $-$ 1)

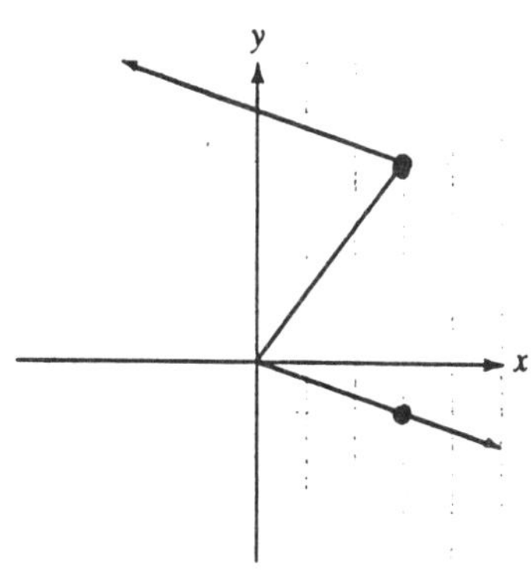

35. No, some one could have a shoe size of 10 when he/she is 25 and 26 years old: (10, 25) and (10, 26)

37. Domain: all real numbers.
Range: all real numbers.

39. Domain: all real numbers.
Range: $y \geq 0$.

41. $f(x) = 4x - 1$
a) $f(1) = 4(1) - 1 = 4 - 1 = 3$
b) $f(-2) = 4(-2) - 1 = -8 - 1 = -9$
c) $f(\frac{1}{4}) = 4(\frac{1}{4}) - 1 = 1 - 1 = 0$
d) $f(50) = 4(50) - 1 = 200 - 1 = 199$

43. $h(t) = 2t^2$
a) $h(0.4) = 2(0.4)^2 = 2(0.16) = 0.32$
b) $h(-2) = 2(-2)^2 = 2(4) = 8$
c) $h(1,000) = 2(1,000)^2$
$= 2(1,000,000) = 2,000,000$
d) $h(\frac{1}{8}) = 2(\frac{1}{8})^2 = 2(\frac{1}{64}) = \frac{2}{64} = \frac{1}{32}$

45. $s(x) = |x - 7|$
a) $s(0) = |(0) - 7| = |-7| = 7$
b) $s(\text{-}7) = |(\text{-}7) - 7| = |\text{-}14| = 14$
c) $s(7) = |(7) - 7| = |0| = 0$
d) $s(8) = |(8) - 7| = |1| = 1$

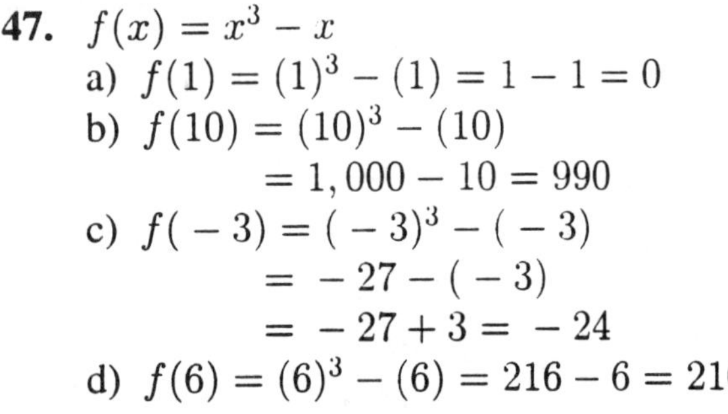

47. $f(x) = x^3 - x$
a) $f(1) = (1)^3 - (1) = 1 - 1 = 0$
b) $f(10) = (10)^3 - (10)$
$= 1,000 - 10 = 990$
c) $f(-3) = (-3)^3 - (-3)$
$= -27 - (-3)$
$= -27 + 3 = -24$
d) $f(6) = (6)^3 - (6) = 216 - 6 = 210$

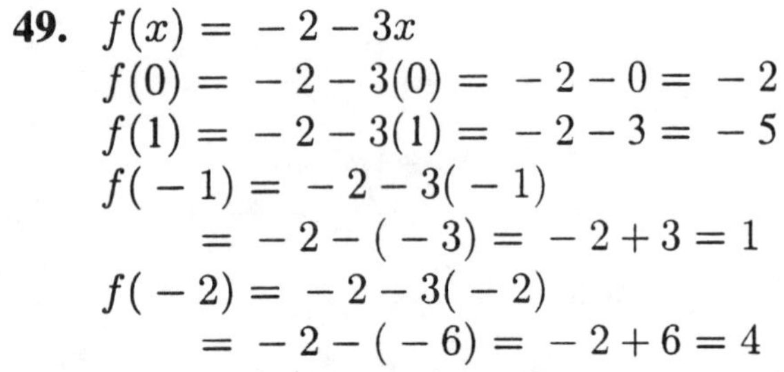

49. $f(x) = -2 - 3x$
$f(0) = -2 - 3(0) = -2 - 0 = -2$
$f(1) = -2 - 3(1) = -2 - 3 = -5$
$f(-1) = -2 - 3(-1)$
$= -2 - (-3) = -2 + 3 = 1$
$f(-2) = -2 - 3(-2)$
$= -2 - (-6) = -2 + 6 = 4$

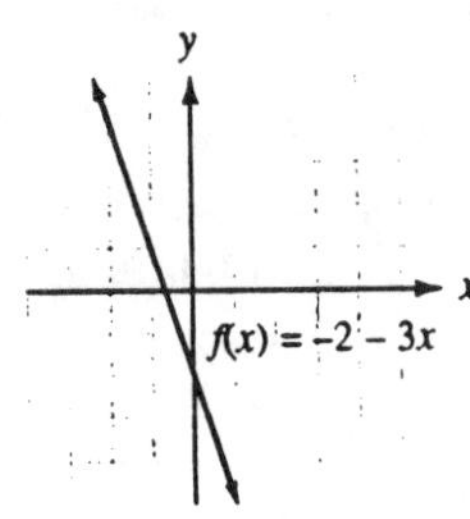

Domain: all real numbers
Range: all real numbers

51. $s(x) = 2 - x^3$
$s(0) = 2 - (0)^3 = 2 - 0 = 2$
$s(1) = 2 - (1)^3 = 2 - 1 = 1$
$s(2) = 2 - (2)^3 = 2 - 8 = -6$
$s(-1) = 2 - (-1)^3 = 2 - (-1)$
$= 2 + 1 = 3$
$s(-2) = 2 - (-2)^3 = 2 - (-8)$
$= 2 + 8 = 10$

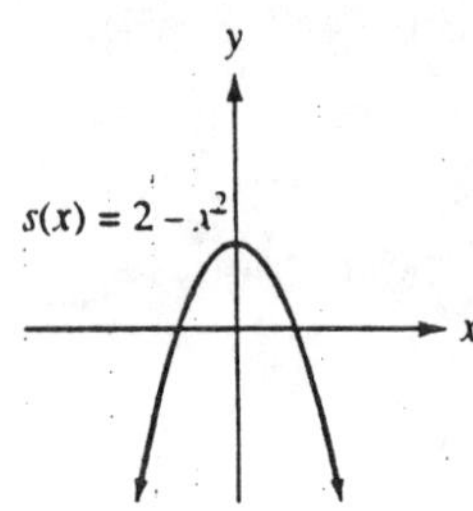

Domain: all real numbers
Range: $y \leq 2$

53. a) If $y = 10$ and $x = 2$.
$y = kx$
$(10) = k(2)$
$\frac{10}{2} = \frac{k2}{2}$
$5 = k$
So, $y = 5x$.

b) When $x = 7$
$y = 5(7) = 35$.

APPLICATIONS

55. Because the beam of light travels at at straight line and the path is in the shape of a V, $y = |x|$ could serve as a model.

57. a) Domain is the number of hours: $0 \leq x \leq 24$.
b) $f(3) = 0.5$
c) $f(6) = 1.5$
d) $f(15) = -1.5$
e) $f(12) = -2.5$. The tide at noon was -2.5 m. This was the lowest tide level.
f) $f(21) = 1.6$

59. $A(r) = 3.14r^2$
$A(5) = 3.14(5)^2 = 3.14(25) \approx 78.5 \text{ ft}^2$
$A(10) = 3.14(10)^2$
$= 3.14(100) \approx 314 \text{ ft}^2$
$A(20) = 3.14(20)^2$
$= 3.14(400) \approx 1256 \text{ ft}^2$

61. Because distance varies directly with gas, use $y = kx$, where $y =$ distance traveled (miles) and $x =$ gallons of gas.
Need to find k.
Let $y = 360$ and $x = 15$.
$360 = k(15)$
$\frac{360}{15} = \frac{k(15)}{15}$
$24 = k$
So, $y = 24x$.
If $x = 7$,
$y = 24(7) = 168$
The car can travel 168 miles on 7 gallons of gas.

63. Because dose varies directly with weight, use $y = kx$, where $y =$ dose (mg) and $x =$ weight (lbs).
Need to find k.
Let $y = 124$ and $x = 20$.
$124 = k(20)$
$\frac{124}{20} = \frac{k(20)}{20}$
$6.2 = k$
So, $y = 6.2x$.
If $x = 28$,
$y = 6.2(28) = 173.6$
A 28-pound child would need a dose of 173.6 mg.

REVIEW

69. For a horizontal line, $m = 0$. Let $x_1 = -3$ and $y_1 = 6$.
$y - y_1 = m(x - x_1)$
$y - (6) = 0(x - (-3))$
$y - 6 = 0$
$y - 6 + 6 = 0 + 6$
$y = 6$

71. profit = revenue − cost
$p = r - c$

73. $-3(2x - 4) = -3(2x) - (-3)(4)$
$= -6x - (-12)$
$= -6x + 12$

75. There are 12 eggs per each dozen. Let $d =$ be the number of dozens of eggs. $12d$ is the number of eggs in d dozens.

CHAPTER 3 REVIEW

1. a)

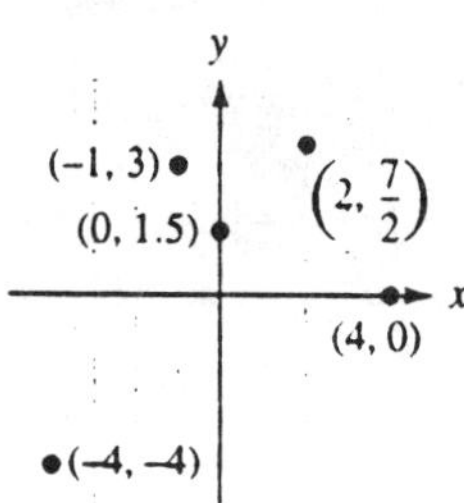

b)

x	y
3	-1
0	0
-3	1

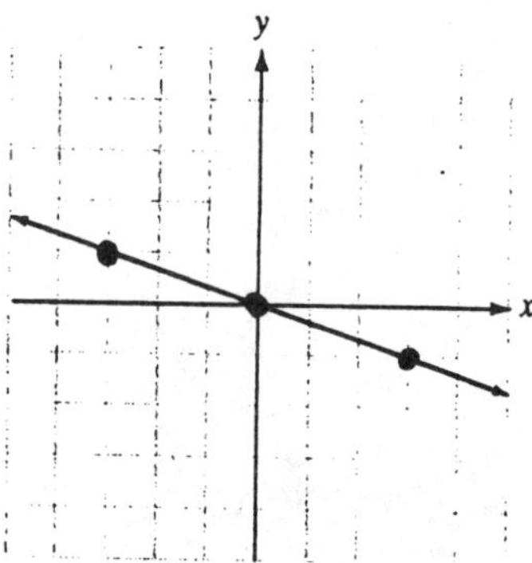

3. a) 2 ft.
b) $3 - 2 = 1$ ft.
c) 6 ft.

5. $y = |2 + x|$
$(5) = |2 + (-3)|$
$5 = |-1|$
$5 = 1$
$(-3, 5)$ is not a solution.

7. a) 9,000
b) It tells us that 40 trees on an acre gives the highest yield of 18,000 oranges.

9. General form of a line is $Ax + By = C$.
So, $A = 5$, $B = 2$, and $C = 10$.

11. $x + 2y = 6$
$x - x + 2y = -x + 6$
$2y = -x + 6$
$\frac{2y}{2} = \frac{-x+6}{2}$
$y = \frac{-x}{2} + \frac{6}{2}$
$y = -\frac{1}{2}x + 3$

If $x = -2$,
$y = -\frac{1}{2}(-2) + 3$
$y = 1 + 3$
$y = 4$ $(-2, 4)$

If $x = 0$,
$y = -\frac{1}{2}(0) + 3$
$y = 0 + 3$
$y = 3$ $(0, 3)$

If $x = 2$,
$y = -\frac{1}{2}(2) + 3$
$y = -1 + 3$
$y = 2$ $(2, 2)$

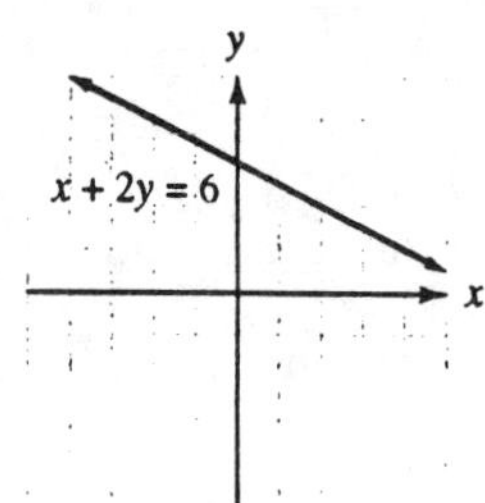

13. a) $y = 4$ is a horizontal line with y-intercept (0, 4).

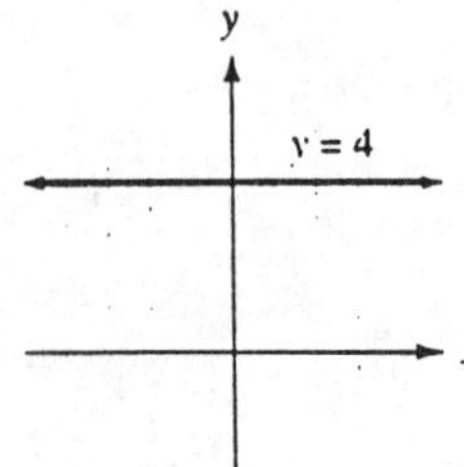

b) $x = -1$ is a vertical line with x-intercept $(-1, 0)$.

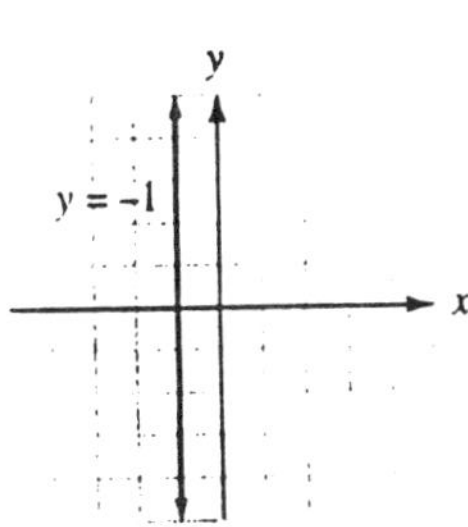

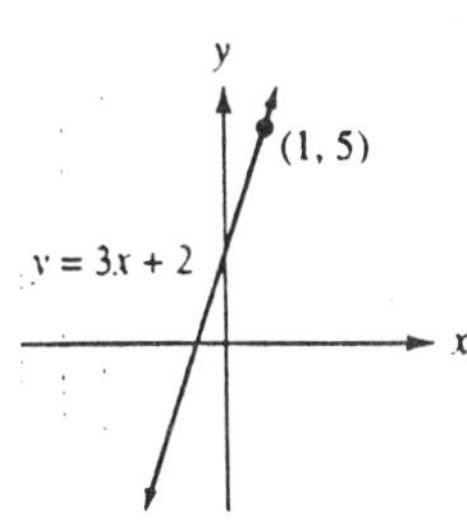

15.

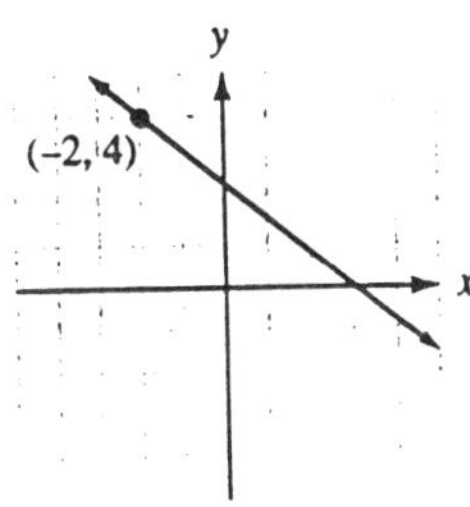

17. a) $y = \frac{3}{4}x - 2$
$m = \frac{3}{4}$ and y-intercept is $(0, -2)$

b) $y = -4x$
$y = -4x + 0$
$m = -4$ and y-intercept is $(0, 0)$

19. a) Let c = # of copies made and w = number of weeks.
Total # copies = 300 times # of weeks plus previous copies.
$c = 300w + 75,000$

b) If $w = 52$,
$c = 300(52) + 75,000$
$c = 15,600 + 75,000$
$c = 90,600$
At the end of the year, the machine would have made 90,600 copies.

21. a) $m = 3$, $x_1 = 1$, and $y_1 = 5$
$y - y_1 = m(x - x_1)$
$y - (5) = (3)(x - (1))$
$y - 5 = 3x - 3(1)$
$y - 5 = 3x - 3$
$y - 5 + 5 = 3x - 3 + 5$
$y = 3x + 2$

b) $m = -\frac{1}{2}$, $x_1 = -4$, and $y_1 = -1$
$y - y_1 = m(x - x_1)$
$y - (-1) = (-\frac{1}{2})(x - (-4))$
$y + 1 = -\frac{1}{2}x - (-\frac{1}{2})(-4)$
$y + 1 = -\frac{1}{2}x - 2$
$y + 1 - 1 = -\frac{1}{2}x - 2 - 1$
$y = -\frac{1}{2}x - 3$

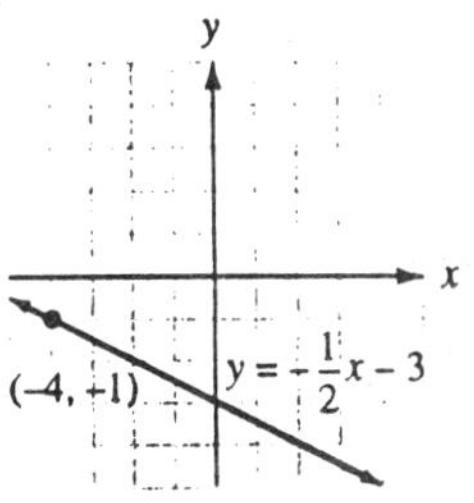

23. The relationship is (years, fees). We were given (2, 380) and (4, 310). In order to determine the equation of the line, need to find the slope.
Let $y_1 = 2$ $f_1 = 380$
$y_2 = 4$ $f_2 = 310$
$m = \frac{f_2 - f_1}{y_2 - y_1} = \frac{310 - 380}{4 - 2}$
$= \frac{-70}{2} = -35$
$f - f_1 = m(y - y_1)$
$f - (380) = (-35)(y - (2))$
$f - 380 = -35y - (-35)(2)$
$f - 380 = -35y - (-70)$
$f - 380 = -35y + 70$
$f - 380 + 380 = -35y + 70 + 380$
$f = -35y + 450$

25. a) Domain: all real numbers
Range: $y \leq 0$
b) Domain: all real numbers
Range: $y \geq 0$

27. $h(x) = 1 - |x|$
$h(0) = 1 - |0| = 1 - 0 = 1$
$h(1) = 1 - |1| = 1 - 1 = 0$
$h(2) = 1 - |2| = 1 - 2 = -1$
$h(-1) = 1 - |-1| = 1 - 1 = 0$
$h(-2) = 1 - |-2| = 1 - 2 = -1$
$h(-3) = 1 - |-3| = 1 - 3 = -2$

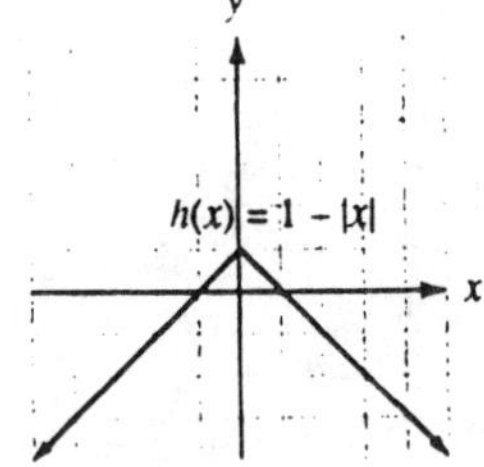

29. $f(r) = 15.7r^2$
$f(8) = 15.7(8)^2$
$= 15.7(64)$
$= 1004.8 \text{ in}^3$

STUDY SET Section 4.1

VOCABULARY

1. system

3. independent

5. inconsistent

CONCEPTS

7. True, since A lies on l_1.

9. False, since A does not lie on l_2.

11. True, since C lies on l_1.

13. a) \$200. Look at the y-value of the cost graph when $x = 30$.
b) \$1200. Look at the y-value of the revenue graph when $x = 30$.
c) Lose \$800 since cost is \$800 more than revenue.

15. One solution; consistent since it has a solution.

17. The method is not accurate enough to find a solution such as $(\frac{3}{5}, -\frac{1}{3})$.

NOTATION

19.
$$\frac{1}{6}x - \frac{1}{3}y = \frac{11}{2}$$
$$6(\frac{1}{6}x - \frac{1}{3}y) = 6(\frac{11}{2})$$
$$6(\frac{1}{6}x) - 6(\frac{1}{3}y) = 6(\frac{11}{2})$$
$$x - 2y = 33$$

PRACTICE

21. Yes, substitute $x = 1$, $y = 1$ into equations:
$$x + y = 2$$
$$(1) + (1) = 2$$
$$2 = 2 \qquad \text{True.}$$

$$2x - y = 1$$
$$2(1) - (1) = 1$$
$$2 - 1 = 1$$
$$1 = 1 \qquad \text{True.}$$

23. Yes, substitute $x = 3$, $y = -2$ into equations:
$$2x + y = 4$$
$$2(3) + (-2) = 4$$
$$6 - 2 = 4$$
$$4 = 4 \qquad \text{True.}$$

$$y = 1 - x$$
$$(-2) = 1 - (3)$$
$$-2 = -2 \qquad \text{True.}$$

25. No, substitute $x = -2$, $y = -4$ into equations:
$$4x + 5y = -23$$
$$4(-2) + 5(-4) = -23$$
$$-8 - 20 = -23$$
$$-28 = -23 \qquad \text{False.}$$

$$-3x + 2y = 0$$
$$2(1) - (1) = 1$$
$$2 - 1 = 1$$
$$1 = 1 \qquad \text{True.}$$

27. No, substitute $x = \frac{1}{2}$, $y = 4$ into equations:
$$2x + y = 4$$
$$2(\frac{1}{2}) + (4) = 4$$
$$1 + 4 = 4$$
$$5 = 4 \qquad \text{False.}$$

$$4x - 11 = 3y$$
$$4(\frac{1}{2}) - 11 = 3(4)$$
$$2 - 11 = 12$$
$$-9 = 12 \qquad \text{False.}$$

29. No, substitute $x = -\frac{2}{5}$, $y = \frac{1}{4}$ into equations:
$$x - 4y = -6$$
$$(-\frac{2}{5}) - 4(\frac{1}{4}) = -6$$
$$-\frac{2}{5} - 1 = -6$$
$$-\frac{2}{5} - \frac{5}{5} = -\frac{30}{5}$$
$$-\frac{7}{5} = -\frac{30}{5} \qquad \text{False.}$$

$8y = 10x + 12$
$8(\frac{1}{4}) = 10(-\frac{2}{5}) + 12$
$2 = -4 + 12$
$2 = 8$ False.

31. Yes, substitute $x = 0.2$, $y = 0.3$ into equations:

$20x + 10y = 7$
$20(0.2) + 10(0.3) = 7$
$4 + 3 = 7$
$7 = 7$ True.

$20y = 15x + 3$
$20(0.3) = 15(0.2) + 3$
$6 = 3 + 3$
$6 = 6$ True.

33. Solve each equation for y.
$y = -x + 2$
$y = x$
(1, 1) is the solution since it is the point that is on both graphs.

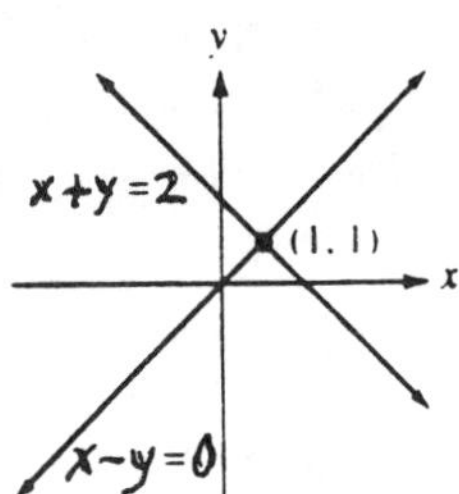

35. Solve each equation for y.
$y = -x + 2$
$y = x - 4$
(3, − 1) is the solution since it is the point that is on both graphs.

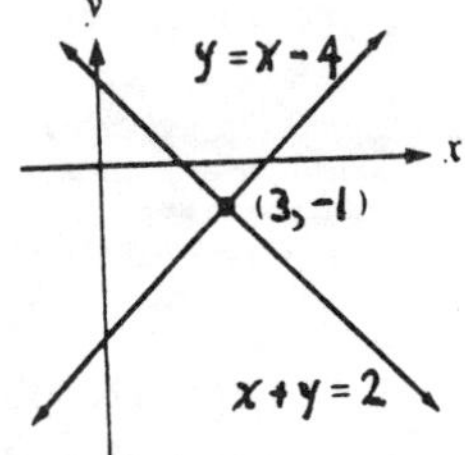

37. Solve each equation for y.
$$y = -\frac{3}{2}x + 4$$
$$y = \frac{2}{3}x + \frac{1}{3}$$
(− 2, − 1) is the solution since it is the point that is on both graphs.

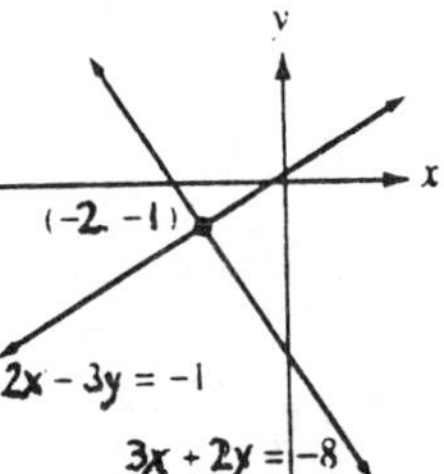

39. Solve each equation for y.
$y = 2x - 4$
$y = 2x - 4$
There are infinite number of solutions since the equations are dependent.

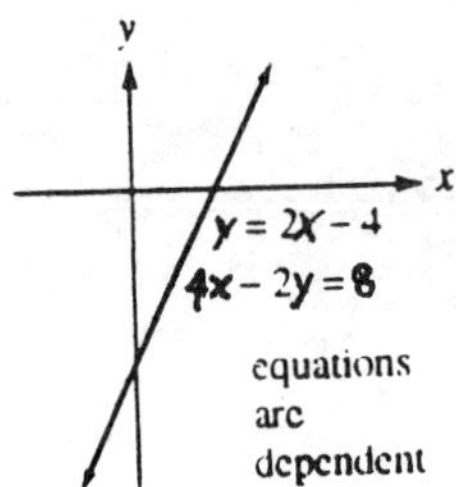

41. Solve each equation for y.
$$y = \frac{2}{3}x + 6$$
$$y = -\frac{3}{2}x - \frac{1}{2}$$
(− 3, 4) is the solution since it is the point that is on both graphs.

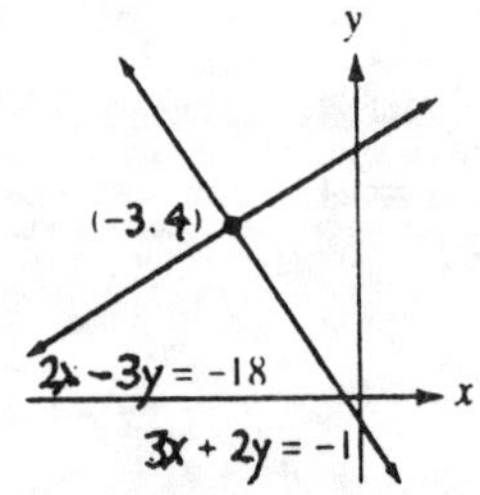

43. Solve the second equation for y.

$$x = 4$$
$$y = 6 - 2x$$

$(4, -2)$ is the solution since it is the point that is on both graphs.

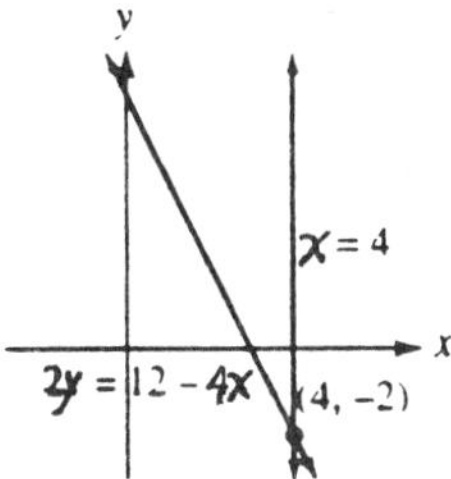

45. Solve each equation for y.

$$y = -\frac{1}{2}x - 2$$
$$y = 2x + 12$$

$(4, -4)$ is the solution since it is the point that is on both graphs.

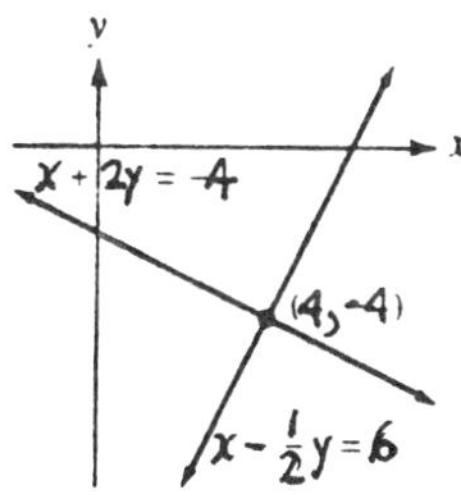

47. Solve each equation for y.

$$y = \frac{3}{4}x + 3$$
$$y = -\frac{1}{4}x - 1$$

$(-4, 0)$ is the solution since it is the point that is on both graphs.

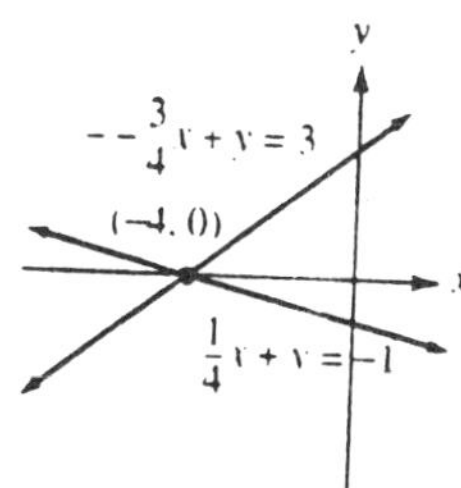

49. Solve each equation for y.

$$y = \frac{3}{2}x + 2$$
$$y = \frac{3}{2}x - 3$$

There are no solutions since the graphs are parallel.

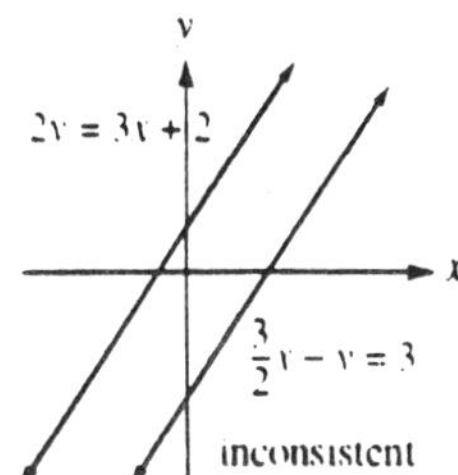

51. Solve each equation for y.

$$y = \frac{2}{3}x - \frac{1}{3}$$
$$y = -\frac{4}{5}x + \frac{13}{5}$$

$(2, 1)$ is the solution since it is the point that is on both graphs.

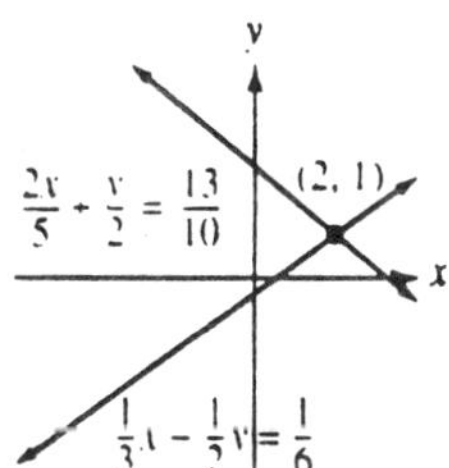

APPLICATIONS

53. a) Donors outnumbered those awaiting a transplant, since the graph of donors is higher than the graph of those awaiting for that year.

b) Look for the intersection of the graphs. the year is 1994; there were about 4,000 people waiting for transplants.

c) The number of people awaiting transplants is greater than the number of donors since the graph is higher than the donors graph in recent years..

55. a) Houston, New Orleans, and St. Augustine.
b) St. Louis, Memphis, and New Orleans
c) New Orleans

57. Solve each equation for y.

$$y = -\frac{1}{2}x + 3$$

$$y = \frac{2}{3}x + \frac{2}{3}$$

The graphs intersect at (2, 2). So, yes, there is a possibility of a collision at (2, 2).

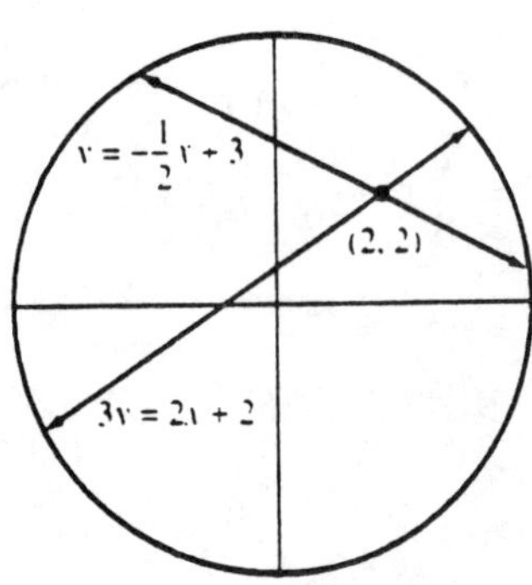

REVIEW

61. Recall, $y = mx + b$, m = slope and b = y-intercept.
slope = -3 and y-intercept = $(0, 4)$

63.
$$f(x) = -4x - x^2$$
$$f(3) = -4(3) - (3)^2$$
$$f(3) = -4(3) - 9$$
$$f(3) = -12 - 9$$
$$f(3) = -21$$

65. Since the y-axis is a vertical line with the x-intercept, (0, 0), the equation of the y-axis is $x = 0$.

67. Recall $y - y_1 = m(x - x_1)$ where x_1 and y_1 are the coordinates of the point and m is the slope. For $y - 2 = 7(x - 5)$, $x_1 = 5$ and $y_1 = 2$. The point is (5, 2).

STUDY SET Section 4.2

VOCABULARY

1. y; terms

3. remove

5. infinite

CONCEPTS

7. a) Each equation contains two variables.

b) $2x + 3y = 12$

$2x + 3(2x + 4) = 12$

The equation now contains one variable.

9. a) $x - 2y = -10$

$x - 2y + 2y = -10 + 2y$

$x = 2y - 10$

b) $x - 2y = -10$

$x - x - 2y = -10 - x$

$-2y = -10 - x$

$\frac{-2y}{-2} = \frac{-10-x}{-2}$

$y = \frac{-10}{-2} - \frac{x}{-2}$

$y = 5 + \frac{x}{2}$

$y = \frac{x}{2} + 5$

c) It was easier to solve for x, because it took only one step.

11. a) Parentheses are needed around $(x - 4)$ in the second line when the substitution $y = x - 4$ is made.

b) $x + 2y = 5$

$x + 2(x - 4) = 5$

$x + 2x - 8 = 5$

$3x - 8 = 5$

$3x - 8 + 8 = 5 + 8$

$3x = 13$

$\frac{3x}{3} = \frac{13}{3}$

$x = \frac{13}{3}$

13. a) Solve each equation for y.

$y = \frac{1}{2}x$

$y = -2x + \frac{5}{3}$

Graph the equations. The solution is difficult to determine because the coordinates are not integers.

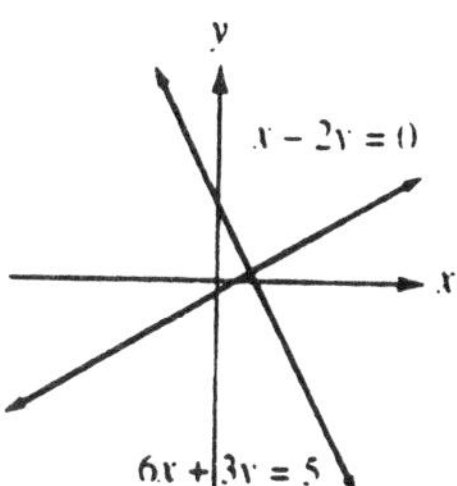

b) $x - 2y = 0$

$6x + 3y = 5$

Use the first equation, $x = 2y$, to substitute into the second equation:

$6(2y) + 3y = 5$

$12y + 3y = 5$

$15y = 5$

$\frac{15y}{15} = \frac{5}{15}$

$y = \frac{5}{15}$

$y = \frac{1}{3}$

Substitute to find x:

$x = 2(\frac{1}{3}) = \frac{2}{3}$

The solution is $\left(\frac{2}{3}, \frac{1}{3}\right)$.

NOTATION

15. $y = 3x$

$x - y = 4$

$x - y = 4$

$x - (3x) = 4$

$-2x = 4$

$\frac{-2x}{-2} = \frac{4}{-2}$

$x = -2$

$y = 3x$

$y = 3(-2)$

$y = -6$

The solution of the system is $(-2, -6)$.

PRACTICE

17. $y = 2x$

$x + y = 6$

Use the first equation, $y = 2x$, to substitute into the second equation:

$x + (2x) = 6$

$3x = 6$

$\frac{3x}{3} = \frac{6}{3}$

$x = 2$

Substitute to find y:

$y = 2(2) = 4$

The solution is $(4, 2)$.

19. $y = 2x - 6$

$2x + y = 6$

Use the first equation, $y = 2x - 6$, to substitute into the second equation:

$2x + (2x - 6) = 6$

$4x - 6 = 6$

$4x - 6 + 6 = 6 + 6$

$4x = 12$

$\frac{4x}{4} = \frac{12}{4}$

$x = 3$

Substitute to find y:

$y = 2(3) - 6 = 6 - 6 = 0$

The solution is $(3, 0)$.

21. $y = 2x + 5$

$x + 2y = -5$

Use the first equation, $y = 2x + 5$, to substitute into the second equation:

$x + 2(2x + 5) = -5$

$x + 4x + 10 = -5$

$5x + 10 = -5$

$5x + 10 - 10 = -5 - 10$

$5x = -15$

$\frac{5x}{5} = \frac{-15}{5}$

$x = -3$

Substitute to find y:

$y = 2(-3) + 5 = -6 + 5 = -1$

The solution is $(-3, -1)$.

23. $2a + 4b = -24$

$a = 20 - 2b$

Substitute the second equation into first equation:

$2(20 - 2b) + 4b = -24$

$40 - 4b + 4b = -24$

$40 = -24$

The system is inconsistent. No solution.

25. $2a = 3b - 13$

$-b = -2a - 7$

Solve the second equation for b.

$b = 2a + 7$.

Substitute it into the first equation:

$2a = 3(2a + 7) - 13$

$2a = 6a + 21 - 13$

$2a = 6a + 8$

$2a - 6a = 6a - 6a + 8$

$-4a = 8$

$\frac{-4a}{-4} = \frac{8}{-4}$

$a = -2$

Substitute to find y:

$b = 2(-2) + 7 = -4 + 7 = 3$

The solution is $(-2, 3)$.

27. $r + 3x = 9$

$3r + 2s = 13$

Solve the first equation for r.

$r + 3s - 3s = 9 - 3s$

$r = 9 - 3s$

Substitute it into the second equation:

$3(9 - 3s) + 2s = 13$

$27 - 9s + 2s = 13$

$27 - 7s = 13$

$27 - 27 - 7s = 13 - 27$

$-7s = -14$

$\frac{-7s}{-7} = \frac{-14}{-7}$

$s = 2$

Substitute to find r:

$r = 9 - 3(2) = 9 - 6 = 3$

The solution is $(3, 2)$.

29. $0.4x + 0.5y = 0.2$
$3x - y = 11$
Solve the second equation for y :
$3x - y = 11$
$3x - 3x - y = 11 - 3x$
$-y = 11 - 3x$
$y = -11 + 3x$
$y = 3x - 11$
Substitute it into the first equation:
$0.4x + 0.5(3x - 11) = 0.2$
$10(0.4x + 0.5(3x - 11)) = 10(0.2)$
$4x + 5(3x - 11) = 2$
$4x + 15x - 55 = 2$
$19x - 55 = 2$
$19x - 55 + 55 = 2 + 55$
$19x = 57$
$\frac{19x}{19} = \frac{57}{19}$
$x = 3$
Substitute to find y:
$y = 3(3) - 11 = 9 - 11 = -2$
The solution is $(3, -2)$.

31. $6x - 3y = 5$
$-2y - x = 0$
Solve the second equation for x :
$-2y - x = 0$
$-2y - x + x = 0 + x$
$-2y = x$
Substitute it into the first equation:
$6(-2y) - 3y = 5$
$-12y - 3y = 5$
$-15y = 5$
$\frac{-15y}{-15} = \frac{5}{-15}$
$y = -\frac{1}{3}$
Substitute to find x:
$x = -2(-\frac{1}{3}) = \frac{2}{3}$
The solution is $\left(\frac{2}{3}, -\frac{1}{3}\right)$.

33. $3x + 4y = -7$
$2y - x = -1$
Solve the second equation for x :
$2y - x = -1$
$2y - 2y - x = -2y - 1$
$-x = -2y - 1$
$x = 2y + 1$
Substitute it into the first equation:
$3(2y + 1) + 4y = -7$
$6y + 3 + 4y = -7$
$10y + 3 = -7$
$10y + 3 - 3 = -7 - 3$
$10y = -10$
$\frac{10y}{10} = \frac{-10}{10}$
$y = -1$
Substitute to find x:
$x = 2(-1) + 1 = -2 + 1 = -1$
The solution is $(-1, -1)$.

35. $9x = 3y + 12$
$4 = 3x - y$
Solve the second equation for y :
$4 = 3x - y$
$4 - 3x = 3x - 3x - y$
$4 - 3x = -y$
$-3x + 4 = -y$
$3x - 4 = y$
Substitute it into the first equation:
$9x = 3(3x - 4) + 12$
$9x = 9x - 12 + 12$
$9x = 9x$
$9x - 9x = 9x - 9x$
$0 = 0$
The equations are dependent. Infinite number of solutions.

37. $0.02x + 0.05y = -0.02$
$4x + 3y = 10$
Solve the second equation for x :
$4x + 3y = 10$
$4x + 3y - 3y = 10 - 3y$
$4x = 10 - 3y$
$\frac{4x}{4} = \frac{10-3y}{4}$
$x = \frac{10}{4} - \frac{3y}{4}$
$x = \frac{5}{2} - \frac{3}{4}y$

Substitute it into the first equation:

$$0.02\left(\frac{5}{2} - \frac{3}{4}y\right) + 0.05y = -0.02$$
$$100\left(0.02\left(\frac{5}{2} - \frac{3}{4}y\right) + 0.05\right) = 100(-0.02)$$
$$2\left(\frac{5}{2} - \frac{3}{4}y\right) + 5y = -2$$
$$5 - \frac{3}{2}y + 5y = -2$$
$$5 - \frac{3}{2}y + \frac{10}{2}y = -2$$
$$5 + \frac{7}{2}y = -2$$
$$5 - 5 + \frac{7}{2}y = -2 - 5$$
$$\frac{7}{2}y = -7$$
$$2(\tfrac{7}{2}y) = 2(-7)$$
$$7y = -14$$
$$\frac{7y}{7} = \frac{-14}{7}$$
$$y = -2$$

Substitute to find x:

$$y = \frac{5}{2} - \frac{3}{4}(-2) = \frac{5}{2} + \frac{3}{2} = \frac{8}{2} = 4$$

The solution is $(4, -2)$.

39. $2a + 3b = 2$
$8a - 3b = 3$

Solve the first equation for b.

$$2a + 3b = 2$$
$$2a - 2a + 3b = 2 - 2a$$
$$3b = 2 - 2a$$
$$\frac{3b}{3} = \frac{2-2a}{3}$$
$$b = \frac{2-2a}{3}$$

Substitute it into the second equation:

$$8a - 3\left(\frac{2-2a}{3}\right) = 3$$
$$8a - (2 - 2a) = 3$$
$$8a - 2 + 2a = 3$$
$$10a - 2 = 3$$
$$10a - 2 + 2 = 3 + 2$$
$$10a = 5$$
$$\frac{10a}{10} = \frac{5}{10}$$
$$a = \frac{1}{2}$$

Substitute to find b:

$$b = \frac{2-2(\frac{1}{2})}{3} = \frac{2-1}{3} = \frac{1}{3}$$

The solution is $\left(\frac{1}{2}, \frac{1}{3}\right)$.

41. $y - x = 3x$
$2(x + y) = 14 - y$

Solve the first equation for y :

$$y - x = 3x$$
$$y - x + x = 3x + x$$
$$y = 4x$$

Substitute it into the second equation:

$$2(x + (4x)) = 14 - (4x)$$
$$2(5x) = 14 - 4x$$
$$10x = 14 - 4x$$
$$10x + 4x = 14 - 4x + 4x$$
$$14x = 14$$
$$\frac{14x}{14} = \frac{14}{14}$$
$$x = 1$$

Substitute to find y:

$$y = 4(1) = 4$$

The solution is $(1, 4)$.

43. $3(x - 1) + 3 = 8 + 2y$
$2(x + 1) = 4 + 3y$

Solve the first equation for x :

$$3(x - 1) + 3 = 8 + 2y$$
$$3x - 3 + 3 = 8 + 2y$$
$$3x = 8 + 2y$$
$$x = \frac{8+2y}{3}$$

Substitute it into the second equation:

$$2\left(\left(\frac{8+2y}{3}\right) + 1\right) = 4 + 3y$$
$$2\left(\frac{8+2y}{3} + \frac{3}{3}\right) = 4 + 3y$$
$$2\left(\frac{8+2y+3}{3}\right) = 4 + 3y$$
$$2\left(\frac{11+2y}{3}\right) = 4 + 3y$$
$$\frac{22+4y}{3} = 4 + 3y$$
$$3\left(\frac{22+4y}{3}\right) = 3(4 + 3y)$$
$$22 + 4y = 12 + 9y$$
$$22 - 12 + 4y - 4y = 12 - 12 + 9y - 4y$$
$$10 = 5y$$
$$\frac{10}{5} = \frac{5y}{5}$$
$$2 = y$$

Substitute to find x:

$$x = \frac{8+2(2)}{3} = \frac{8+4}{3} = \frac{12}{3} = 4$$

The solution is $(4, 2)$.

45. $\frac{1}{2}x + \frac{1}{2}y = -1$

$\frac{1}{3}x - \frac{1}{2}y = -4$

Solve the first equation for y :

$\frac{1}{2}x + \frac{1}{2}y = -1$

$2(\frac{1}{2}x + \frac{1}{2}y) = 2(-1)$

$x + y = -2$

$x - x + y = -2 - x$

$y = -2 - x$

Substitute it into the second equation:

$\frac{1}{3}x - \frac{1}{2}(-2 - x) = -4$

$\frac{1}{3}x + 1 + \frac{1}{2}x = -4$

$6(\frac{1}{3}x + 1 + \frac{1}{2}x) = 6(-4)$

$2x + 6 + 3x = -24$

$5x + 6 = -24$

$5x + 6 - 6 = -24 - 6$

$5x = -30$

$\frac{5x}{5} = \frac{-30}{5}$

$x = -6$

Substitute to find y:

$y = -2 - (-6) = -2 + 6 = 4$

The solution is $(-6, 4)$.

47. $5x = \frac{1}{2}y - 1$

$\frac{1}{4}y = 10x - 1$

Solve the second equation for y :

$\frac{1}{4}y = 10x - 1$

$4(\frac{1}{4}y) = 4(10x - 1)$

$y = 40x - 4$

Substitute it into the first equation:

$5x = \frac{1}{2}(40x - 4) - 1$

$5x = 20x - 2 - 1$

$5x = 20x - 3$

$5x - 20x = 20x - 20x - 3$

$-15x = -3$

$\frac{-15x}{-15} = \frac{-3}{-15}$

$x = \frac{-3}{-15}$

$x = \frac{1}{5}$

Substitute to find y:

$y = 40(\frac{1}{5}) - 4 = 8 - 4 = 4$

The solution is $\left(\frac{1}{5}, 4\right)$.

49. $\frac{6x-1}{3} - \frac{5}{3} = \frac{3y+1}{2}$

$\frac{1+5y}{4} + \frac{x+3}{4} = \frac{17}{2}$

Solve the first equation for y :

$6\left(\frac{6x-1}{3} - \frac{5}{3}\right) = 6\left(\frac{3y+1}{2}\right)$

$2(6x-1) - 2(5) = 3(3y+1)$

$12x - 2 - 10 = 9y + 3$

$12x - 12 = 9y + 3$

$12x - 12 - 3 = 9y + 3 - 3$

$12x - 15 = 9y$

$\frac{12x-15}{9} = \frac{9y}{9}$

$\frac{12x-15}{9} = y$

$\frac{4x-5}{3} = y$

Before substituting, we need to simplify.

$\frac{1+5y}{4} + \frac{x+3}{4} = \frac{17}{2}$

$4\left(\frac{1+5y}{4} + \frac{x+3}{4}\right) = 4\left(\frac{17}{2}\right)$

$1 + 5y + x + 3 = 34$

$5y + x + 4 = 34$

Now, substitute into the first equation:

$5\left(\frac{4x-5}{3}\right) + x + 4 = 34$

$\frac{20x-25}{3} + x + 4 = 34$

$\frac{20x-25}{3} + x + 4 - 4 = 34 - 4$

$\frac{20x-25}{3} + x = 30$

$\frac{20x-25}{3} + \frac{3x}{3} = 30$

$\frac{20x-25+3x}{3} = 30$

$\frac{23x-25}{3} = 30$

$3\left(\frac{23x-25}{3}\right) = 3(30)$

$23x - 25 = 90$

$23x - 25 + 25 = 90 + 25$

$23x = 115$

$\frac{23x}{23} = \frac{115}{23}$

$x = 5$

Substitute to find y:

$y = \frac{4(5)-5}{3} = \frac{20-5}{3} = \frac{15}{3} = 5$

The solution is $(5, 5)$.

APPLICATIONS

51. Melon, because it's the same price as hash browns.

REVIEW

57. Recall, $y = mx + b$, $m =$ slope and $b = y$-intercept.
slope $= -\frac{5}{8}$ and
y-intercept $= (0, -12)$

59. Solve for y.

$$2x - 3y = 18$$
$$2x - 2x - 3y = -2x + 18$$
$$-3y = -2x + 18$$
$$\frac{-3y}{-3} = \frac{-2x+18}{-3}$$
$$y = \frac{-2x}{-3} + \frac{18}{-3}$$
$$y = \frac{2}{3}x - 6$$

y-intercept is $(0, -6)$.

61. No, since the circle does not pass the vertical line test. The graph of a function must pass the vertical line test.

63. $s = kt$

STUDY SET Section 4.3

VOCABULARY

1. coefficient

3. general

CONCEPTS

5. Like terms are not aligned. Needs to be

$$\begin{aligned} 2x - 5y &= -3 \\ 3x - 2y &= 10 \end{aligned}$$

7. Multiply the equation by 15 which is the LCM of 3 & 5.

9. a) Solve both equations for y.

$y = -2x + 1$

$y = \frac{3}{2} - 6$

From the graph, the intersection is (2, -3).

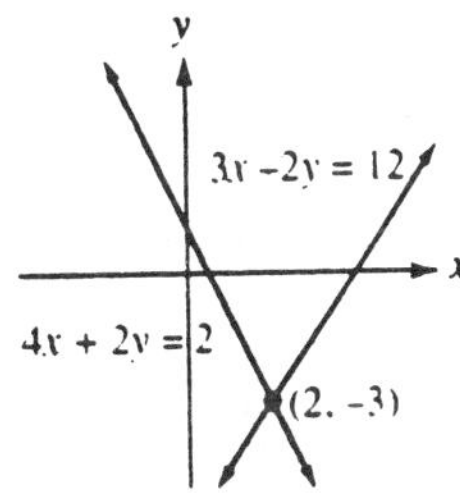

b) $4x + 2y = 2$

$3x - 2y = 12$

Solve the first equation for x :

$4x + 2y = 2$

$4x - 4x + 2y = -4x + 2$

$2y = -4x + 2$

$\frac{2y}{2} = \frac{-4x+2}{2}$

$y = -2x + 1$

Substitute it into the second equation:

$3x - 2(-2x + 1) = 12$

$3x + 4x - 2 = 12$

$7x - 2 = 12$

$7x - 2 + 2 = 12 + 2$

$7x = 14$

$\frac{7x}{7} = \frac{14}{7}$

$x = 2$

Substitute to find y:

$y = -2(2) + 1 = -4 + 1 = -3$

The solution is $(2, -3)$.

c) $4x + 2y = 2$

$3x - 2y = 12$

$7x \quad = 14$

$\frac{7x}{7} = \frac{14}{7}$

$x = 2$

To get y, we substitute 2 into either equation:

$4(2) + 2y = 2$

$8 + 2y = 2$

$8 - 8 + 2y = 2 - 8$

$2y = -6$

$\frac{2y}{2} = \frac{-6}{2}$

$y = -3$

The solution is (2, -3).

NOTATION

11. $x + y = 5$

$x - y = -3$

$2x \quad = 2$

$\frac{2x}{2} = \frac{2}{2}$

$x = 1$

$(1) + y = 5$

$1 - 1 + y = 5 - 1$

$y = 4$

The solution is (1, 4)

PRACTICE

13. $x - y = -5$

$x + y = 1$

$2x \quad = 6$

$\frac{2x}{2} = \frac{6}{2}$

$x = 3$

$(3) + y = 1$

$3 - 3 + y = 1 - 3$

$y = -2$

The solution is (3, -2)

15. $$\begin{aligned} 2r + s &= \text{-}1 \\ \text{-}2r + s &= 3 \\ \hline 2s &= 2 \end{aligned}$$
$\frac{2s}{2} = \frac{2}{2}$
$s = 1$
$2r + x = -1$
$2r + (1) = -1$
$2r + 1 - 1 = -1 - 1$
$2r = -2$
$\frac{2r}{2} = \frac{-2}{2}$
$r = -1$
The solution is (− 1, 1)

17. $$\begin{aligned} 2x + y &= \text{-}2 \\ \text{-}2x - 3y &= \text{-}6 \\ \hline -2y &= 8 \end{aligned}$$
$\frac{-2y}{-2} = \frac{-8}{-2}$
$y = 4$
$2x + y = -2$
$2x + (4) = -2$
$2x + 4 - 4 = -2 - 4$
$2x = -6$
$\frac{2x}{2} = \frac{-6}{2}$
$x = -3$
The solution is (− 3, 4)

19. $$\begin{aligned} 4x + 3y &= 24 \\ 4x - 3y &= \text{-}24 \\ \hline 8x \quad &= 0 \end{aligned}$$
$\frac{8x}{8} = \frac{0}{8}$
$x = 0$
$4x + 3y = 24$
$4(0) + 3y = 24$
$0 + 3y = 24$
$3y = 24$
$\frac{3y}{3} = \frac{24}{3}$
$x = 8$
The solution is (0, 8)

21. $$\begin{aligned} x + y &= 5 \\ x + 2y &= 8 \\ \hline \end{aligned}$$

Multiply both sides of the first equation by -1.
$$\begin{aligned} \text{-}x - y &= \text{-}5 \\ x + 2y &= 8 \\ \hline y &= 3 \end{aligned}$$
$x + 2y = 8$
$x + 2(3) = -5$
$x + 6 = 8$
$x + 6 - 6 = 8 - 6$
$x = 2$
The solution is (2, 3).

23. $$\begin{aligned} 2x + y &= 4 \\ 2x + 3y &= 0 \\ \hline \end{aligned}$$

Multiply both sides of the first equation by -1.
$$\begin{aligned} \text{-}2x - y &= \text{-}4 \\ 2x + 3y &= 0 \\ \hline 2y &= \text{-}4 \end{aligned}$$
$\frac{2y}{2} = \frac{-4}{2}$
$y = -2$
$2x + y = 4$
$2x + (-2) = 4$
$2x - 2 = 4$
$2x - 2 + 2 = 4 + 2$
$2x = 6$
$\frac{2x}{2} = \frac{6}{2}$
$x = 3$
The solution is (3, − 2).

25. $3x - 5y = -29$
$3x + 4y = 34$

Multiply both sides of the first equation by -1.

$-3x + 5y = 29$
$3x + 4y = 34$
$9y = 63$
$\frac{9y}{9} = \frac{63}{9}$
$y = 7$
$3x - 5y = -29$
$3x - 5(7) = -29$
$3x - 35 = -29$
$3x - 35 + 35 = -29 + 35$
$3x = 6$
$\frac{3x}{3} = \frac{6}{3}$
$x = 2$

The solution is (2, 7).

27. $2a - 3b = -6$
$2(a + 4) = 3b$

First, we put the second equation into the correct form:

$2(a + 4) = 3b$
$2(a + 4) = 3b$
$2a + 8 = 3b$
$2a - 3b + 8 - 8 = 3b - 3b - 8$
$2a - 3b = -8$

Multiply both sides of this equation by -1.

$-2a + 3b = 8$

We have

$2a - 3b = -6$
$-2a + 3b = 8$
$0 = 2$

There is no solution. The system is inconsistent.

29. $-2(x + 1) = 3(y - 2)$
$3(y + 2) = 6 - 2(x - 2)$

We first work with each equation separately to get them in the proper form.

First equation:

$-2(x + 1) = 3(y - 2)$
$-2x - 2 = 3y - 6$
$-2x - 2 + 2 = 3y - 6 + 2$
$-2x = 3y - 4$
$-2x - 3y = 3y - 3y - 4$
$-2x - 3y = -4$

Second equation:

$3(y + 2) = 6 - 2(x - 2)$
$3y + 6 = 6 - 2x + 4$
$3y + 6 = -2x + 10$
$3y + 6 - 6 = -2x + 10 - 6$
$3y = -2x + 4$
$2x + 3y = -2x + 2x + 4$
$2x + 3y = 4$

So, we have the system

$-2x - 3y = -4$
$2x + 3y = 4$
$0 = 0$

The equations are dependent. There are infinitely many solutions.

31. $4(x + 1) = 17 - 3(y - 1)$
$2(x + 2) + 3(y - 1) = 9$

We first work with each equation separately to get them in the proper form.

First equation:

$4(x + 1) = 17 - 3(y - 1)$
$4x + 4 = 17 - 3y + 3$
$4x + 4 = 20 - 3y$
$4x + 3y + 4 = 20 - 3y + 3y$
$4x + 3y + 4 = 20$
$4x + 3y + 4 - 4 = 20 - 4$
$4x + 3y = 16$

Second equation:

$2(x + 2) + 3(y - 1) = 9$
$2x + 4 + 3y - 3 = 9$
$2x + 3y + 1 = 9$
$2x + 3y + 1 - 1 = 9 - 1$
$2x + 3y = 8$

So, we have the system

$4x + 3y = 16$
$2x + 3y = 8$

Multiply both sides of the second equation by -2.

$$4x + 3y = 16$$
$$-3x - 6y = -16$$
$$-3y = 0$$
$$\frac{-3y}{-3} = \frac{0}{-3}$$
$$y = 0$$
$$4x + 3y = 16$$
$$4x + 3(0) = 16$$
$$4x + 0 = 16$$
$$4x = 16$$
$$\frac{4x}{4} = \frac{16}{4}$$
$$x = 4$$

The solution is $(4, 0)$.

33. $2x + y = 10$
$0.1x + 0.2y = 1.0$

Multiply both sides of the second equation by -20.

$$2x + y = 10$$
$$-2x - 4y = -20$$
$$-3y = -10$$
$$\frac{-3y}{-3} = \frac{-10}{-3}$$
$$y = \frac{10}{3}$$
$$2x + y = 10$$
$$2x + (\tfrac{10}{3}) = 10$$
$$2x + \frac{10}{3} - \frac{10}{3} = 10 - \frac{10}{3}$$
$$2x = 10 - \frac{10}{3}$$
$$2x = \frac{30}{3} - \frac{10}{3}$$
$$2x = \frac{20}{3}$$
$$\frac{2x}{2} = \frac{20}{3} \div 2$$
$$x = \frac{10}{3}$$

The solution is $(\frac{10}{3}, \frac{10}{3})$.

35. $2x - y = 16$
$0.03x + 0.02y = 0.03$

Multiply both sides of the second equation by 100 and the first equation by 2.

$$4x - 2y = 32$$
$$3x + 2y = 3$$
$$7x = 35$$
$$\frac{7x}{7} = \frac{35}{7}$$
$$x = 5$$
$$3(5) + 2y = 3$$
$$15 + 2y = 3$$
$$15 - 15 + 2y = 3 - 15$$
$$2y = -12$$
$$\frac{2y}{2} = \frac{-12}{2}$$
$$y = -6$$

The solution is $(5, -6)$.

37. $6x + 3y = 0$
$5y = 2x + 12$

First, we put the second equation into the correct form:

$$5y = 2x + 12$$
$$-2x + 5y = 2x - 2x + 12$$
$$-2x + 5y = 12$$

Multiply both sides of this equation by 3.

$$-6x + 15y = 36$$

We have

$$6x + 3y = 0$$
$$-6x + 15y = 36$$
$$18y = 36$$
$$\frac{18y}{18} = \frac{36}{18}$$
$$y = 2$$
$$6x + 3y = 0$$
$$6x + 3(2) = 0$$
$$6x + 6 = 0$$
$$6x + 6 - 6 = 0 - 6$$
$$6x = -6$$
$$\frac{6x}{6} = \frac{-6}{6}$$
$$y = -1$$

The solution is $(-1, 2)$.

39. $8x - 4y = 18$
$3x - 2y = 8$

Multiply both sides of the second equation by -2.

$8x - 4y = 18$
$-6x + 4y = -16$
$2x \quad = 2$
$\frac{2x}{2} = \frac{2}{2}$
$x = 1$
$8(1) - 4y = 18$
$8 - 4y = 18$
$8 - 8 - 4y = 18 - 8$
$-4y = 10$
$\frac{-4y}{-4} = \frac{10}{-4}$
$y = -\frac{5}{2}$

The solution is $(1, -\frac{5}{2})$.

41. $\frac{3}{5}s + \frac{4}{5}t = 1$
$-\frac{1}{4}s + \frac{3}{8}t = 1$

Multiply both sides of the first equation by 5 and the second equation by 8 to clear the fractions.

$5(\frac{3}{5}s + \frac{4}{5}t) = 5(1)$
$3s + 4t = 5$
$8(-\frac{1}{4}s + \frac{3}{8}t) = 8(1)$
$-2s + 3t = 8$

Now, we have

$3s + 4t = 5$
$-2s + 3t = 8$

Multiply both sides of the first equation by 2 and the second equation by 3 to set up the equations.

$6s + 8t = 10$
$-6s + 9t = 24$
$17t = 34$
$\frac{17t}{17} = \frac{34}{17}$
$t = 2$
$6s + 8t = 10$
$6s + 8(2) = 10$
$6s + 16 = 10$
$6s + 16 - 16 = 10 - 16$
$6s = -6$
$\frac{6s}{6} = \frac{-6}{6}$
$s = -1$

The solution is $(-1, 2)$.

43. $\frac{3}{5}x + y = 1$
$\frac{4}{5}x - y = -1$

Multiply both sides of the first and second equations by 5 to clear the fractions.

$5(\frac{3}{5}x + y) = 5(1)$
$3x + 5y = 5$
$5(\frac{4}{5}x - y) = 5(-1)$
$4x - 5y = -5$

Now, we have

$3x + 5y = 5$
$4x - 5y = -5$
$7x \quad = 0$
$\frac{7x}{7} = \frac{0}{7}$
$x = 0$
$3x + 5y = 5$
$3(0) + 5y = 5$
$0 + 5y = 5$
$5y = 5$
$\frac{5y}{5} = \frac{5}{5}$
$y = 1$

The solution is $(0, 1)$.

45. $\frac{x}{2} - \frac{y}{3} = -2$
$\frac{2x-3}{2} + \frac{6y+1}{3} = \frac{17}{6}$

Multiply both sides of the first and second equations by 6 to clear the fractions and put in proper form.

$6(\frac{x}{2} - \frac{y}{3}) = 6(-2)$
$3x - 2y = -12$
$6\left(\frac{2x-3}{2} + \frac{6y+1}{3}\right) = 6\left(\frac{17}{6}\right)$
$3(2x-3) + 2(6y+1) = 17$
$6x - 9 + 12y + 2 = 17$
$6x + 12y - 7 = 17$

$6x + 12y - 7 + 7 = 17 + 7$
$6x + 12y = 24$
Now, we have
$3x - 2y = -12$
$6x + 12y = 24$

Multiply both sides of the first equation by -2 to set up the equations.
$-6x + 4y = 24$
$6x + 12y = 24$
$16y = 48$
$\frac{16y}{16} = \frac{48}{16}$
$y = 3$
$3x - 2y = -12$
$3x - 2(3) = -12$
$3x - 6 = -12$
$3x - 6 + 6 = -12 + 6$
$3x = -6$
$\frac{3x}{3} = \frac{-6}{3}$
$x = -2$
The solution is $(-2, 3)$.

47. $\frac{x-3}{2} + \frac{y+5}{3} = \frac{11}{6}$
$\frac{x+3}{3} - \frac{5}{12} = \frac{y+3}{4}$

Multiply both sides of the first equation by 6 and the second equation by 12 to clear the fractions and put in proper form.
$6\left(\frac{x-3}{2} + \frac{y+5}{3}\right) = 6\left(\frac{11}{6}\right)$
$3(x-3) + 2(y+5) = 11$
$3x - 9 + 2y + 10 = 11$
$3x + 2y + 1 = 11$
$3x + 2y + 1 - 1 = 11 - 1$
$3x + 2y = 10$
$12\left(\frac{x+3}{3} - \frac{5}{12}\right) = 12\left(\frac{y+3}{4}\right)$
$4(x+3) - 5 = 3(y+3)$
$4x + 12 - 5 = 3y + 9$
$4x + 7 = 3y + 9$
$4x - 3y + 7 - 7 = 3y - 3y + 9 - 7$
$4x - 3y = 2$
Now, we have
$3x + 2y = 10$
$4x - 3y = 2$

Multiply both sides of the first equation by 3 and the second equation by 2 to set up the equations.
$9x + 6y = 30$
$8x - 6y = 4$
$17x = 34$
$\frac{17x}{17} = \frac{34}{17}$
$x = 2$
$3x + 2y = 10$
$3(2) + 2y = 10$
$6 + 2y = 10$
$6 - 6 + 2y = 10 - 6$
$2y = 4$
$\frac{2y}{2} = \frac{4}{2}$
$x = 2$
The solution is $(2, 2)$.

REVIEW

51. $8(3x - 5) - 12 = 4(2x + 3)$
$24x - 40 - 12 = 8x + 12$
$24x - 52 = 8x + 12$
$24x - 8x - 52 = 8x - 8x + 12$
$16x - 52 = 12$
$16x - 52 + 52 = 12 + 52$
$16x = 64$
$\frac{16x}{16} = \frac{64}{16}$
$x = 4$

53. $x - x = 0$

55. The formula for the area of a triangle is $A = \frac{1}{2}bh$. We have $b = 4$ ft. and $h = 3.75$ ft.
$A = \frac{1}{2}(4)(3.75)$
$A = 2(3.75)$
$A = 7.5 \text{ ft}^2$

57. Key word: less
Translation: subtract
$x - 10$

STUDY SET Section 4.4
VOCABULARY

1. variable

3. system

CONCEPTS

5. speed of canoe upstream
$= $ speed in still water $-$ speed of current
$s = x - c$
speed of canoe downstream
$= $ speed in still water $+$ speed of current
$s = x + c$

7. a) $\$y$ was deposited in USA Savings.
b) The City Bank account earned 5% interest.
c) For City Bank,
$x(0.05)(1) = 0.05x$
For USA Savings,
$y(0.11)(1) = 0.11y$

9. a) Two equations are needed to solve a system with two variables.
b) Graphing, Substitution, or Addition

NOTATION

11. Area of a rectangle $=$ length • width
$A = l \cdot w$

13. Distance traveled $=$ rate • time
$d = rt$

15. Let $l =$ length of pool and
$w =$ width of pool.
$2l + 2w = 90$

PRACTICE

17. Let x denote the smaller integer and y denote the larger integer.
The first sentence gives us $y = 2x$.
The second sentence gives us
$x + y = 96$.
We have $x + y = 96$
$2x - y = 0$
$3x \quad = 96$
$\frac{3x}{3} = \frac{96}{3}$
$x = 32$
$y = 2x$
$y = 2(32)$
$y = 64$
The integers are 32 and 64.

19. Let x be one integer and y be the other integer. The first sentence tells us
$3x + y = 29$.
The second sentence tells us
$x + 2y = 18$.
Multiply the first equation by -2:
$-2(3x + y) = -2(29)$
$-6x - 2y = -58$
Now, we have
$-6x - 2y = -58$
$x + 2y = 18$
$-5x = -40$
$\frac{-5x}{-5} = \frac{-40}{-5}$
$x = 8$
$(8) + 2y = 18$
$8 - 8 + 2y = 18 - 8$
$2y = 10$
$\frac{2y}{2} = \frac{10}{2}$
$y = 5$
The two integers are 8 and 5.

APPLICATIONS

21. **Analyze:** We know that if we add the upper part and lower part we get 51 and that the upper part is 7 feet shorter than the lower part.
Form: Let $u =$ length of upper part and $l =$ length of lower part.
lower part $+$ upper part $= 51$
$l + u = 51$
upper part $=$ lower part $- 7$
$u = l - 7$
Solve: We have $l + u = 51$
$u = l - 7$
Substitute the second equation into the first:
$l + u = 51$
$l + (l - 7) = 51$
$2l - 7 = 51$
$2l - 7 + 7 = 51 + 7$

$2l = 58$
$\frac{2l}{2} = \frac{58}{2}$
$l = 29$
$u = l - 7$
$u = (29) - 7$
$u = 22$

State: The upper part is 22 feet and the lower part is 29 feet.

Check: $(22) + (29) = 51$
$51 = 51$
$(22) = (29) - 7$
$22 = 22$

23. **Analyze:** We know that the sum of the salaries is \$371,500. Also, the President's salary is \$28,500 more than the Vice President's salary.

Form: Let $P =$ the President's salary and $V =$ Vice President's salary.

Pres. Salary + V.P. Salary = 371,500
$P + V = 371,500$
Pres. Salary = V.P. Salary + 28,500
$P = V + 28,500$

Solve: We have $P + V = 371,500$
$P = V + 28,500$

Substitute the second equation into the first:

$P + V = 371,500$
$(V + 28,500) + V = 371,500$
$2V + 28,500 = 371,500$
$2V + 28500 - 28500 = 371500 - 28500$
$2V = 343,000$
$\frac{2V}{2} = \frac{343,000}{2}$
$V = 171,500$
$P = V + 28,500$
$P = (171,500) + 28,500$
$P = 200,000$

State: The President's salary is \$200,000 and the Vice President's salary is \$171,500.

Check:
$(200,000) + (171,500) = 371,500$
$371,500 = 371,500$
$(200,000) = (171,500) + 28,500$
$200,000 = 200,000$

25. **Analyze:** The two items that were purchased are latex paint and paint brushes. Eight gallons of latex paint and three brushes costs \$135. Six gallons of paint and 2 brushes costs \$100.

Form: Let p represent the cost of a gallon of latex paint and b represent the cost of a brush.

8 gallons of paint + 3 brushes costs \$135
$8p + 3b = 135$
6 gallons of paint + 2 brushes costs \$100
$6p + 2b = 100$

Solve: We have $8p + 3b = 135$
$6p + 2b = 100$

Multiply the first equation by 2 and the second equation by -3.

$2(8p + 3b) = 2(135)$
$16p + 6b = 270$
$-3(6p + 2b) = -3(100)$
$-18p - 6p = -300$

Now, we have $16p + 6b = 270$
$-18p - 6p = -300$
$-2p \qquad = -30$

$\frac{-2p}{-2} = \frac{-30}{-2}$
$p = 15$
$6p + 2b = 100$
$6(15) + 2b = 100$
$90 + 2b = 100$
$90 - 90 + 2b = 100 - 90$
$2b = 10$
$\frac{2b}{2} = \frac{10}{2}$
$b = 5$

State: The paint costs \$15 per gallon and the brushes are \$5 each.

Check: $8(15) + 3(5) = 135$
$120 + 15 = 135$
$135 = 135$
$6p + 2b = 100$
$6(15) + 2(5) = 100$
$90 + 10 = 100$
$100 = 100$

27. Analyze: We know that the total number of people is 190 and the total of the money taken in is $720.

Form: Let x represent the number of non-seniors and y represent the number of seniors.

# of non-seniors + # of seniors is 190

$x + y = 190$

4•# of non-seniors + 3•# of seniors is 720

$4x + 3y = 720$

Solve: We have $x + y = 190$

$4x + 3y = 720$

Multiply the first equation by -3

$-3(x + y) = -3(190)$

$-3x - 3y = -570$

Now, we have $-3x - 3y = -570$

$4x + 3y = 720$

$x \quad = 150$

$x + y = 190$

$(150) + y = 190$

$150 - 150 + y = 190 - 150$

$y = 40$

State: There were 40 senior citizens at the movie.

Check: $(150) + (40) = 190$

$190 = 190$

$4(150) + 3(40) = 720$

$600 + 120 = 720$

$720 = 720$

29. Analyze: We know that the sum of the two angles is 180°. We need to find the two angles by using the relationship given. Angle 1 is the larger angle.

Form: Let x be angle 1 and y be angle 2.

angle 1 + angle 2 is 180

$x + y = 180$

2 • angle 2 − 15° is angle 1

$2y - 15 = x$

Solve: We have $x + y = 180$

$2y - 15 = x$

Substitute the second equation into the first:

$y + x = 180$

$y + (2y - 15) = 180$

$3y - 15 = 180$

$3y - 15 + 15 = 180 + 15$

$3y = 195$

$\frac{3y}{3} = \frac{195}{3}$

$y = 65$

$2y - 15 = x$

$2(65) - 15 = x$

$130 - 15 = x$

$115 = x$

State: Angle 1 is 115° and angle 2 is 65°.

Check: $x + y = 180$

$(115) + (65) = 180$

$180 = 180$

$2(65) - 15 = (115)$

$130 - 15 = 115$

$115 = 115$

31. Analyze: We need to find the area of the rectangular screen, so we will use $A = l•w$. We know the perimeter is 332 ft and $P = 2l + 2w$.

Form: Let l = length of screen and w = width of screen.

width is length $- 26$

$w = l - 26$

perimeter is 2 • length + 2 • width

$332 = 2l + 2w$

Solve: We have $w = l - 26$

$332 = 2l + 2w$

Substitute the first equation into the second:

$332 = 2l + 2w$

$332 = 2l + 2(l - 26)$

$332 = 2l + 2l - 52$

$332 = 4l - 52$

$332 + 52 = 4l - 52 + 52$

$384 = 4l$

$\frac{384}{4} = \frac{4l}{4}$

$96 = l$

$w = l - 26$

$w = (96) - 26$

$w = 70$

Now, use $A = l•w$.

$A = 96(70)$

$A = 6720$

State: The area of the screen is 6720 ft^2.

Check: $(70) = (96) - 26$
$70 = 70$
$332 = 2(70) + 2(96)$
$332 = 140 + 192$
$332 = 332$

33. a) The break point is the number of tires that will cost equal amounts to make on either machine. We must write equations for each machine. Let y = cost to make tires and x = number of tires.

Machine 1:
$y = 1000 + 15x$

Machine 2:
$y = 3000 + 10x$

When costs are equal , we have
$1000 + 15x = 3000 + 10x$
$1000 + 15x - 10x = 3000 + 10x - 10x$
$1000 + 5x = 3000$
$1000 - 1000 + 5x = 3000 - 1000$
$5x = 2000$
$\frac{5x}{5} = \frac{2000}{5}$
$x = 400$

The break point is 400 tires.

b) The intersection is at (400, 1,000).

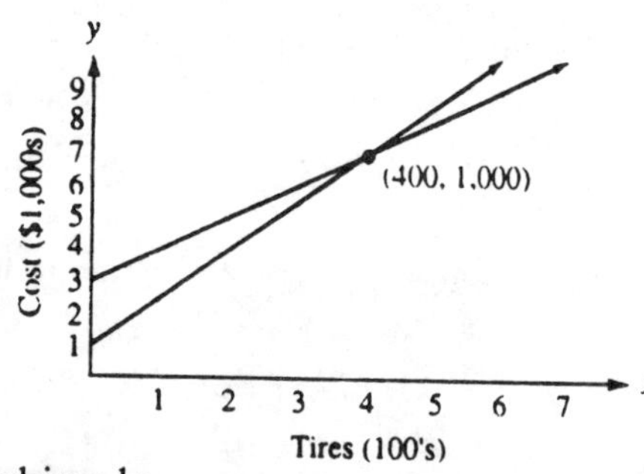

c) Machine 1:
$y = 1000 + 15x$
$y = 1000 + 15(500)$
$y = 1000 + 7500$
$y = 8500$

Machine 2:
$y = 3000 + 10x$
$y = 3000 + 10(500)$
$y = 3000 + 5000$
$y = 8000$

The second mold should be used since the cost is less.

35. Analyze: We know there are tow loans that total \$5000. We know one is for 5% interest and the other for 7% interest. We know the total interest after 1 year is \$310.

Form: Let n represent the loan to the nursing student and b represent the loan to the business student.

nursing loan + business loan is \$5000
$n + b = 5000$

nursing loan•(.05) + business loan•(.07) is \$310
$0.05n + 0.07b = 310$

Solve: We have $n + b = 5000$
$0.05n + 0.07b = 310$

Solve the first equation for n.
$n + b = 5000$
$n + b - b = 5000 - b$
$n = 5000 - b$

Substitute this equation into the second:
$0.05n + 0.07b = 310$
$0.05(5000 - b)n + 0.07b = 310$
$250 - 0.05b + 0.07b = 310$
$250 + 0.02b = 310$
$250 - 250 + 0.02b = 310 - 250$
$0.02b = 60$
$\frac{0.02b}{0.02} = \frac{60}{0.02}$
$b = 3000$
$n = 5000 - b$
$n = 5000 - (3000)$
$n = 2000$

State: The nursing loan is for \$2000 and the business loan is for \$3000.

Check: $(2000) + (3000) = 5000$
$5000 = 5000$
$0.05(2000) + 0.07(3000) = 310$
$100 + 210 = 310$
$310 = 310$

37. **Analyze:** Traveling with the current, the ship travels faster. Traveling against the current, the ship travels slower.
Form: Let s represent the speed of the boat and c represent the speed of the current. Use $d = rt$.
With current:
$$300 = (s + c)\cdot 10$$
Against current:
$$300 = (s - c)\cdot 15$$
Solve: We have $300 = (s + c)\cdot 10$
$$300 = (s - c)\cdot 15$$
Divide both sides of the first equation by 10 to get:
$$30 = s + c$$
Divide both sides of the second equation by 15 to get:
$$20 = s - c$$
Now, we have
$$\begin{array}{r} 30 = s + c \\ 20 = s - c \\ \hline 50 = 2s \end{array}$$
$$\frac{50}{2} = \frac{2s}{2}$$
$$25 = s$$
$$s + c = 30$$
$$(25) + c = 30$$
$$25 - 25 + c = 30 - 25$$
$$c = 5$$
State: The speed of the current is 5 miles per hour.
Check: $30 = (5) + (25)$
$$30 = 30$$
$$20 = (25) - (5)$$
$$20 = 20$$

39. **Analyze:** When flying downwind, the total speed is the speed of the plane in still air plus the speed of the wind. When flying upwind the total speed is the speed of the plane in still air minus the speed of the wind.
Form: Use $d = rt$. Let $p =$ speed of the plane is still air and $w =$ the speed of the wind.
Downwind:
$$600 = (p + w)\cdot 2$$
Upwind:
$$600 = (p - w)\cdot 3$$
Solve: We have $600 = (p + w)\cdot 2$
$$600 = (p - w)\cdot 3$$
Divide both sides of the first equation by 2 to get:
$$300 = p + w$$
Divide both sides of the second equation by 3 to get:
$$200 = p - w$$
Now, we have
$$\begin{array}{r} 300 = p + w \\ 200 = p - w \\ \hline 500 = 2p \end{array}$$
$$\frac{500}{2} = \frac{2p}{2}$$
$$250 = p$$
$$p + w = 300$$
$$(250) + c = 300$$
$$250 - 250 + c = 300 - 250$$
$$c = 50$$
State: The speed of the wind is 50 miles per hour.
Check: $600 = ((250) + (50))\cdot 2$
$$600 = (300)\cdot 2$$
$$600 = 600$$
$$600 = ((250) - (50))\cdot 3$$
$$600 = (200)\cdot 3$$
$$600 = 600$$

41. **Analyze:** Some 6% solution must be added to the 2% solution to make a 3% solution.
Form: Let x represent the amount of the 6% solution and y represent the amount of 2% solution.
amt. of 6% sol + amt. of 2% sol. is 16
$$x + y = 16$$
0.06(amt. of 6%) + 0.02(amt. of 2%) is 0.03•16
$$0.06x + 0.02 = 0.03(16)$$
$$0.06x + 0.02 = 0.48$$
Solve: We have $x + y = 16$
$$0.06x + 0.02y = 0.48$$
Multiply second equation by 100.
$$100(0.06x + 0.02y) = 100(0.48)$$
$$6x + 2y = 48$$
Multiply the first equation by -2.
$$-2(x + y) = -2(16)$$
$$-2x - 2y = -32$$

Now, we have $6x + 2y = 48$
$$\begin{array}{r} -2x - 2y = -32 \\ \hline 4x \qquad = 16 \end{array}$$
$\frac{4x}{4} = \frac{16}{4}$
$x = 4$
$x + y = 16$
$(4) + y = 16$
$4 - 4 + y = 16 - 4$
$y = 12$

State: The biologist should mix 4 liters of 6% solution and 12 liters of 2% solution.

Check: $(4) + (12) = 16$
$16 = 16$
$0.06(4) + 0.02(12) = 0.48$
$0.24 + 0.24 = 0.48$

43. **Analyze:** The peanuts and cashews must be mixed to make 48 pounds of mixed nuts.

Form: Let x be the number of pounds of peanuts and y be the number of pounds of cashews needed to make 48 pounds of mixed nuts.

We have

Weight:

lbs. of peanuts + # lbs. of cashews is 48
$x + y = 48$

Value:

3(# lbs. of peanuts) + 6(# lbs of cashews) is 4(# lbs of mixed)
$3x + 6y = 4(48)$
$3x + 6y = 192$

Solve: We have $x + y = 48$
$3x + 6y = 192$

Multiply the first equation by -3.
$-3(x + y) = -3(48)$
$-3x - 3y = -144$

Now, we have $-3x - 3y = -144$
$$\begin{array}{r} 3x + 6y = 192 \\ \hline 3y = 48 \end{array}$$
$\frac{3y}{3} = \frac{48}{3}$
$y = 16$
$x + y = 48$
$x + (16) = 48$
$x + 16 - 16 = 48 - 16$
$x = 32$

State: The merchant should use 32 pounds of peanuts and 16 pounds of cashews.

Check: $(16) + (32) = 48$
$48 = 48$
$3(32) + 6(16) = 192$
$96 + 76 = 192$
$192 = 192$

45. **Analyze:** The sale price is \$384 and the discount is 40%.

Form: We have
retail − discount = sale
$r - d = 384$
discount = discount rate • retail
$d = 0.4r$

Solve: We have $r - d = 384$
$d = 0.4r$

Substitute the second equation into the first.
$r - d = 384$
$r - (0.4r) = 384$
$0.6r = 384$
$\frac{0.6r}{0.6} = \frac{384}{0.6}$
$r = 640$
$d = 0.4r$
$d = 0.4(640)$
$d = 256$

State: The original retail price was \$640.

Check: $(640) - (256) = 384$
$384 = 384$

REVIEW

49. $x < 4$ <======)-------
4

51. $-1 < x \leq 2$ ----(======]----
-1 2

53. $8•8•8•c = 8^3c$

55. $a•a•b•b = a^2b^2$

STUDY SET Section 4.5

VOCABULARY

1. inequality

3. boundary

CONCEPTS

5. $5x - 3y \geq 0$

a) $(1, 1)$
$5(1) - 3(1) \geq 0$
$5 - 3 \geq 0$
$2 \geq 0$
Yes, (1, 1) is a solution.

b) $(-2, -3)$
$5(-2) - 3(-3) \geq 0$
$-10 + 9 \geq 0$
$-1 \geq 0$
No, $(-2, -3)$ is not a solution.

c) $(0, 0)$
$5(0) - 3(0) \geq 0$
$0 - 0 \geq 0$
$0 \geq 0$
Yes, (0, 0) is a solution.

d) $(\frac{1}{5}, \frac{4}{3})$
$5(\frac{1}{5}) - 3(\frac{4}{3}) \geq 0$
$1 \quad 4 \geq 0$
$-3 \geq 0$
No, $(\frac{1}{5}, \frac{4}{3})$ is not a solution.

7. a) No, does not include the boundary line.
b) Yes, does include the boundary line.

9. a) No, it is on the boundary, which is not included.
b) Yes, it is in the shaded region.
c) Yes, it is in the shaded region.
d) No, it is on the wrong side of the line.

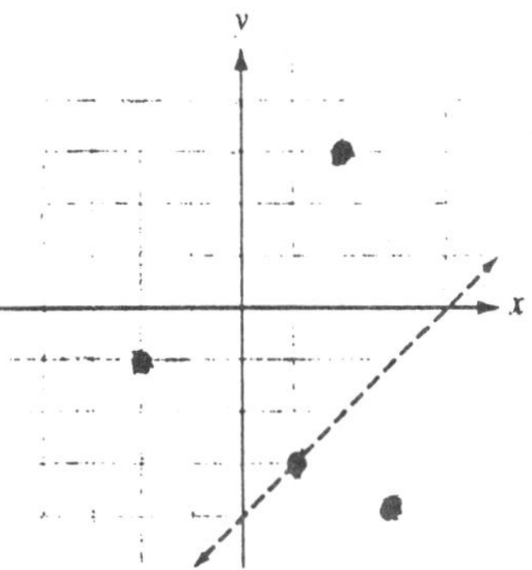

11. The test point should be on one side of the boundary. For this graph, the origin in on the boundary.

PRACTICE

13. Graph the line $y = x + 2$. Use a solid line because of the " $\leq$ ".
Use (0, 0) as the test point.
$(0) \leq (0) + 2$
$0 \leq 2$
Shade the region with the point (0,0).

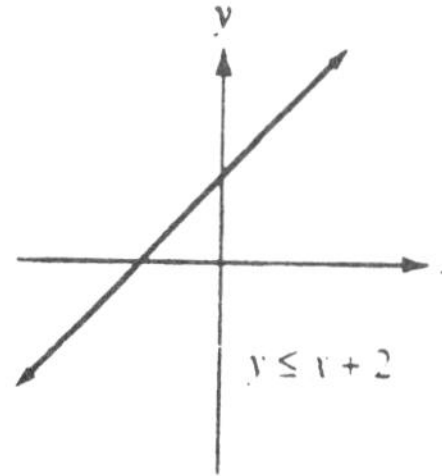

15. Graph the line $y = 2x - 4$. Use a dotted line because of the " $>$ ".
Use (0, 0) as the test point.

$(0) > 2(0) - 4$
$0 > 0 - 4$
$0 > -4$

Shade the region with the point (0,0).

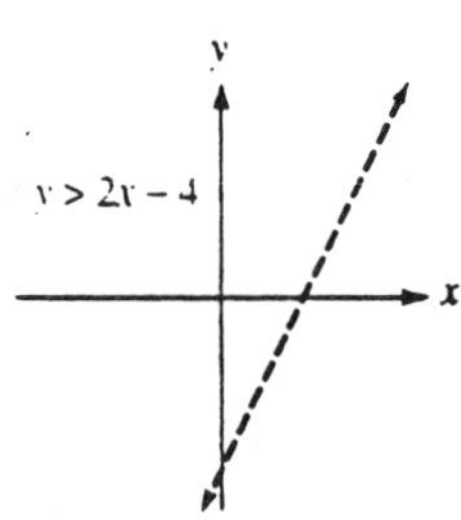

17. $x - 2y \geq 4$
$x - x - 2y \geq -x + 4$
$-2y \geq -x + 4$
$\frac{-2y}{-2} \leq \frac{-x+4}{-2}$
$y \leq \frac{x}{2} - 2$

Graph the line $y = \frac{x}{2} - 2$. Use a solid line because of the " $\leq$ ".
Use (0, 0) as the test point.

$(0) - 2(0) \geq 4$
$0 - 0 \geq 4$
$0 \geq 4$

Shade the region without the point (0,0).

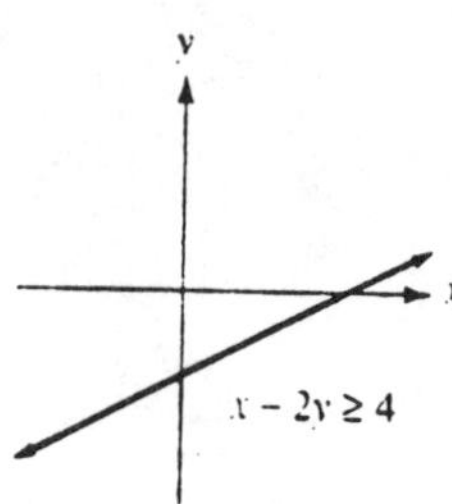

19. Graph the line $y = 4x$. Use a solid line because of the " $\leq$ ".
Use (2, 0) as the test point.

$(0) \leq 4(2)$
$0 \leq 8$

Shade the region with the point (2,0).

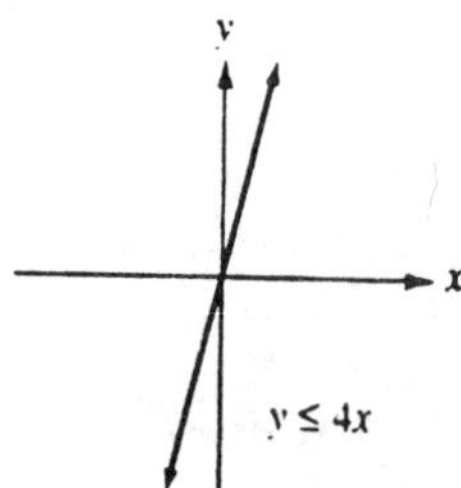

21. Graph the line $y = 3 - x$. Use a solid line because of the " $\geq$ ".
Use (0, 0) as the test point.

$(0) \geq 3 - (0)$
$0 \geq 3$

Shade the region without the point (0,0).

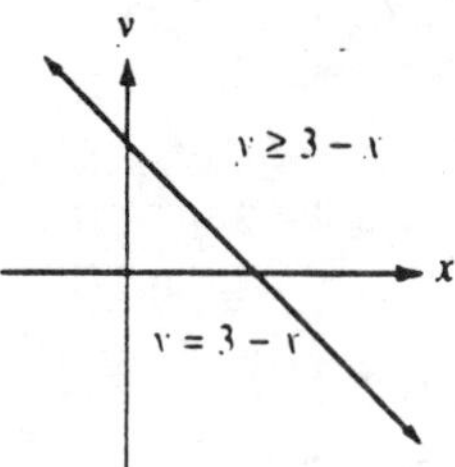

23. Graph the line $y = 2 - 3x$. Use a dotted line because of the " $<$ ".
Use (0, 0) as the test point.

$(0) < 2 - 3(0)$
$0 < 2 - 0$
$0 < 2$

Shade the region with the point (0,0).

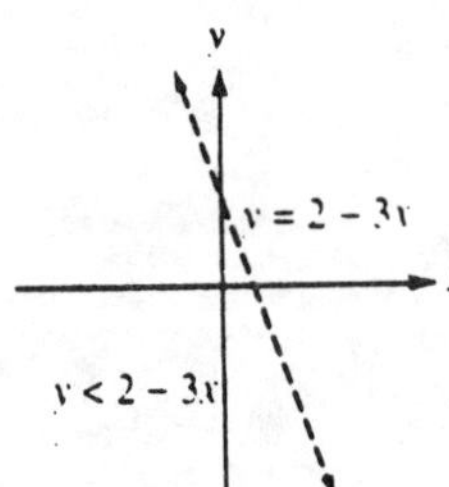

25. Graph the line $y = 2x$. Use a solid line because of the " $\geq$ ".
Use (2, 0) as the test point.
$(0) \geq 2(2)$
$0 \geq 4$
Shade the region without the point (2,0).

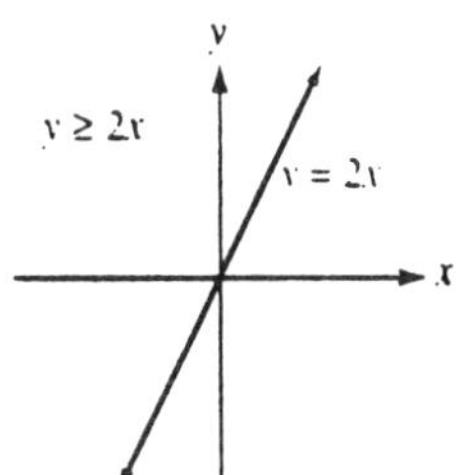

27. $2y - x < 8$
$2y - x + x < x + 8$
$2y < x + 8$
$\frac{2y}{2} < \frac{x+8}{2}$
$y < \frac{x}{2} + 4$
Graph the line $y = \frac{x}{2} + 4$. Use a dotted line because of the " $<$ ".
Use (0, 0) as the test point.
$(0) < \frac{(0)}{2} + 4$
$0 < 0 + 4$
$0 < 4$
Shade the region with the point (0,0).

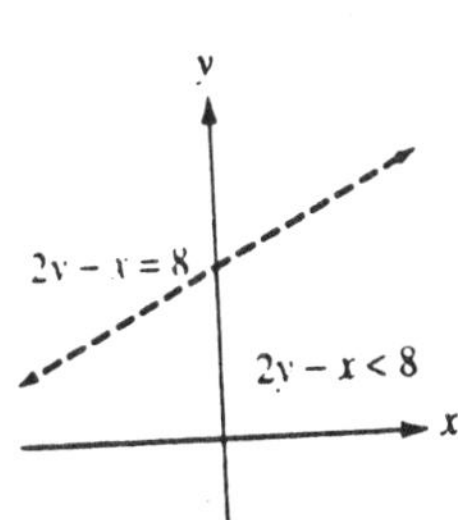

29. $y - x \geq 0$
$y - x + x \geq 0 + x$
$y \geq x$
Graph the line $y = x$. Use a solid line because of the " $\geq$ ".
Use (2, 0) as the test point.
$(0) \geq (2)$
Shade the region without the point (2,0).

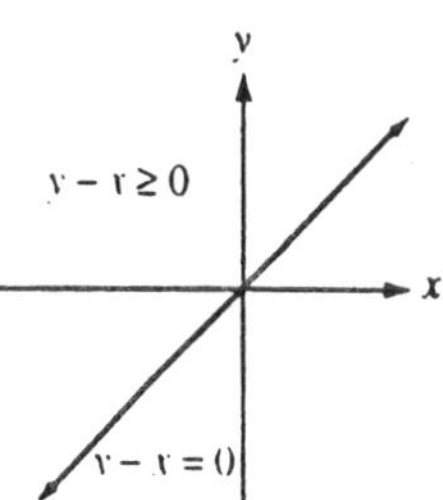

31. $2x + y > 2$
$2x - 2x + y > -2x + 2$
$y > -2x + 2$
Graph the line $y = -2x + 2$. Use a dotted line because of the " $>$ ".
Use (0, 0) as the test point.
$2(0) + (0) > 2$
$0 + 0 > 2$
$0 > 2$
Shade the region without the point (0,0).

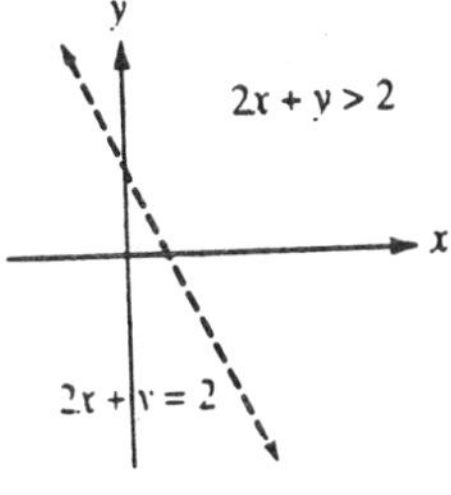

33. $3x - 4y > 12$
$3x - 3x - 4y > -3x + 12$
$-4y > -3x + 12$
$\frac{-4y}{-4} < \frac{-3x+12}{-4}$
$y < \frac{3x}{4} - 3$
Graph the line $y = \frac{3x}{4} - 3$. Use a dotted line because of the " $<$ ".
Use (0, 0) as the test point.
$3(0) - 4(0) > 12$
$0 - 0 > 12$
$0 > 12$
Shade the region without the point (0,0).

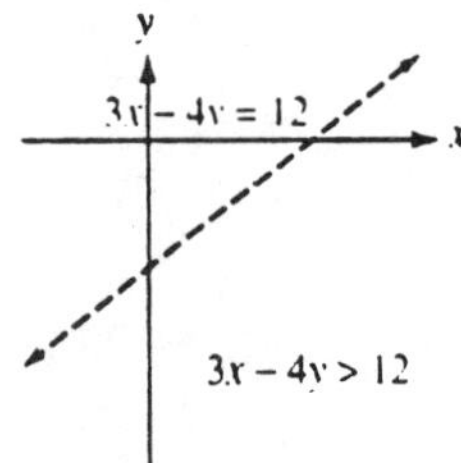

35. $5x + 4y \geq 20$
$5x - 5x + 4y \geq -5x + 20$
$4y \geq -5x + 20$
$\frac{4y}{4} \geq \frac{-5x+20}{4}$
$y \geq -\frac{5x}{4} + 5$
Graph the line $y = -\frac{5x}{4} + 5$. Use a solid line because of the " $\geq$ ".
Use (0, 0) as the test point.
$5(0) + 4(0) \geq 20$
$0 + 0 \geq 20$
$0 \geq 20$
Shade the region without the point (0,0).

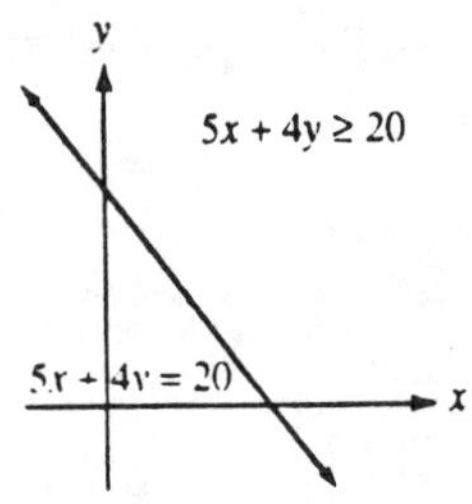

37. Graph the vertical line $x = 2$. Use a dotted line because of the " $<$ ".
Use (0, 0) as the test point.
$(0) < 2$
Shade the region with the point (0,0).

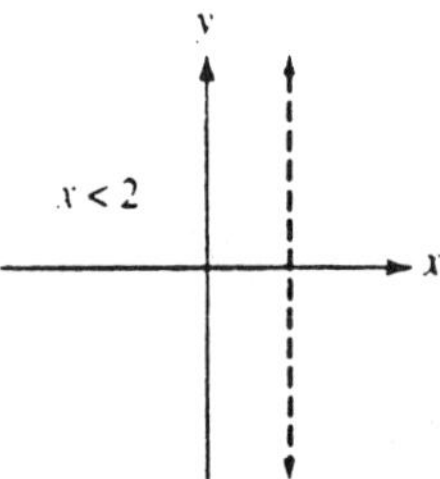

39. Graph the horizontal line $y = 1$. Use a solid line because of the " $\leq$ ".
Use (0, 0) as the test point.
$(0) \leq 1$
Shade the region without the point (0,0).

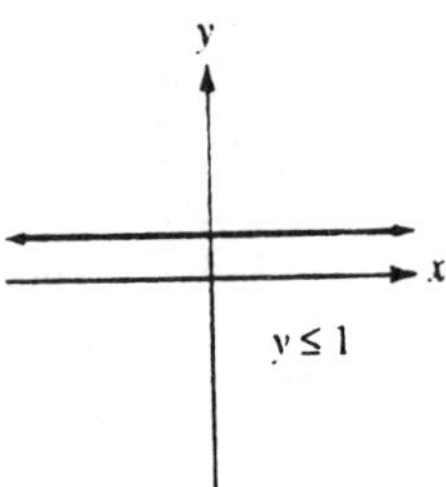

41. $3(x+y)+x<6$
$3x+3y+x<6$
$4x+3y<6$
$4x-4x+3y<-4x+6$
$3y<-4x+6$
$\frac{3y}{3}<\frac{-4x+6}{3}$
$y<\frac{-4x}{3}+2$
Graph the line $y=-\frac{4x}{3}+2$. Use a dotted line because of the " $<$ ".
Use (0, 0) as the test point.
$3((0)+(0))+(0)<6$
$3(0)+0<6$
$0+0<6$
$0<6$
Shade the region with the point (0,0).

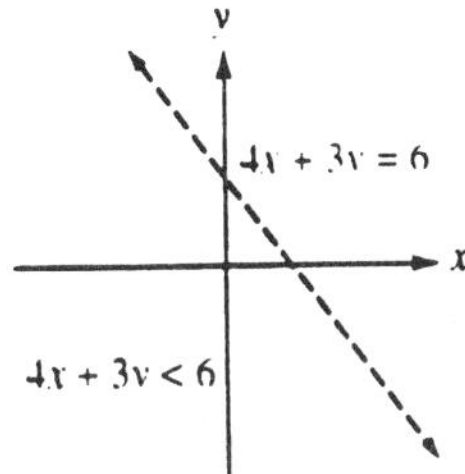

43. $4x-3(x+2y)\geq -6y$
$4x-3x-6y\geq -6y$
$x-6y\geq -6y$
$x-6y+6y\geq -6y+6y$
$x\geq 0$
Graph the vertical line $x=0$. Use a solid line because of the " $\geq$ ".
Use (2, 0) as the test point.
$4(2)-3((2)+2(0))\geq -6(0)$
$8-3(2+0)\geq 0$
$8-3(2)\geq 0$
$8-6\geq 0$
$2\geq 0$
Shade the region with the point (2,0).

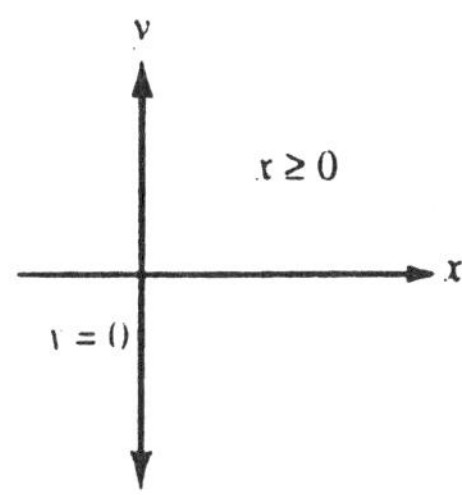

45. We let x represent the number of cakes and y represent the number of pies. We have
3(# of cakes) + 4(# of pies)
is less than or equal to \$120
$3x+4y\leq 120$
$3x-3x+4y\leq -3x+120$
$4y\leq -3x+120$
$\frac{4y}{4}\leq\frac{-3x+120}{4}$
$y\leq -\frac{3x}{4}+30$
Graph the line $y=-\frac{3x}{4}+30$. Use a solid line because of the " $\leq$ ".
Use (10, 10) as the test point.
$3(10)+4(10)\leq 120$
$30+40\leq 120$
$70\leq 120$
Shade the region with the point (10, 10). Also, shade above the x-axis and to the right of the y-axis. Negative numbers do not make sense in the problem.
Other combinations are (20, 10) and (10, 20).

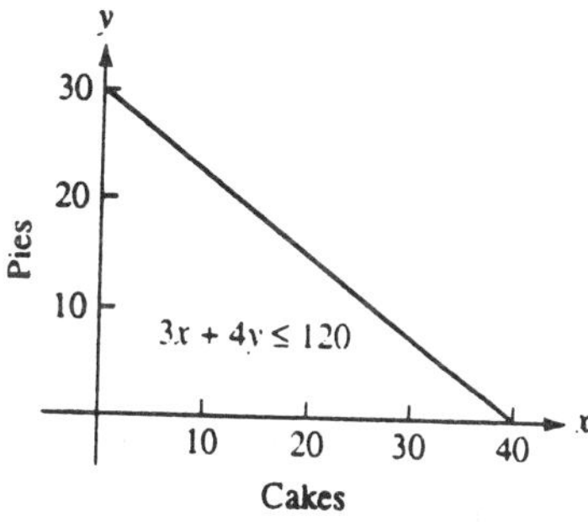

47. We let x represent the number of leather jackets and y represent the number of nylon jackets. We have
100(# of leather) + 88(# of nylon)
is greater than or equal to 4400.

$$100x + 88y \geq 4400$$
$$100x - 100x + 88y \geq -100x + 4400$$
$$88y \geq -100x + 4400$$
$$\frac{88y}{88} \geq \frac{-100x+4400}{88}$$
$$y \geq -\frac{100x}{88} + 50$$

Graph the line $y = -\frac{100x}{88} + 50$. Use a solid line because of the " $\geq$ ".
Use (10, 10) as the test point.

$$100(10) + 88(10) \geq 4400$$
$$1000 + 880 \geq 4400$$
$$1880 \geq 4400$$

Shade the region with the point (10, 10).
Combinations that work are (50, 50), (30, 40), and (40, 40).

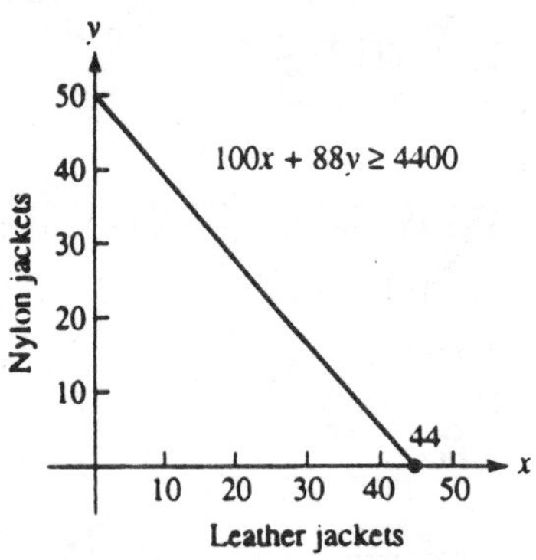

49. We let x represent the shares of Robotronic and y represent the shares of Macrohard.
Value:
40(# of Robo.) + 50(# of Macro.)
is less than or equal to 8000.

$$40x + 50y \leq 800$$
$$40x - 40x + 50y \leq -40x + 800$$
$$50y \leq -40x + 800$$
$$\frac{50y}{50} \leq \frac{-40x+800}{50}$$
$$y \leq -\frac{4x}{5} + 16$$

Graph the line $y = -\frac{4x}{5} + 16$. Use a solid line because of the " $\leq$ ".
Use (40, 40) as the test point.

$$40(40) + 50(40) \leq 8000$$
$$1600 + 2000 \leq 8000$$
$$3600 \leq 8000$$

Shade the region with the point (40, 40). Also, shade above the x-axis and to the right of the y-axis. Negative numbers do not make sense in the problem.
Combinations that work are (80, 40), (80, 80), and (120, 40).

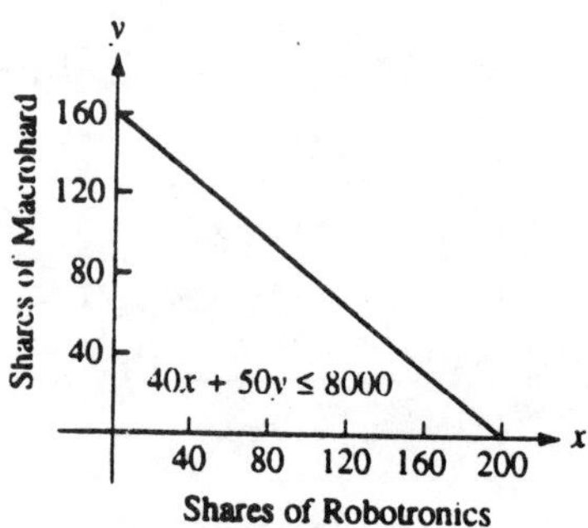

REVIEW

53. $g(x) = 3x^2 - 4x + 3$
$$g(2) = 3(2)^2 - 4(2) + 3$$
$$= 3(4) - 4(2) + 3$$
$$= 12 - 8 + 3$$
$$= 7$$

55. True. $\frac{1}{2}x = \frac{x}{2}$

57. distance = rate • time
$d = r \cdot t$

59. $A = P + Prt$
$$A - P = P - P + Prt$$
$$A - P = Prt$$
$$\frac{A-P}{Pr} = \frac{Prt}{Pr}$$
$$\frac{A-P}{Pr} = t$$

STUDY SET Section 4.6

VOCABULARY

1. inequality

3. doubly shaded

5. a) True
b) False
c) False
d) True
e) True
f) True

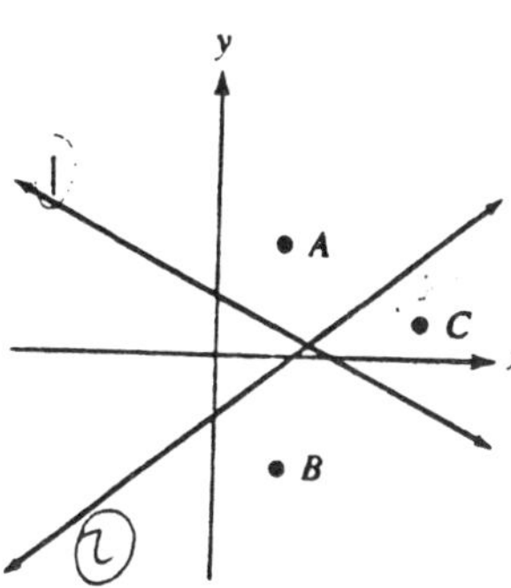

7. a) Yes
b) No.
c) Yes.

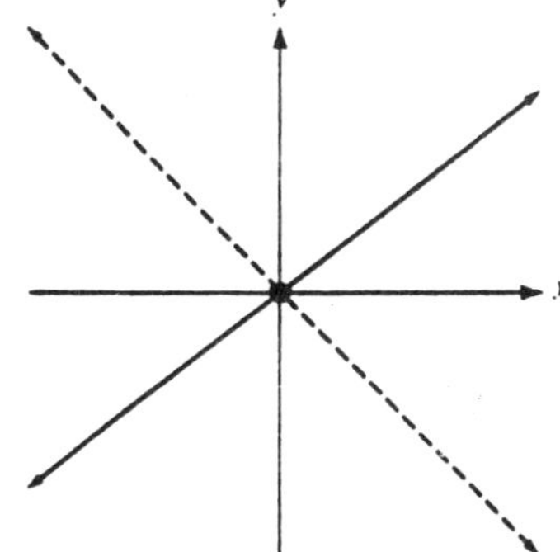

9. *ABC*

PRACTICE

11. First inequality:
$x+2y\leq 3$
$x-x+2y\leq -x+3$
$2y\leq -x+3$
$\frac{2y}{2}\leq\frac{-x+3}{2}$
$y\leq -\frac{x}{2}+\frac{3}{2}$
Second inequality:
$2x-y\geq 1$
$2x-2x-y\geq -2x+1$
$-y\geq -2x+1$
$-1(-y)\leq -1(-2x+1)$
$y\leq 2x-1$
Graph the line $y=-\frac{x}{2}+\frac{3}{2}$. Use a solid line because of the " $\leq$ ". Graph the line $y=2x-1$. Use a solid line because of the " $\leq$ ".
Use (0, 0) as the test point for both inequalities.
$x+2y\leq 3$
$(0)+2(0)\leq 3$
$0+0\leq 3$
$0\leq 3$
For the first inequality, shade the region with the point $(0,0)$.
$2x-y\geq 1$
$2(0)-(0)\geq 1$
$0-0\geq 1$
$0\geq 1$
For the second inequality, shade the region without the point $(0,0)$.

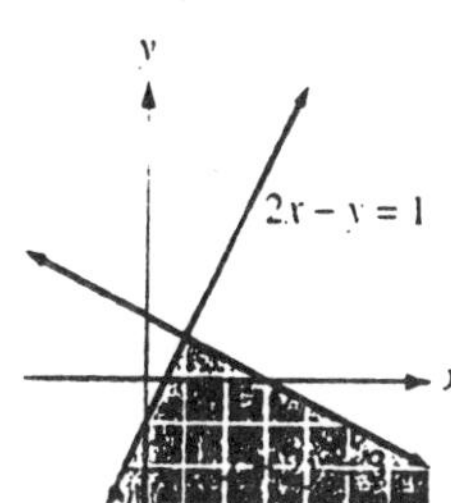

13. First inequality:
$x + y < -1$
$x - x + y < -x - 1$
$y < -x - 1$
Second inequality:
$x - y > -1$
$x - x - y > -x - 1$
$-y > -x - 1$
$-1(-y) < -1(-x - 1)$
$y < x + 1$
Graph the line $y = -x - 1$. Use a dotted line because of the " $<$ ".
Graph the line $y = x + 1$. Use a dotted line because of the " $<$ ".
Use (0, 0) as the test point for both inequalities.

$x + y < -1$
$(0) + (0) < -1$
$0 + 0 < -1$
$0 < -1$

For the first inequality, shade the region without the point $(0, 0)$.

$x - y > -1$
$(0) - (0) > -1$
$0 - 0 > -1$
$0 > -1$

For the second inequality, shade the region with the point $(0, 0)$.

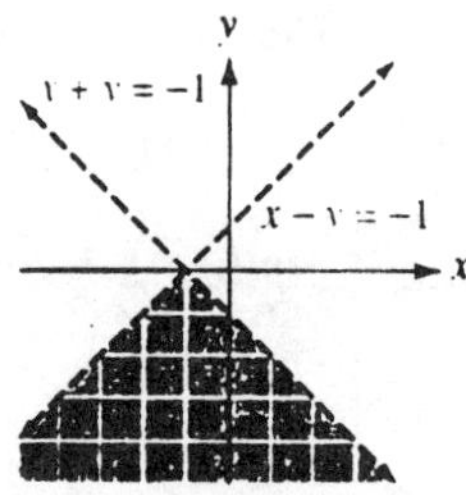

15. Graph the vertical line $x = 2$. Use a dotted line because of the " $>$ ".
Graph the horizontal line $y = 3$. Use a solid line because of the " $\leq$ ".
Use (0, 0) as the test point for both inequalities.

$x > 2$
$(0) > 2$

For the first inequality, shade the region without the point $(0, 0)$.

$y \leq 3$
$(0) \leq 3$

For the second inequality, shade the region with the point $(0, 0)$.

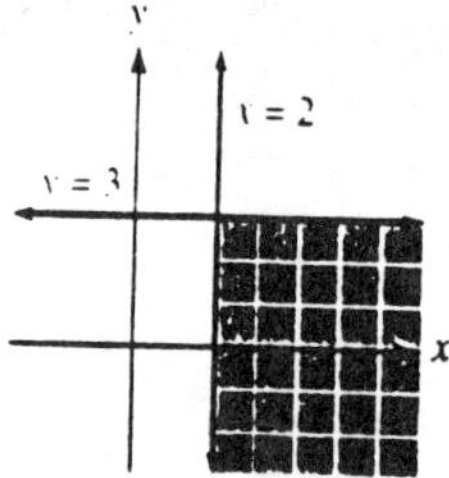

17. First inequality:
$2x - 3y < 0$
$2x - 2x - 3y < -2x + 0$
$-3y < -2x$
$\frac{-3y}{-3} > \frac{-2x}{-3}$
$y > \frac{2x}{3}$
Second inequality:
$y > x - 1$
Graph the line $y = \frac{2x}{3}$. Use a dotted line because of the " $>$ ". Graph the line $y = x - 1$. Use a dotted line because of the " $>$ ".
Use (2, 0) as the test point for both inequalities.

$2x - 3y < 0$
$2(2) - 3(0) < 0$
$4 + 0 < 0$
$4 < 0$

For the first inequality, shade the region without the point $(2, 0)$.

$y > x - 1$
$(0) > (2) - 1$
$0 > 1$

For the second inequality, shade the region without the point $(2, 0)$.

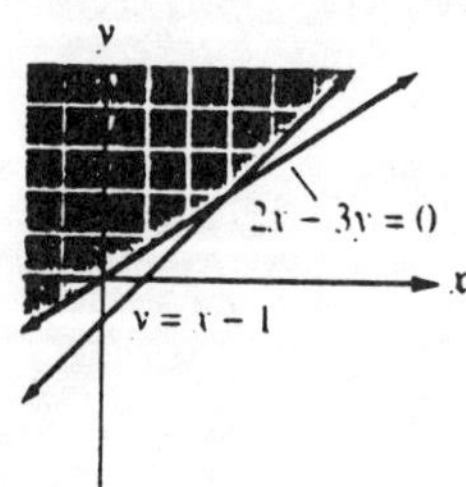

19. First inequality:
$y < -x + 1$
Second inequality:
$y > -x + 3$
Graph the line $y = -x + 1$. Use a dotted line because of the " $<$ ". Graph the line $y = -x + 3$. Use a dotted line because of the " $>$ ".
Use (0, 0) as the test point for both inequalities.

$y < -x + 1$
$(0) < -(0) + 1$
$0 < 1$

For the first inequality, shade the region with the point $(0, 0)$.

$y > -x + 3$
$(0) > -(0)3$
$0 > 3$

For the second inequality, shade the region without the point $(0, 0)$.
There is no solution because the shaded regions of the inequalities do not overlap.

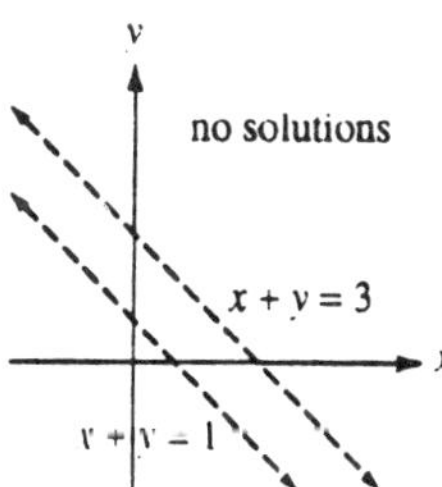

21. First inequality:
$x > 0$
Second inequality:
$y > 0$
Graph the vertical line $x = 0$. Use a dotted line because of the " $>$ ". Graph the horizontal line $y = 0$. Use a dotted line because of the " $>$ ".
Use (1, 1) as the test point for both inequalities.

$x > 0$
$(1) > 0$

For the first inequality, shade the region with the point $(1, 1)$.

$y > 0$
$(1) > 0$

For the second inequality, shade the region with the point $(1, 1)$.

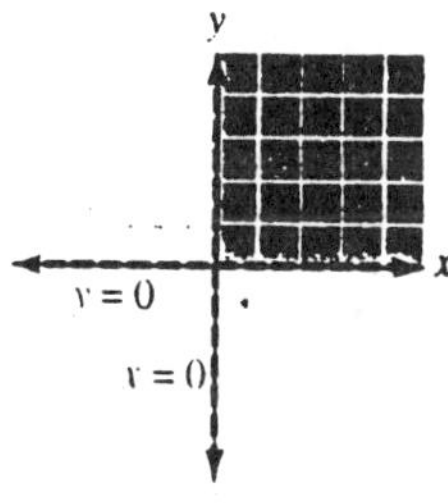

23. First inequality:

$3x + 4y > -7$

$3x - 3x + 4y < -3x - 7$

$4y < -3x - 7$

$\frac{4y}{4} > \frac{-3x-7}{4}$

$y > -\frac{3x}{4} - \frac{7}{4}$

Second inequality:

$2x - 3y \geq 1$

$2x - 2x - 3y \geq -2x + 1$

$-3y \geq -2x + 1$

$\frac{-3y}{-3} \leq \frac{-2x+1}{-3}$

$y \leq \frac{2x}{3} - \frac{1}{3}$

Graph the line $y = -\frac{3x}{4} - \frac{7}{4}$. Use a dotted line because of the " $>$ ". Graph the line $y = \frac{2x}{3} - \frac{1}{3}$. Use a solid line because of the " $\leq$ ".

Use (0, 0) as the test point for both inequalities.

$3x + 4y > -7$

$3(0) + 4(0) > -7$

$0 + 0 > -7$

$0 > -7$

For the first inequality, shade the region with the point $(0, 0)$.

$2x - 3y \geq 1$

$2(0) - 3(0) \geq 1$

$0 - 0 \geq 1$

$0 \geq 1$

For the second inequality, shade the region without the point $(0, 0)$.

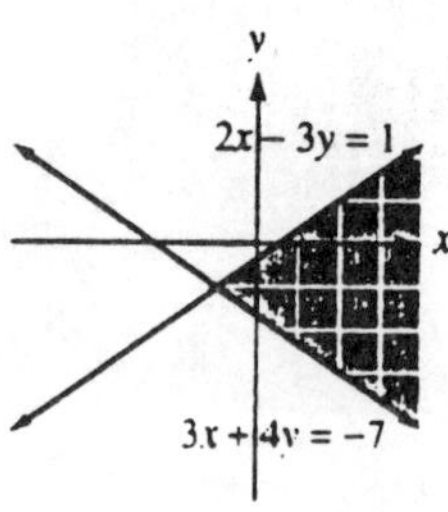

25. First inequality:

$2x + y < 7$

$2x - 2x + y < -2x + 7$

$y < -2x + 7$

Second inequality:

$y > 2(1 - x)$

$y > 2 - 2x$

$y > -2x + 2$

Graph the line $y = -2x + 7$. Use a dotted line because of the " $<$ ". Graph the line $y = -2x + 2$. Use a dotted line because of the " $<$ ".

Use (0, 0) as the test point for both inequalities.

$2x + y < 7$

$2(0) + (0) < 7$

$0 + 0 < 7$

$0 < 7$

For the first inequality, shade the region with the point $(0, 0)$.

$y > 2(1 - x)$

$(0) > 2(1 - (0))$

$0 > 2(1)$

$0 > 2$

For the second inequality, shade the region without the point $(0, 0)$.

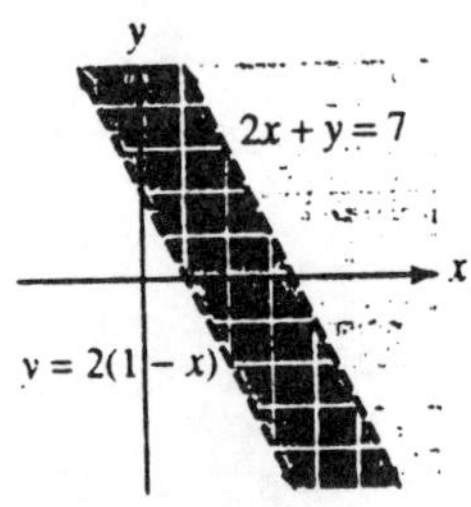

27. First inequality:
$2x - 4y > -6$
$2x - 2x - 4y > -2x - 6$
$-4y > -2x - 6$
$\frac{-4y}{-4} < \frac{-2x-6}{-4}$
$y < \frac{x}{2} + \frac{3}{2}$
Second inequality:
$3x + y \geq 5$
$3x - 3x + y \geq -3x + 5$
$y \geq -3x + 5$
Graph the line $y = \frac{x}{2} + \frac{3}{2}$. Use a dotted line because of the " $<$ ". Graph the line $y = -3x + 5$. Use a solid line because of the " $\geq$ ".
Use (0, 0) as the test point for both inequalities.

$2x - 4y > -6$
$2(0) - 4(0) > -6$
$0 + 0 > -6$
$0 > -6$

For the first inequality, shade the region with the point $(0, 0)$.

$3x + y \geq 5$
$3(0) + (0) \geq 5$
$0 + 0 \geq 5$
$0 \geq 5$

For the second inequality, shade the region without the point $(0, 0)$.

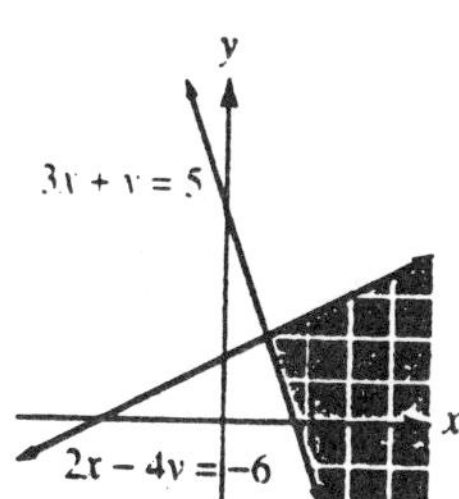

29. First inequality:
$3x - y \leq -4$
$3x - 3x - y \leq -3x - 4$
$-y \leq -3x - 4$
$-1(-y) \geq -1(-3x - 4)$
$y \geq 3x + 4$
Second inequality:
$3y > -2(x + 5)$
$3y > -2x - 10$
$3y > -2x - 10$
$\frac{3y}{3} > \frac{-2x-10}{3}$
$y > -\frac{2x}{3} - \frac{10}{3}$
Graph the line $y = 3x + 4$. Use a solid line because of the " $\geq$ ". Graph the line $y = -\frac{2x}{3} - \frac{10}{3}$. Use a dotted line because of the " $>$ ".
Use (0, 0) as the test point for both inequalities.

$3x - y \leq -4$
$3(0) - (0) \leq -4$
$0 + 0 \leq -4$
$0 \leq -4$

For the first inequality, shade the region without the point $(0, 0)$.

$3y > -2(x + 5)$
$3(0) > -2((0) + 5)$
$0 \geq -2(5)$
$0 \geq -10$

For the second inequality, shade the region with the point $(0, 0)$.

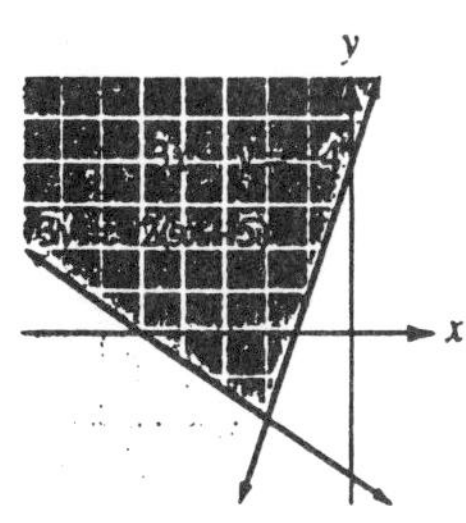

31. First inequality:

$\frac{x}{2} + \frac{y}{3} \geq 2$

$6(\frac{x}{2} + \frac{y}{3}) \geq 6(2)$

$3x + 2y \geq 12$

$3x - 3x + 2y \geq -3x + 12$

$2y \geq -3x + 12$

$\frac{2y}{2} \geq \frac{-3x+12}{2}$

$y \geq -\frac{3x}{2} + 6$

Second inequality:

$\frac{x}{2} - \frac{y}{2} < -1$

$2(\frac{x}{2} - \frac{y}{2}) < 2(-1)$

$x - y < -2$

$x - x - y < -x - 2$

$-y < -x - 2$

$-1(-y) > -1(-x - 2)$

$y > x + 2$

Graph the line $y = -\frac{3x}{2} + 6$. Use a solid line because of the " $\geq$ ". Graph the line $y = x + 2$. Use a dotted line because of the " $>$ ".

Use (0, 0) as the test point for both inequalities.

$\frac{x}{2} + \frac{y}{3} \geq 2$

$\frac{(0)}{2} + \frac{(0)}{3} \geq 2$

$0 + 0 \geq 2$

$0 \geq 2$

For the first inequality, shade the region without the point $(0, 0)$.

$\frac{x}{2} - \frac{y}{2} < -1$

$\frac{(0)}{2} - \frac{(0)}{2} < -1$

$0 - 0 < -1$

$0 < -1$

For the second inequality, shade the region without the point $(0, 0)$.

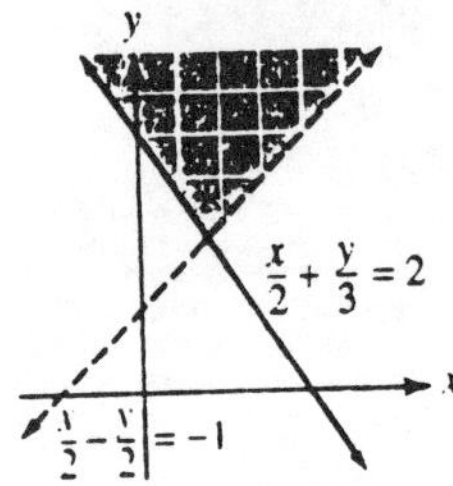

33. First inequality:

$x \geq 0$

Second inequality:

$y \geq 0$

Third inequality:

$x + y \leq 3$

$x - x - y \leq -x + 3$

$y \leq -x + 3$

Graph the vertical line $x = 0$.

Use a solid line because of the " $\geq$ ".

Graph the horizontal line $y = 0$. Use a solid line because of the " $\geq$ ". Graph the line $y = -x + 3$. Use a solid line because of the " $\leq$ ".

Use (1, 1) as the test point for the three inequalities.

$x \geq 0$

$(1) \geq 0$

For the first inequality, shade the region with the point $(1, 1)$.

$y \geq 0$

$(1) \geq 0$

For the second inequality, shade the region with the point $(1, 1)$.

$x + y \leq 3$

$(1) + (1) \leq 3$

$2 \leq 3$

For the third inequality, shade the region with the point $(1, 1)$.

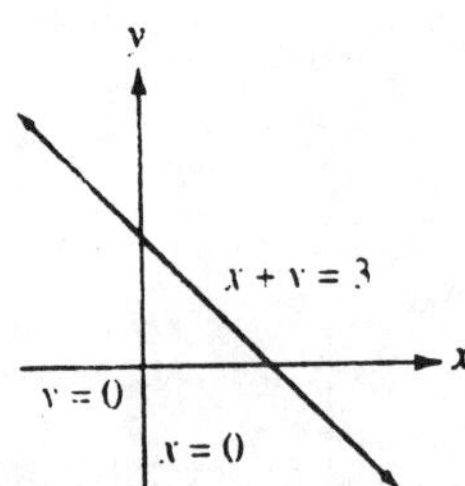

35. **Analyze:** Let x represents \$10 CD's and y represents \$15 CD's. The customer can spend at least \$30 and no more than \$60 on a combination CD's.

Form: Value:

10•(# of \$10 CD's) + 15•(# of \$15 CD's) is greater than or equal to 30.

$10x + 15y \geq 30$

10•(# of \$10 CD's) + 15•(# of \$15 CD's) is less than or equal to 60.

$10x + 15y \leq 60$

Solve: We have $10x + 15y \geq 30$

$10x + 15y \leq 60$

First inequality:

$10x + 15y \geq 30$

$10x - 10x + 15y \geq -10x + 30$

$15y \geq -10x + 30$

$\frac{15y}{15} \geq \frac{-10x+30}{15}$

$y \geq -\frac{2x}{3} + 2$

Second inequality:

$10x + 15y \leq 60$

$10x - 10x + 15y \leq -10x + 60$

$15y \leq -10x + 60$

$\frac{15y}{15} \geq \frac{-10x+60}{15}$

$y \geq -\frac{2x}{3} + 4$

Graph the line $y = -\frac{2x}{3} + 2$. Use a solid line because of the " $\geq$ ". Graph the line $y = -\frac{2x}{3} + 4$. Use a solid line because of the " $\geq$ ".

Use (0, 0) as the test point for both inequalities.

$10x + 15y \geq 30$

$10(0) + 15(0) \geq 30$

$0 + 0 \geq 30$

$0 \geq 30$

For the first inequality, shade the region without the point $(0, 0)$.

$10x + 15y \leq 60$

$10(0) + 15(0) \leq 60$

$0 + 0 \leq 60$

$0 \leq 60$

For the second inequality, shade the region with the point $(0, 0)$.

Two coordinates that are in the shaded region are (1, 2) and (4, 1).

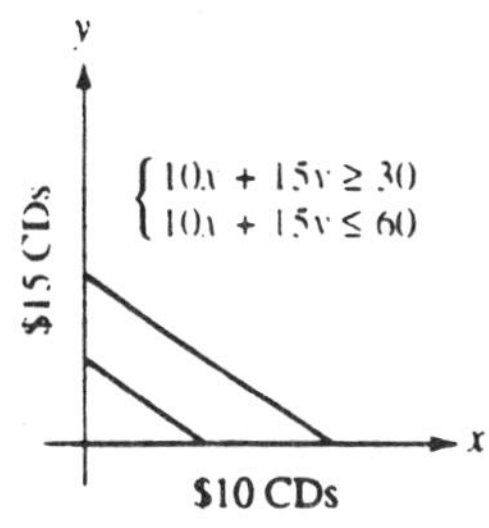

State: The customer can buy 1 \$10 CD and 2 \$15 CD's or 4 \$10 CD and 1 \$15 CD'

Check: For (1, 2):

$10x + 15y \geq 30$

$10(1) + 15(2) \geq 30$

$10 + 30 \geq 30$

$40 \geq 30$

$10x + 15y \leq 60$

$10(1) + 15(2) \leq 60$

$10 + 30 \leq 60$

$40 \leq 60$

For (4, 1):

$10x + 15y \geq 30$

$10(4) + 15(1) \geq 30$

$40 + 15 \geq 30$

$55 \geq 30$

$10(4) + 15(1) \leq 60$

$40 + 15 \leq 60$

$55 \leq 60$

37. **Analyze:** Given that x represents desk chairs and y represents side chairs. We know that we want more side chairs than desk chairs. Each desk chair costs \$150 and each side chair costs \$100. The total is no more than \$900.

Form:

# of side chairs is more than # of desk.

$y > x$

Value:

150•(# of desk) + 100•(# of side) is less than or equal to 900.

$150x + 100y \leq 900$

Solve: We have $y > x$

$150x + 100y \leq 900$

First inequality:

$y > x$

Second inequality:

$150x + 100y \leq 900$

$150x - 150x + 100y \leq -150x + 900$

$100y \leq -150x + 900$

$\frac{100y}{100} \leq \frac{-150x+900}{100}$

$y \leq -\frac{3x}{2} + 9$

Graph the line $y = x$. Use a dotted line because of the " $>$ ". Graph the line $y = -\frac{3x}{2} + 9$. Use a solid line because of the " $\geq$ ".

Use (2, 0) as the test point for both inequalities.

$y > x$

$(0) > (2)$

For the first inequality, shade the region without the point $(2, 0)$.

$150x + 100y \leq 900$

$150(2) + 100(0) \leq 900$

$300 + 0 \leq 900$

$300 \leq 900$

For the second inequality, shade the region with the point $(2, 0)$.

Two coordinates that are in the shaded region are (2, 4) and (1, 5).

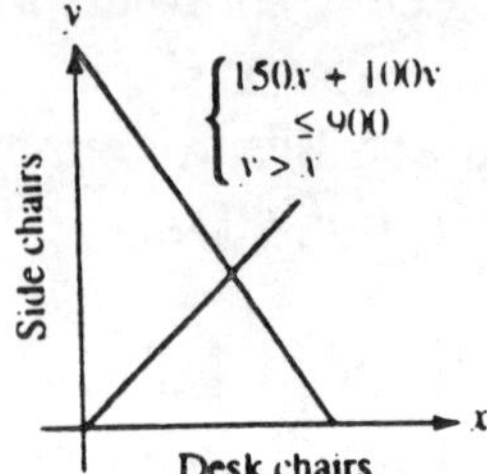

State: Best Furniture can order 2 desk chairs and 4 side chairs or 1 desk chair and 5 side chairs.

Check: For (2, 4):

$y > x$

$(4) > (2)$

$150x + 100y \leq 900$

$150(2) + 100(4) \leq 900$

$300 + 400 \leq 900$

$700 \leq 900$

For $(1, 5)$:

$y > x$

$(5) > (1)$

$150x + 100y \leq 900$

$150(1) + 100(5) \leq 900$

$150 + 500 \leq 900$

$650 \leq 900$

39. Let y represent the North and South directions and x represent the East and West directions.

East of First Street translates to $x > 1$.

North of Second Avenue translates to $y > 2$.

South of Sixth Avenue translates to $y < 6$.

West of Fifth Street translates to $x < 5$.

REVIEW

45. For $x = 8$:

$y = 2x^2 = 2(8)^2 = 2(64) = 128$

For $x = -2$:

$y = 2x^2 = 2(-2)^2 = 2(4) = 8$

47. $f(x) = 4 + x^3$

$f(0) = 4 + (0)^3 = 4 + 0 = 4$

$f(0) = 4 + (\text{-}3)^3 = 4 + (\text{-}27) = \text{-}23$

CHAPTER REVIEW

1. a) $3(2) - 2(-3) = 12$

$6 - (-6) = 12$

$6 + 6 = 12$

$12 = 12$

$2(2) + 3(-3) = -5$

$4 + (-9) = -5$

$4 - 9 = -5$

$-5 = -5$

Yes, (2, -3) is a solution of the system.

b) $4\left(\frac{7}{2}\right) - 6\left(-\frac{2}{3}\right) = 18$

$14 - (-4) = 18$

$14 + 4 = 18$

$18 = 18$

$\frac{\left(\frac{7}{2}\right)}{3} + \frac{\left(-\frac{2}{3}\right)}{2} = \frac{5}{6}$

$\frac{7}{6} + \left(-\frac{1}{3}\right) = \frac{5}{6}$

$\frac{7}{6} - \frac{1}{3} = \frac{5}{6}$

$\frac{7}{6} - \frac{2}{6} = \frac{5}{6}$

$\frac{5}{6} = \frac{5}{6}$

Yes, $\left(\frac{7}{2}, -\frac{2}{3}\right)$ is a solution of the system.

3. a) $x = y$

$5x - 4y = 3$

Substitute the first equation into the second equation.

$5x - 4(x) = 3$

$x = 3$

Find y.

$(3) = y$

The solution is (3, 3).

b) $3x = 15 - y$

$7y + 3x = 15$

Solve the first equation for y.

$3x = 15 - y$

$3x - 15 = 15 - 15 - y$

$3x - 15 = -y$

$-1(3x - 15) = -1(-y)$

$-3x + 15 = y$

Substitute this equation into the second equation.

$7(-3x + 15) + 3x = 15$

$-21x + 105 + 3x = 15$

$-18x + 105 = 15$

$-18x + 105 - 105 = 15 - 105$

$-18x = -90$

$\frac{-18x}{-18} = \frac{-90}{-18}$

$x = 5$

Find y.

$y = 15 - 3x$

$y = 15 - 3(5)$

$y = 15 - 15$

$y = 0$

The solution is $(5, 0)$.

c) $0.2x + 0.2y = 0.6$

$3x = 2 - y$

Solve the second equation for y.

$3x = 2 - y$

$3x - 2 = 2 - 2 - y$

$3x - 2 = -y$

$-1(3x - 2) = -1(-y)$

$-3x + 2 = y$

Substitute this equation into the first equation.

$0.2x + 0.2(-3x + 2) = 0.6$

$0.2x - 0.6x + 0.4 = 0.6$

$-0.4x + 0.4 = 0.6$

$-0.4x + 0.4 - 0.4 = 0.6 - 0.4$

$-0.4x = 0.2$

$\frac{-0.4x}{-0.4} - \frac{0.2}{-0.4}$

$x = -\frac{1}{2}$

Find y.

$-3(-\frac{1}{2}) + 2 = y$

$\frac{3}{2} + 2 = y$

$\frac{3}{2} + \frac{4}{2} = y$

$\frac{7}{2} = y$

The solution is $\left(-\frac{1}{2}, \frac{3}{2}\right)$.

d) $6(r + 2) = s - 1$

$5(s - 1) = r + 2$

Solve the first equation for s.

$6(r + 2) = s - 1$

$6r + 12 = s - 1$

$6r + 12 + 1 = s - 1 + 1$

$6r + 13 = s$

Substitute this equation into the second equation.

$5(s-1)=r+2$

$5((6r+13)-1)=r+2$

$5(6r+12)=r+2$

$30r+60=r+2$

$30r-r+60=r-r+2$

$29r+60=2$

$29r+60-60=2-60$

$29r=-58$

$\frac{29r}{29}=\frac{-58}{29}$

$r=-2$

Find s.

$6r+13=s$

$6(-2)+13=s$

$-12+13=s$

$1=s$

The solution is $(1, -2)$.

e) $9x+3y=5$

$3x+y=\dfrac{5}{3}$

Solve the second equation for y.

$3x+y=\frac{5}{3}$

$3x-3x+y=-3x+\frac{5}{3}$

$y=-3x+\frac{5}{3}$

Substitute this equation into the first equation.

$9x+3\left(-3x+\frac{5}{3}\right)=5$

$9x-9x+5=5$

$5=5$

The equations are dependent; there are infinitely many solutions.

f) $\dfrac{x}{6}+\dfrac{y}{10}=3$

$\dfrac{5x}{16}-\dfrac{3y}{16}=\dfrac{15}{8}$

Multiply the first equation by 30 and the second equation by 16 to clear the denominators.

$30\left(\frac{x}{6}+\frac{y}{10}\right)=30(3)$

$5x+3y=90$

$16\left(\frac{5x}{16}-\frac{3y}{16}\right)=16\left(\frac{15}{8}\right)$

$5x-3y=30$

Now, we have $5x+3y=90$

$5x-3y=30$

Solve the second equation for y.

$5x-3y=30$

$5x-5x-3y=-5x+30$

$-3y=-5x+30$

$\frac{-3y}{-3}=\frac{-5x+30}{-3}$

$y=\frac{5x}{3}-10$

Substitute this equation into the first equation.

$5x+3y=90$

$5x+3\left(\frac{5x}{3}-10\right)=90$

$5x+5x-30=90$

$10x-30=90$

$10x-30+30=90+30$

$10x=120$

$\frac{10x}{10}=\frac{120}{10}$

$x=12$

Find y.

$y=\frac{5(12)}{3}-10$

$y=\frac{60}{3}-10$

$y=20-10$

$y=10$

The solution is $(12, 10)$.

5. a) $2x+y=1$

$5x-y=20$

$7x\quad=21$

$\frac{7x}{7}=\frac{21}{7}$

$x=3$

Find y.

$2x+y=1$

$2(3)+y=1$

$6+y=1$

$6-6+y=1-6$

$y=-5$

The solution is $(3, -5)$.

b) $x + 8y = 7$
$\underline{x - 4y = 1}$

Multiply the second equation by -1.

$$\begin{array}{r} x + 8y = 7 \\ \underline{-x + 4y = -1} \\ 12y = 6 \end{array}$$

$$\frac{12y}{12} = \frac{6}{12}$$
$$y = \frac{1}{2}$$

Find x.

$$x + 8y = 7$$
$$x + 8(\tfrac{1}{2}) = 7$$
$$x + 4 = 7$$
$$x + 4 - 4 = 7 - 4$$
$$x = 3$$

The solution is $\left(3,\ \frac{1}{2}\right)$.

c) $5a + b = 2$
$\underline{3a + 2b = 11}$

Multiply the first equation by -2.

$$-2(5a + b) = -2(2)$$
$$-10a - 2b = -4$$

Now, we have

$$\begin{array}{r} -10a - 2b = -4 \\ \underline{3a + 2b = 11} \\ -7a \quad = 7 \end{array}$$

$$\frac{-7a}{-7} = \frac{7}{-7}$$
$$a = -1$$

Find b.

$$5a + b = 2$$
$$5(-1) + b = 2$$
$$-5 + b = 2$$
$$-5 + 5 + b = 2 + 5$$
$$b = 7$$

The solution is $(-1,\ 7)$.

d) $11x + 3y = 27$
$\underline{8x + 4y = 36}$

Multiply the first equation by -4 and the second equation by 3.

$$-4(11x + 3y) = -4(27)$$
$$-44x - 12y = -108$$
$$3(8x + 4y) = 3(36)$$
$$24x + 12y = 108$$

Now, we have

$$\begin{array}{r} -44x - 12y = -108 \\ \underline{24x + 12y = 108} \\ -20x \quad = 0 \end{array}$$

$$\frac{-20x}{-20} = \frac{0}{-20}$$
$$x = 0$$

Find y.

$$11x + 3y = 27$$
$$11(0) + 3y = 27$$
$$0 + 3y = 27$$
$$\frac{3y}{3} = \frac{27}{3}$$
$$y = 9$$

The solution is $(0,\ 9)$.

e) $9x + 3y = 15$
$\underline{3x = 5 - y}$

Put the second equation into proper form.

$$3x = 5 - y$$
$$3x + y = 5 \quad y + y$$
$$3x + y = 5$$

Multiply this equation by -3.

$$-3(3x + y) = -3(5)$$
$$-9x - 3y = -15$$

Now, we have

$$\begin{array}{r} 9x + 3y = 15 \\ \underline{-9x - 3y = -15} \\ 0 \quad = 0 \end{array}$$

The equations are dependent. There are infinitely many solutions.

f) $\dfrac{x}{3} + \dfrac{y+2}{2} = 1$

$\dfrac{x+8}{8} + \dfrac{y-3}{3} = 0$

Put both equations into proper form. Multiply the first equation by 6 and the second equation by 24 to get rid of the denominator.

$$6\left(\tfrac{x}{3} + \tfrac{y+2}{2}\right) = 6(1)$$
$$2x + 3(y+2) = 6$$
$$2x + 3y + 6 = 6$$
$$2x + 3y + 6 - 6 = 6 - 6$$
$$2x + 3y = 0$$
$$24\left(\tfrac{x+8}{8} + \tfrac{y-3}{3}\right) = 24(0)$$
$$3(x+8) + 8(y-3) = 0$$
$$3x + 24 + 8y - 24 = 0$$
$$3x + 8y = 0$$

Now, we have

$$2x + 3y = 0$$
$$3x + 8y = 0$$

Multiply the first equation by -3 and the second equation by 2.

$$-3(2x + 3y) = -3(0)$$
$$-6x - 9y = 0$$
$$2(3x + 8y) = 2(0)$$
$$6x + 16y = 0$$

Now, we're ready to add equations.

$$-6x - 9y = 0$$
$$6x + 16y = 0$$
$$7y \;\; = 0$$

$$\tfrac{7y}{7} = \tfrac{0}{7}$$
$$y = 0$$

Find x.

$$2x + 3y = 0$$
$$2x + 3(0) = 0$$
$$2x + 0 = 0$$
$$2x = 0$$
$$\tfrac{2x}{2} = \tfrac{0}{2}$$
$$x = 0$$

The solution is $(0,\ 0)$.

g) $0.02x + 0.05y = 0$

$0.3x - 0.2y = -1.9$

Multiply the first equation by 100 and the second equation by 10 to get rid of the decimals.

$$100(0.02x + 0.05y) = 100(0)$$
$$2x + 5y = 0$$
$$10(0.3x - 0.2y) = 10(-1.9)$$
$$3x - 2y = -19$$

We have $2x + 5y = 0$

$3x - 2y = -19$

Multiply the first equation by -3 and the second equation by 2.

$$-3(2x + 5y) = -3(0)$$
$$-6x - 15y = 0$$
$$2(3x - 2y) = 2(-19)$$
$$6x - 4y = -38$$

Now, we're ready to add equations.

$$-6x - 15y = 0$$
$$6x - 4y = -38$$
$$-19y \;\; = -38$$

$$\tfrac{-19y}{-19} = \tfrac{-38}{-19}$$
$$y = 2$$

Find x.

$$2x + 5y = 0$$
$$2x + 5(2) = 0$$
$$2x + 10 = 0$$
$$2x + 10 - 10 = 0 - 10$$
$$2x = -10$$
$$\tfrac{2x}{2} = \tfrac{-10}{2}$$
$$x = -5$$

The solution is $(-5,\ 2)$.

h) $-\frac{1}{4}x = 1 - \frac{2}{3}y$

$6(x-3y)+2y=5$

Multiply the first equation by 12 to get rid of the denominator and put the second equation in proper form.

$12\left(-\frac{1}{4}x\right) = 12\left(1-\frac{2}{3}y\right)$

$-3x = 12 - 8y$

$-3x + 8y = 12 - 8y + 8y$

$-3x + 8y = 12$

$6(x-3y)+2y=5$

$6x - 18y + 2y = 5$

$6x - 16y = 5$

We have $-3x + 8y = 12$

$6x - 16y = 5$

Multiply the first equation by 2.

$2(-3x+8y) = 2(12)$

$-6x + 16y = 24$

Now, we're ready to add equations.

$-6x + 16y = 24$

$6x - 16y = 5$

$0 \quad = 29$

The system is inconsistent. There are no solutions.

7. **Analyze:** Some people are dying from heart disease; others are dying from stroke. The total number of deaths was 900,000. More people were dying from heart disease.

Form: Let x represent the number of people dying from strokes and y represent the number of people dying from heart disease.

from heart disease is 5 times # from stroke.

$y = 5x$

from heart disease + # from stroke is 900,000.

$x + y = 900,000$

Solve: $y = 5x$

$x + y = 900,000$

Substitute the first equation into the second equation.

$x + y = 900,000$

$x + (5x) = 900,000$

$6x = 900,000$

$\frac{6x}{6} = \frac{900,000}{6}$

$x = 150,000$

Find y.

$x + y = 900,000$

$(150,000) + y = 900,000$

$150,000 - 150,000 + y$

$= 900,00 - 150,00$

$y = 750,000$

State: There were 150,00 people who died from stroke and 750,000 who died from heart disease.

Check: $y = 5x$

$(750,000) = 5(150,000)$

$750,000 = 750,000$

9. **Analyze:** We need to find the area of a rectangle. We know the perimeter is 420 yds. We also know that the width is three-fourths of the length.

Form: Let l represent the length of the rectangle and w represent the width of the rectangle.

2•(length) + 2•(width) is 420.

$2l + 2w = 420$

width is three-fourths the length

$w = \frac{3}{4}l$

Solve: $2l + 2w = 420$

$w = \frac{3}{4}l$

Substitute the second equation into the first equation.

$2l + 2w = 420$

$2l + 2(\frac{3}{4}l) = 420$

$2l + \frac{3}{2}l = 420$

$\frac{4}{2}l + \frac{3}{2}l = 420$

$\frac{7}{2}l = 420$

$2\left(\frac{7}{2}l\right) = 2(420)$

$7l = 840$

$\frac{7l}{7} = \frac{840}{7}$

$l = 120$

Find w.

$w = \frac{3}{4}l$

$w = \frac{3}{4}(120)$

$w = 90$

Area: $A = l \cdot w$

$A = (120)(90)$

$A = 10800$

State: The area of the rectangle is 10,800 ft^2.

Check: $2l + 2w = 420$

$2(120) + 2(90) = 420$

$240 + 180 = 420$

$420 = 420$

11. Analyze: The merchant needs to mix some of the gummy bears with some of the gummy worms. We need to find the amounts of each candy that is used.

Form: Let x represent the number of pounds of gummy bears and y represent the number of gummy worms.

Weight:

lbs bears + # lbs. worms is 30.

$x + y = 30$

Value:

1.5(# lbs bears) + 3(# lbs worms) is 2.10

$1.5x + 3y = 2.10(30)$

Solve: $x + y = 30$

$1.5x + 3y = 63$

Multiply the second equation by 10 to get rid of the decimals.

$10(1.5x + 3y) = 10(63)$

$15x + 30y = 630$

Solve the first equation for x to substitute into the second equation.

$x + y = 30$

$x + y - y = 30 - y$

$x = 30 - y$

Substitute this equation into the second equation.

$15x + 30y = 630$

$15(30 - y) + 30y = 630$

$450 - 15y + 30y = 630$

$450 + 15y = 630$

$450 - 450 + 15y = 630 - 450$

$15y = 180$

$\frac{15y}{15} = \frac{180}{15}$

$y = 12$

Find x.

$x + y = 30$

$x + (12) = 30$

$x + 12 - 12 = 30 - 12$

$x = 18$

State: The merchant should use 12 pounds of worms (yuck!) and 18 pounds of bears to make the mixture.

Check: $1.5x + 3y = 2.10$

$1.5(18) + 3(12) = 63$

$27 + 36 = 63$

$63 = 63$

13. Analyze: We need to find the price of a bottle of lens cleaner and the price of a bottle of soaking solution.

Form: Let x represent the price of a bottle of lens cleaner and y represent the price of a bottle of soaking solution.

Value:

2($ of cleaner) + 3($ of sol.) is 29.40.

$2x + 3y = 29.40$

3($ of cleaner) + 2($ of sol.) is 28.60

$3x + 2y = 28.60$

Solve: $2x + 3y = 29.40$

$3x + 2y = 28.60$

Multiply the first equation by -3 and the second equation by 2 in order to set up the x-terms to be eliminated.

$-3(2x + 3y) = -3(29.40)$

$-6x - 9y = -88.20$

$2(3x + 2y) = 2(28.60)$

$6x + 4y = 57.20$

Now, we have

$$\begin{array}{r} -6x - 9y = -88.20 \\ 6x + 4y = 57.20 \\ \hline -5y = -31 \end{array}$$

$\frac{-5y}{-5} = \frac{-31}{-5}$

$y = 6.2$

Find x.

$3x + 2y = 28.60$
$3x + 2(6.2) = 28.60$
$3x + 12.4 = 28.60$
$3x + 12.4 - 12.4 = 28.60 - 12.4$
$3x = 16.2$
$\frac{3x}{3} = \frac{16.2}{3}$
$x = 5.4$

State: The lens cleaner costs $5.40 per bottle and the soaking solution costs $6.20 per bottle.

Check: $2x + 3y = 29.40$
$2(5.4) + 3(6.2) = 29.40$
$10.8 + 18.6 = 29.40$
$29.40 = 29.40$

15. **Analyze:** Some of the 40% antifreeze solution must be mixed with some of the 70% solution to make a 50% solution. The total number of gallons of the 50% solution is 20.

Form: Let x represent the number of gallons of 40% solution and y represent the number of gallons of the 70% solution.

Amount of solution:
gallons of 40% sol. + gallons of 70% sol. is 20.

$x + y = 20$

Amount of antifreeze:
0.4(gallons of 40% sol.)
+ 0.7(gallons of 70% sol.)
is 0.5(20)

$0.4x + 0.7y = 10$

Solve: $x + y = 20$
$0.4x + 0.7y = 10$

Solve the first equation for y to substitute into the second equation.

$x + y = 20$
$x - x + y = -x + 20$
$y = -x + 20$

Substitute this equation into the second equation.

$0.4x + 0.7y = 10$
$0.4x + 0.7(-x + 20) = 10$
$0.4x - 0.7x + 14 = 10$
$-0.3x + 14 = 10$
$-0.3x + 14 - 14 = 10 - 14$
$-0.3x = -4$
$\frac{-0.3x}{-0.3} = \frac{-4}{-0.3}$
$x = \frac{40}{3}$

Find y.

$x + y = 20$
$\frac{40}{3} + y = 20$
$\frac{40}{3} - \frac{40}{3} + y = 20 - \frac{40}{3}$
$y = \frac{60}{3} - \frac{40}{3}$
$y = \frac{20}{3}$

$\frac{40}{3} = 13\frac{1}{3}$ and $\frac{20}{3} = 6\frac{2}{3}$

State: The mechanic should mix $13\frac{1}{3}$ gallons of 40% solution with $6\frac{2}{3}$ gallons of 70%.

Check: $0.4x + 0.7y = 10$
$0.4\left(\frac{40}{3}\right) + 0.7\left(\frac{20}{3}\right) = 10$
$\frac{4}{10}\left(\frac{40}{3}\right) + \frac{7}{10}\left(\frac{20}{3}\right) = 10$
$\frac{16}{3} + \frac{14}{3} = 10$
$\frac{30}{3} = 10$
$10 = 10$

17. a) $x - y < 5$
$x - x - y < -x + 5$
$-y < -x + 5$
$-1(-y) > -1(-x + 5)$
$y > x - 5$

Graph the line $y = x - 5$. Use a dotted line because of the " $>$ ".
Use (0, 0) as the test point.

$y > x - 5$
$(0) > (0) - 5$
$0 > -5$

Shade the region with the point (0,0).

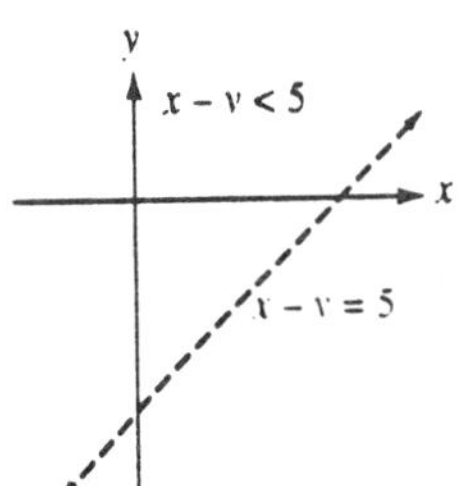

b) $2x - 3y \geq 6$

$2x - 2x - 3y \geq -2x + 6$

$-3y \geq -2x + 6$

$\frac{-3y}{-3} \leq \frac{-2x+6}{-3}$

$y \leq \frac{2x}{3} - 2$

Graph the line $y = \frac{2x}{3} - 2$. Use a solid line because of the " $\leq$ ".

Use (0, 0) as the test point.

$y \leq \frac{2x}{3} - 2$

$(0) \leq \frac{2(0)x}{3} - 2$

$0 \leq 0 - 2$

$0 \leq -2$

Shade the region without the point (0,0).

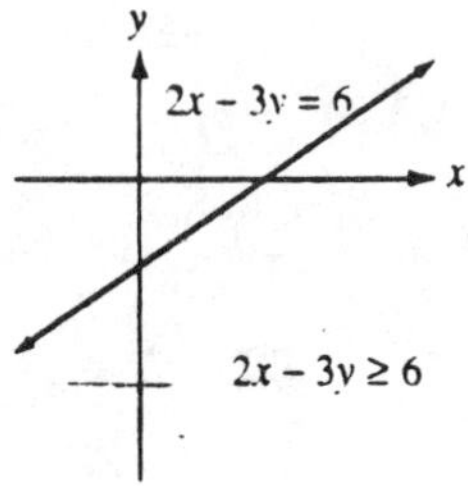

c) $y \leq -2x$

Graph the line $y = -2x$. Use a solid line because of the " $\leq$ ".

Use (1, 1) as the test point.

$y \leq -2x$

$(1) \leq -2(1)$

$1 \leq -2$

Shade the region without the point (1, 1).

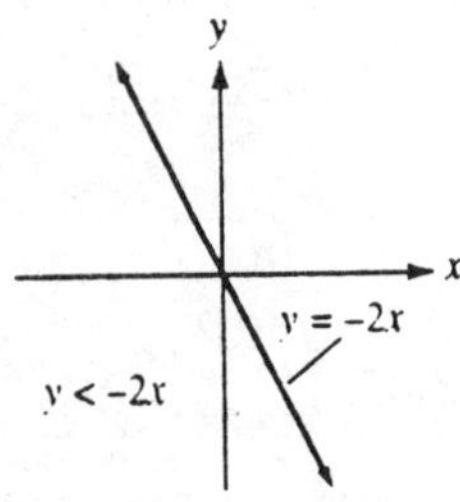

d) $y < -4$

Graph the horizontal line $y = -4$. Use a dotted line because of the " $<$ ".

Use (0, 0) as the test point.

$y < -4$

$(0) < -4$

$0 < -4$

Shade the region without the point (1, 1).

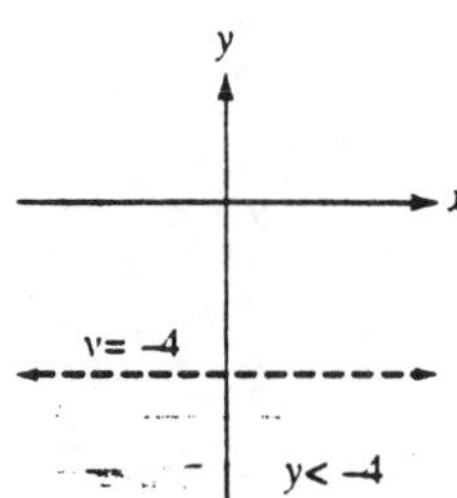

19. Analyze: The student can work a total of 30 hours a week. She can work either 3 hour shifts or 5 hour shifts.

Form: Let x represent the number of 3 hour shifts and y represent the number of 5 hour shifts.

Hours worked:

3(# 3-hour shifts) + 5(# 5-hour shifts) is less than or equal to 30.

$3x + 5y \leq 30$

Solve: $3x + 5y \leq 30$

$3x - 3x + 5y \leq -3x + 30$

$5y \leq -3x + 30$

$\frac{5y}{5} \leq \frac{-3x+30}{5}$

$y \leq -\frac{3x}{5} + 6$

Graph the line $y = -\frac{3x}{5} + 6$.

Use a solid line because of the " $\leq$ ". Use (0, 0) as the test point.

$3x + 5y \leq 30$

$3(0) + 5(0) \leq 30$

$0 \leq 30$

Shade the region without the point $(0, 0)$.

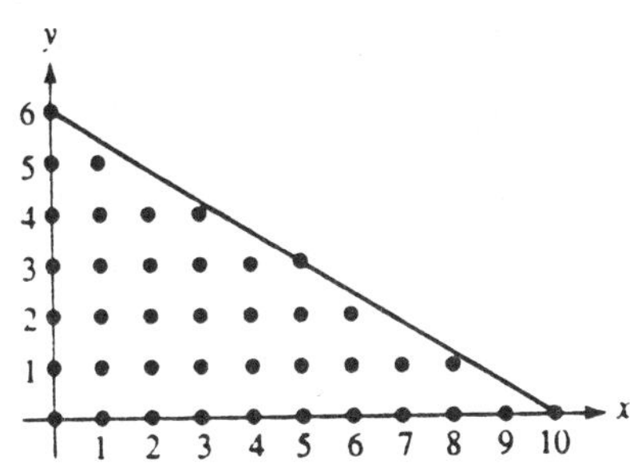

Possible ordered pairs:
(2, 4), (5, 3), or (6, 2)

21. **Analyze:** She wants to spend between \$40 and \$60. She will by T-shirts and pants.
Form: Let x represent the number of T-shirts and y represent the number of pants.
Value:
10(# of T-shirts) + 20(# of pants)
is more than or equal to 40.
$10x + 20y \geq 40$
10(# of T-shirts) + 20(# of pants)
is less than or equal to 60.
$10x + 20y \leq 60$
Solve: $10x + 20y \geq 40$
$10x + 20y \leq 60$
First inequality:
$10x + 20y \geq 40$
$10x - 10x + 20y \geq -10x + 40$
$20y \geq -10x + 40$
$\frac{20y}{20} \geq \frac{-10x+40}{20}$
$y \geq -\frac{1}{2}x + 2$
Second inequality:
$10x + 20y \leq 60$
$10x - 10x + 20y \leq -10x + 60$
$20y \leq -10x + 60$
$\frac{20y}{20} \leq \frac{-10x+60}{20}$
$y \leq -\frac{1}{2}x + 3$
Graph the line $y = -\frac{1}{2}x + 2$. Use a solid line because of the " $\geq$ ". Graph the line $y = -\frac{1}{2}x + 3$. Use a solid line because of the " $\leq$ ".

Use (0, 0) as the test point for both inequalities.
$10x + 20y \geq 40$
$10(0) + 20(0) \geq 40$
$0 + 0 \geq 40$
$0 \geq 40$
For the first inequality, shade the region without the point $(0, 0)$.
$10x + 20y \leq 60$
$10(0) + 20(0) \leq 60$
$0 + 0 \leq 60$
$0 \leq 60$
For the second inequality, shade the region with the point $(0, 0)$.

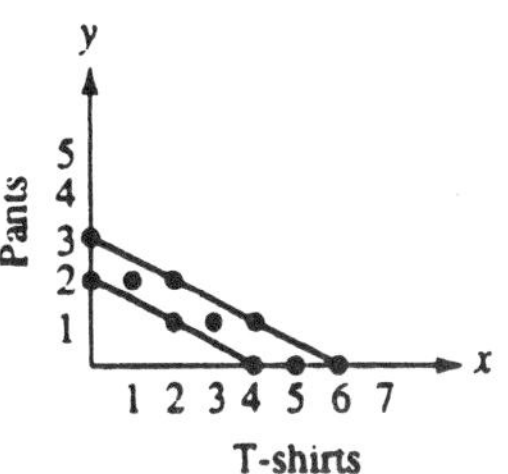

Two possible solutions:
(3, 1)--3 T-shirts and 1 pair of pants.
(1, 2)--1 T-shirt and 2 pairs of pants.

STUDY SET Section 5.1

VOCABULARY

1. base; exponent

3. power

CONCEPTS

5. $(3x)^4$ means $(3x)(3x)(3x)(3x)$

7. $x^m x^n = x^{m+n}$

9. $\left(\frac{a}{b}\right)^n = \frac{a^n}{b^n}$

11. $\frac{x^m}{x^n} = x^{m-n}$

13. $(xy) = (xy)^1$

15. a) $x^2 + x^2 = 2x^2$
b) $x^2 - x^2 = 0$
c) $x^2 \bullet x^2 = x^4$
d) $\frac{x^2}{x^2} = x^{2-2} = x^0 = 1$

17. a) $6x^3 + 2x^2$ Because the exponents are not the same, we cannot simplify.
b) $6x^3 - 2x^2$ Because the exponents are not the same, we cannot simplify.
c) $6x^3 \bullet 2x^2 = 6(2)x^3 \bullet x^2$
$= 12x^{3+2} = 12x^5$

d) $\frac{6x^3}{2x^2} = \frac{6}{2}\frac{x^3}{x^2} = 3x^{3-2} = 3x^1 = 3x$

19. $A = l \bullet w$.
Each side has length a^5.
So, $A = a^5 a^5 = a^{5+5} = a^{10}$
The area is a^{10} mi^2.

21. $V = l \bullet w \bullet h$
$V = (6x^2)(3x^3)(2x^2)$
$V = 6(3)(2)(x^2)(x^3)(x^2)$
$V = 36x^{2+3+2}$
$V = 36x^7$
Volume is $36x^7$ m^3.

23. Evaluate $x = 1$:
$3^x = 3^1 = 3$
Evaluate $x = 2$:
$3^x = 3^2 = 9$
Evaluate $x = 3$:
$3^x = 3^3 = 27$
Evaluate $x = 4$:
$3^x = 3^4 = 81$

25. Evaluate $x = 1$:
$(-4)^x = (-4)^1 = -4$
Evaluate $x = 2$:
$(-4)^x = (-4)^2 = 16$
Evaluate $x = 3$:
$(-4)^x = (-4)^3 = -64$
Evaluate $x = 4$:
$(-4)^x = (-4)^4 = 256$

27. Letting each grid represent 1 unit, the corresponding y-values are found for each x-value from the graph.

x	y
1	2
2	4
3	27
4	16

NOTATION

29. $(x^4x^2)^3 = (x^{4+2})^3 = (x^6)^3$

$= x^{6 \bullet 3} = x^{18}$

PRACTICE

31. 4^3 has base 4 and exponent 3.

33. x^5 has base x and exponent 5.

35. $(-3x)^2$ has base $-3x$ and exponent 2.

37. $-\frac{1}{3}y^6$ has base y and exponent 6. $-\frac{1}{3}$ is a coefficient.

39. $5^3 = 5 \bullet 5 \bullet 5$

41. $x^6 = x \bullet x \bullet x \bullet x \bullet x \bullet x$

43. $-\frac{3}{4}x^5 = -\frac{3}{4}x{\bullet}x{\bullet}x{\bullet}x{\bullet}x$

45. $\left(\frac{1}{3}t\right)^3 = \left(\frac{1}{3}t\right)\left(\frac{1}{3}t\right)\left(\frac{1}{3}t\right)$

47. $4t(4t)(4t)(4t) = (4t)^4$

49. $-4{\bullet}t{\bullet}t{\bullet}t = -4t^3$

51. $2(5^4 - 4^3) = 2(625 - 64)$

$= 2(561) = 1,122$

53. $-5^2(3^4 + 4^3) = -5^2(81 + 64)$

$= -5^2(145) = -25(145) = -3,625$

55. $x^4x^3 = x^{4+3} = x^7$

57. $a^3aa = a^{3+1+5} = a^9$

59. $y^3(y^2y^4) = y^3(y^{2+4}) = y^3(y^6)$

$= y^{3+6} = y^9$

61. $4x^2(3x^5) = 4(3)x^2x^5 = 12x^{2+5} = 12x^7$

63. $(-y^2)(4y^3) = (-1)(4)y^2y^3$

$= -4y^{2+3} = -4y^5$

65. $(3^2)^4 = 3^{2{\bullet}4} = 3^8$

67. $(y^5)^3 = y^{5{\bullet}3} = y^{15}$

69. $(x^2x^3)^5 = (x^{2+3})^5 = (x^5)^5 = x^{5{\bullet}5} = x^{25}$

71. $(3zz^2z^3)^5 = (3z^{1+2+3})^5 = (3z^6)^5$

$= 3^5z^{6{\bullet}5} = 243z^{30}$

73. $(x^5)^2(x^7)^3 = x^{5{\bullet}2}x^{7{\bullet}3} = x^{10}x^{21}$

$= x^{10+21} = x^{31}$

75. $(xy)^3 = x^3y^3$

77. $(r^3s^2)^2 = r^{3{\bullet}2}s^{2{\bullet}2} = r^6s^4$

79. $(4ab^2)^2 = 4^2a^2b^{2{\bullet}2} = 16a^2b^4$

81. $(-2r^2s^3)^3 = (-2)^3(r^2)^3(s^3)^3$

$= (-8)r^{2{\bullet}3}s^{3{\bullet}3} = -8r^6s^9$

83. $\left(\frac{a}{b}\right)^3 = \frac{a^3}{b^3}$

85. $\left(\frac{x^2}{y^3}\right)^5 = \frac{(x^2)^5}{(y^3)^5} = \frac{x^{2+5}}{y^{3+5}} = \frac{x^7}{y^8}$

87. $(-2a)^5 = (-2)^5(a)^5 = -32a^5$

89. $\left(\frac{b^2}{3a}\right)^3 = \frac{(b^2)^3}{(3a)^3} = \frac{b^{2{\bullet}3}}{3^3a^3} = \frac{b^6}{27a^3}$

91. $\frac{x^5}{x^3} = x^{5-3} = x^2$

93. $\frac{y^3y^4}{yy^2} = \frac{y^{3+4}}{y^{1+2}} = \frac{y^7}{y^3} = y^{7-3} = y^4$

95. $\frac{12a^2a^3a^4}{4(a^4)^2} = \frac{12a^{2+3+4}}{4a^{4{\bullet}2}} = \frac{12a^9}{4a^6}$

$= \frac{12}{4}\frac{a^9}{a^6} = 3a^{9-6} = 3a^3$

97. $\frac{(ab^2)^3}{(ab)^2} = \frac{a^3b^{2{\bullet}3}}{a^2b^2} = \frac{a^3b^6}{a^2b^2}$

$= a^{3-2}b^{6-2} = ab^4$

99. $\frac{20(r^4s^3)^4}{6(rs^3)^3} = \frac{20r^{4{\bullet}4}s^{3{\bullet}4}}{6r^3s^{3{\bullet}3}} = \frac{20r^{16}s^{12}}{6r^3s^9}$

$= \frac{20}{6}\frac{r^{16}}{r^3}\frac{s^{12}}{s^9} = \frac{10}{3}r^{16-3}s^{12-9} = \frac{10}{3}r^{13}s^3$

101. $\left(\frac{y^3y}{2yy^2}\right)^3 = \left(\frac{y^{3+1}}{2y^{1+2}}\right)^3 = \left(\frac{y^4}{2y^3}\right)^3$

$= \left(\frac{1}{2}y^{4-3}\right)^3 = \left(\frac{1}{2}y\right)^3 = (\frac{1}{2})^3y^3$

$= \frac{1}{8}y^3 = \frac{y^3}{8}$

APPLICATIONS

103. a) Area of square is $A = s^2$.
Since height is $5x$, each side has length $5x$.
$A = 5x \cdot 5x = 5(5)x \cdot x = 25x^2$
The area of the square is $25x^2$ ft^2.
b) Area of circle is $A = \pi r^2$.
Since distance from waist to feet is $3x$, the radius is $3x$.
$A = \pi(3x)^2 = \pi \cdot 3^2 x^2 = \pi \cdot 9x^2 = 9\pi x^2$
The are of the circle is $9\pi x^2$ ft^2.

105. a)

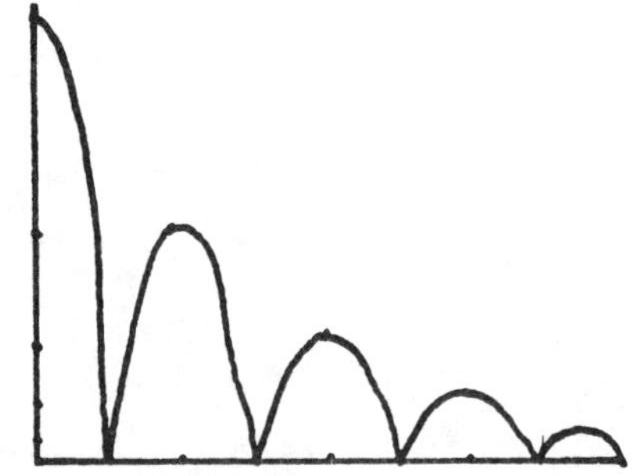

b) $32(\frac{1}{2})$ is the height on the first bounce because it rebounds to half of the original height of 32 ft. The height of the second bounce is half of first bounce:
$\frac{1}{2}(32(\frac{1}{2})) = 32(\frac{1}{2})^{1+1} = 32(\frac{1}{2})^2$. The third bounce is half of the second:
$\frac{1}{2}(32(\frac{1}{2})^2) = 32(\frac{1}{2})^{2+1} = 32(\frac{1}{2})^3$. Notice that the exponent matches the bounce.
So, the height of the fourth bounce is $32(\frac{1}{2})^4$. Calculating these:
$32(\frac{1}{2}) = 16$ ft. on the 1st bounce,
$32(\frac{1}{2})^2 = 32(\frac{1}{4}) = 8$ ft. on the 2nd,
$32(\frac{1}{2})^3 = 32(\frac{1}{8}) = 4$ ft. on the 3rd, and
$32(\frac{1}{2})^4 = 32(\frac{1}{16}) = 2$ ft. on the 4th.

107. Each pack has 12 pencils. Since there are 12 packs in a box, we have
$12 \cdot 12 = 12^{1+1} = 12^2$ pencils in a box.
Since there are 12 boxes in a carton, we have $12(12^2) = 12^{2+1} = 12^3$ pencils in a carton. Since there are 12 cartons in a case, we have $12(12^3) = 12^{3+1} = 12^4$ pencils in a case.

109. An investment doubles every 7 years. Since $28 \div 7 = 4$, the money will double 4 times. Thus, after 7 years we have $2 \cdot 1000 = 2000$. After 7 more years, we have $2 \cdot 2000 = 4000$. After 7 more years, we have $2 \cdot 4000 = 8000$. After the last 7 years, we have $2 \cdot 8000 = 16000$. Thus, the investment is worth \$16,000 after 28 years.

REVIEW

115. $y = 2x - 1$. The graph of this equation has slope 2 and y-intercept $(0, -1)$.

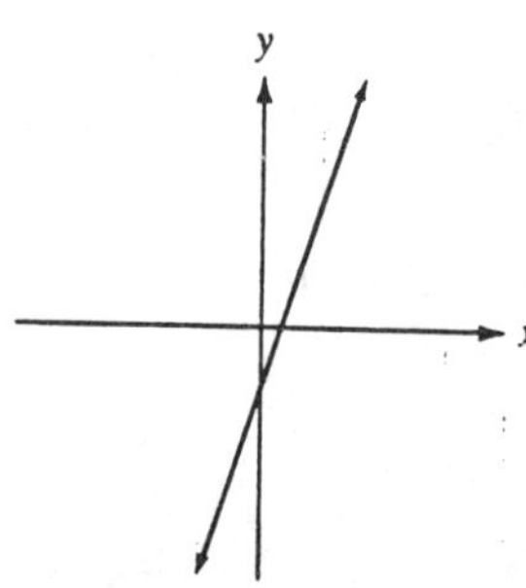

117. $y = 3$. The graph of this line will be a horizontal line with y-intercept $(0, 3)$.

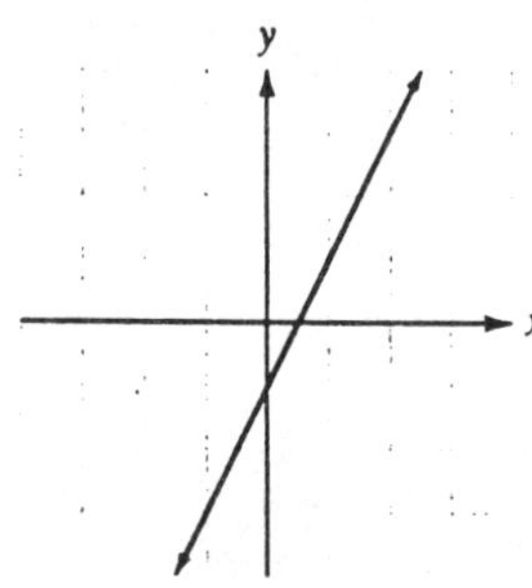

STUDY SET Section 5.2

VOCABULARY

1. base; exponent

3. negative

CONCEPTS

5. a) $\frac{6^4}{6^4} = 6^{4-4} = 6^0$
   b) $\frac{6^4}{6^4} = \frac{6•6•6•6}{6•6•6•6} = 1$
   c) So, we define 6^0 to be 1, and in general, if x is any nonzero number, then $x^0 = 1$.

7. Evaluate $x = 2$:
   $3^x = 3^2 = 9$
   Evaluate $x = 1$:
   $3^x = 3^1 = 3$
   Evaluate $x = 0$
   $3^x = 3^0 = 1$
   Evaluate $x = -1$:
   $3^x = 3^{-1} = \frac{1}{3}$
   Evaluate $x = -2$:
   $3^x = 3^{-2} = \frac{1}{3^2} = \frac{1}{9}$

9. Evaluate $x = 2$:
   $(-9)^x = (-9)^2 = 81$
   Evaluate $x = 2$:
   $(-9)^x = (-9)^1 = -9$
   Evaluate $x - 0$:
   $(-9)^x = (-9)^0 = 1$
   Evaluate $x = -1$:
   $(-9)^x = (-9)^{-1} = \frac{1}{-9} = -\frac{1}{9}$
   Evaluate $x = -2$:
   $(-9)^x = (-9)^{-2} = \frac{1}{(-9)^2} = \frac{1}{81}$

11.

x	y	*power*
2	4	2^2
1	2	2^1
0	1	2^0
-1	$\frac{1}{2}$	2^{-1}
-2	$\frac{1}{4}$	2^{-2}

NOTATION

13. $(y^5y^3)^{-5} = (y^{5+3})^{-5} = (y^8)^{-5}$
    $= y^{8•(-5)} = y^{-40} = \frac{1}{y^{40}}$

15. For $3x^{-2}$, x is the base and -2 is the exponent. 3 is the coefficient.

17. a) For -4^2, 4 is the base and 2 is the exponent. $-4^2 = -16$.
    b) For 4^{-2}, 4 is the base and -2 is the exponent. $4^{-2} = \frac{1}{4^2} = \frac{1}{16}$.
    c) For -4^{-2}, 4 is the base and -2 is the exponent. $-4^{-2} = -\frac{1}{4^2} = -\frac{1}{16}$.

PRACTICE

19. $7^0 = 1$

21. $\left(\frac{1}{4}\right)^0 = 1$

23. $2x^0 = 2•1 = 2$

25. $(-x)^0 = 1$

27. $\left(\frac{a^2b^3}{ab^4}\right)^0 = 1$

29. $\frac{5}{2x^0} = \frac{5}{2•1} = \frac{5}{2}$

31. $12^{-2} = \frac{1}{12^2} = \frac{1}{144}$

33. $(-4)^{-1} = \frac{1}{-4} = -\frac{1}{4}$

35. $\frac{1}{5^{-3}} = \frac{1}{\frac{1}{5^3}} = 1 \div \frac{1}{5^3} = 1•\frac{5^3}{1} = 5^3 = 125$

37. $\frac{2^{-4}}{3^{-1}} = \frac{\frac{1}{2^4}}{\frac{1}{3^1}} = \frac{1}{2^4} \div \frac{1}{3^1} = \frac{1}{2^4}•\frac{3^1}{1}$
    $= \frac{3^1}{2^4} = \frac{3}{16}$

39. $-4^{-3} = -\frac{1}{4^3} = -\frac{1}{64}$

41. $-(-4)^{-3} = -\frac{1}{(-4)^3}$
    $= -\frac{1}{-64} = \frac{1}{64}$

43. $x^{-2} = \frac{1}{x^2}$

45. $-b^{-5} = -\frac{1}{b^5}$

47. $(2y)^{-4} = \frac{1}{(2y)^4} = \frac{1}{2^4y^4} = \frac{1}{16y^4}$

49. $(ab^2)^{-3} = \frac{1}{(ab^2)^3} = \frac{1}{a^3b^{2\cdot3}} = \frac{1}{a^3b^6}$

51. $2^5 \cdot 2^{-2} = \frac{2^5}{2^2} = 2^{5-2} = 2^3 = 8$

53. $4^{-3} \cdot 4^{-2} \cdot 4^5 = 4^{(-3)+(-2)+5}$

$= 4^0 = 1$

55. $\left(\frac{7}{8}\right)^{-1} = \frac{1}{\frac{7}{8}} = 1 \div \frac{7}{8} = 1 \cdot \frac{8}{7} = \frac{8}{7}$

57. $\frac{3^5 \cdot 3^{-2}}{3^3} = \frac{3^{5+(-2)}}{3^3} = \frac{3^3}{3^3} = 3^{3-3}$

$= 3^0 = 1$

59. $\frac{y^4}{y^5} = y^{4-5} = y^{-1} = \frac{1}{y}$

61. $\frac{(r^2)^3}{(r^3)^4} = \frac{r^{2\cdot3}}{r^{3\cdot4}} = \frac{r^6}{r^{12}} = r^{6-12}$

$= r^{-6} = \frac{1}{r^6}$

63. $\frac{y^4y^3}{y^4y^{-2}} = \frac{y^{4+3}}{y^{4+(-2)}} = \frac{y^7}{y^2} = y^{7-2} = y^5$

65. $\frac{10a^4a^{-2}}{5a^2a^0} = \frac{10}{5}\frac{a^{4+(-2)}}{a^{2+0}} = 2\frac{a^2}{a^2}$

$= 2a^{2-2} = 2a^0 = 2 \cdot 1 = 2$

67. $(ab^2)^{-2} = \frac{1}{(ab^2)^2} = \frac{1}{a^2b^{2\cdot2}} = \frac{1}{a^2b^4}$

69. $(x^2y)^{-3} = \frac{1}{(x^2y)^3} = \frac{1}{x^{2\cdot3}y^3} = \frac{1}{x^6y^3}$

71. $(x^{-4}x^3)^3 = (x^{-4+3})^3 = (x^{-1})^3$

$= x^{-1\cdot3} = x^{-3} = \frac{1}{x^3}$

73. $(a^{-2}b^3)^{-4} = \frac{1}{(a^{-2}b^3)^4} = \frac{1}{a^{-2\cdot4}b^{3\cdot4}}$

$= \frac{1}{a^{-8}b^{12}} = \frac{a^8}{b12}$

75. $(-2x^3y^{-2})^{-5} = \frac{1}{(-2x^3y^{-2})^5}$

$= \frac{1}{(-2)^5x^{3\cdot5}y^{-2\cdot5}} = \frac{1}{-32x^{15}y^{-10}}$

$= \frac{y^{10}}{-32x^{15}} = -\frac{y^{10}}{32x^{15}}$

77. $\left(\frac{a^3}{a^{-4}}\right)^2 = \left(a^{3-(-4)}\right)^2 = (a^7)^2$

$= a^{7\cdot2} = a^{14}$

79. $\left(\frac{b^5}{b^{-2}}\right)^{-2} = \left(b^{5-(-2)}\right)^{-2} = (b^7)^{-2}$

$= b^{7\cdot(-2)} = b^{-14} = \frac{1}{b^{14}}$

81. $\left(\frac{4x^2}{3x^{-5}}\right)^4 = \left(\frac{4}{3}\frac{x^2}{x^{-5}}\right)^4 = \left(\frac{4}{3}x^{2-(-5)}\right)^4$

$= \left(\frac{4}{3}x^7\right)^4 = \left(\frac{4}{3}\right)^4 x^{7\cdot4} = \frac{4^4}{3^4}x^{28} = \frac{256x^{28}}{81}$

83. $\left(\frac{12y^3z^{-2}}{3y^{-4}z^3}\right)^2 = \left(\frac{12}{3}y^{3-(-4)}z^{-2-3}\right)^2$

$= (4y^7z^{-5})^2 = 4^2y^{7\cdot2}z^{-5\cdot2}$

$= 16y^{14}z^{-10} = \frac{16y^{14}}{z^{10}}$

85. $x^{2m}x^m = x^{2m+m} = x^{3m}$

87. $u^{2m}u^{-3m} = u^{2m-3m} = u^{-m} = \frac{1}{u^m}$

89. $y^{3m+2}y^{-m} = y^{3m+2+(-m)} = y^{2m+2}$

91. $\frac{y^{3m}}{y^{2m}} = y^{3m-2m} = y^m$

93. $\frac{x^{3n}}{x^{6n}} = x^{3n-6n} = x^{-3n} = \frac{1}{x^{3n}}$

95. $(x^{m+1})^2 = x^{(m+1)\bullet 2} = x^{2m+2}$

97. 661.2759 can be described as
$6 \times 100 = 6 \times 10^2$
$6 \times 10 = 6 \times 10^1$
$1 \times 1 = 1 \times 10^0$
$2 \times 0.1 = 2 \times \frac{1}{10} = 2 \times 10^{-1}$
$7 \times 0.01 = 7 \times \frac{1}{100} = 7 \times 10^{-2}$
$5 \times 0.001 = 2 \times \frac{1}{1000} = 2 \times 10^{-3}$
$9 \times 0.0001 = 9 \times \frac{1}{10000} = 9 \times 10^{-4}$

99. The formula is $P = A(1 + ni)^{-1}$.
From the information:
$A = \$100,000$; $P =$ what we want to find; $n = 40$ years; and $i = 8\% = 0.08$.
$P = 100,000(1 + 40\bullet 0.08)^{-1}$
$P = 100,000(1 + 3.2)^{-1}$
$P = 100,000(4.2)^{-1}$
$P = \frac{100,000}{4.2}$
$P = 23,809.52$
So, the couple should invest \$23,809.52.

REVIEW

103. $2x + y > 4$ matches graph c. The " > " indicates the graph will have a dotted line.

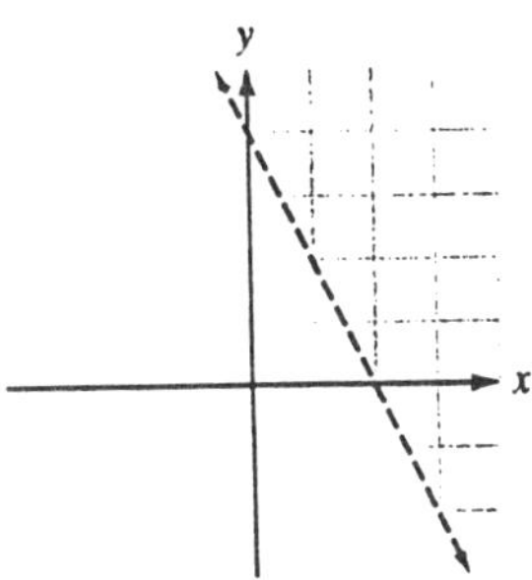

105. $2x + y \geq 4$ matches graph c. The " $\geq$ " indicates the graph will have a solid line.

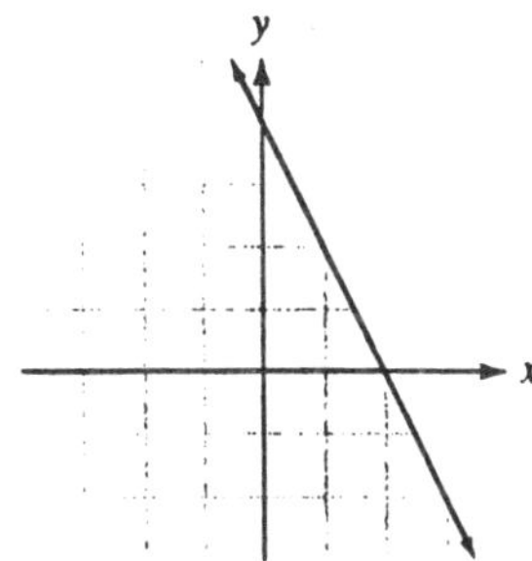

STUDY SET Section 5.3

VOCABULARY

1. scientific

CONCEPTS

3. $2.5 \times 10^2 = 250$

5. $2.5 \times 10^{-5} = 0.000025$

7. $387,000 = 3.87 \times 10^5$

9. $0.00387 = 3.87 \times 10^{-3}$

11. When multiplying a decimal by 10^5, the decimal point moves 5 places to the right.

13. Dividing a decimal by 10^4 is equivalent to multiplying it by 10^{-4}.

NOTATION

15. $63.7 \times 10^5 = 6.37 \times 10^1 \times 10^5$
$= 6.37 \times 10^{1+5}$
$= 6.37 \times 10^6$

PRACTICE

17. $23,000 = 2.3 \times 10^4$

19. $1,700,000 = 1.7 \times 10^6$

21. $0.062 = 6.2 \times 10^{-2}$

23. $0.0000051 = 5.1 \times 10^{-6}$

25. $42.5 \times 10^2 = 4.25 \times 10^3$

27. $0.25 \times 10^{-2} = 2.5 \times 10^{-3}$

29. $2.3 \times 10^2 = 230$

31. $8.12 \times 10^5 = 812,000$

33. $1.15 \times 10^{-3} = 0.0115$

35. $9.76 \times 10^{-4} = 0.000976$

37. $5 \times 10^6 = 25,000,000$

39. $0.51 \times 10^{-3} = 0.00051$

41. $25,700,000,000,000 = 2.57 \times 10^{13}$mi

43. $6.38 \times 10^7 = 63,800,000 \text{ mi}^2$

45. $0.00622 = 6.22 \times 10^{-3}$

47. $(3.4 \times 10^2)(2.1 \times 10^3)$
$= (3.4)(2.1) \times 10^2 10^3$
$= 7.14 \times 10^5$
$= 714,000$

49. $\frac{9.3\times10^2}{3.1\times10^{-2}} = \frac{9.3}{3.1} \times \frac{10^2}{10^{-2}}$

$= 3 \times 10^{2-(-2)}$
$= 3 \times 10^4$
$= 30,000$

51. $\frac{96{,}000}{(12{,}000)(0.00004)}$

$= \frac{9.6\times10^4}{(1.2\times10^4)(4\times10^{-5})}$

$= \frac{9.6\times10^4}{(1.2\times10^4)(4\times10^{-5})}$

$= \frac{9.6\times10^4}{(1.2)(4)\times10^4 10^{-5}}$

$= \frac{9.6\times10^4}{4.8\times10^{4+(-5)}}$

$= \frac{9.6\times10^4}{4.8\times10^{-1}}$

$= \frac{9.6}{4.8} \times \frac{10^4}{10^{-1}}$

$= 2 \times 10^{4-(-1)}$

$= 2 \times 10^5$

$= 200,000$

53. $9.038030748 \times 10^{15}$

55. $1.881676423 \times 10^{12}$

APPLICATIONS

57. gamma ray (8.9×10^{-14}) shortest
x-ray (2.3×10^{-11})
ultraviolet (6.1×10^{-8})
visible light (9.3×10^{-6})
Infrared (3.7×10^{-5})
microwave (1.1×10^{-2})
radio wave (3.0×10^{2}) longest

59. $(1.7 \times 10^{-24}) \bullet 1{,}000{,}000$
$= (1.7 \times 10^{-24}) \bullet (1 \times 10^{6})$
$= (1.7)(1) \times 10^{-24} \times 10^{6}$
$= 1.7 \times 10^{-24+6}$
$= 1.7 \times 10^{-18}$ grams

61. $(5.87 \times 10^{12}) \bullet 5280$
$= (5.87 \times 10^{12}) \bullet (5.28 \times 10^{3})$
$= (5.87)(5.28) \times 10^{12} 10^{3}$
$= 30.9936 \times 10^{12+3}$
$= 30.9936 \times 10^{15}$
$= (3.09936 \times 10^{1}) \times 10^{15}$
$= 3.09936 \times 10^{1+15}$
$= 3.09936 \times 10^{16}$ ft.

63. What is 4% of 3.79×10^{12}?
$x = 4\%$ of 3.79×10^{12}
$x = 0.04 \bullet 3.79 \times 10^{12}$
$x = (4 \times 10^{-2})(3.79 \times 10^{12})$
$x = (4)(3.79) \times 10^{-2} 10^{12}$
$x = 15.16 \times 10^{-2+12}$
$x = (1.516 \times 10^{1}) \times 10^{10}$
$x = 1.516 \times 10^{1+10}$
$x = 1.516 \times 10^{11}$ dollars

65. a) 1.7×10^{6} ; $1,700,000$
b) smallest: 1.5×10^{6}; $1,500,000$
largest: 2.05×10^{6}; $2,050,000$

REVIEW

69. If $y = -1$,
$-5y^{55} = -5(-1)^{55} = -5(-1) = 5$

71. commutative property of addition

73. $3(x-4) - 6 = 0$
$3x - 3(4) - 6 = 0$
$3x - 12 - 6 = 0$
$3x - 18 = 0$
$3x - 18 + 18 = 0 + 18$
$3x = 18$
$\frac{3x}{3} = \frac{18}{3}$
$x = 6$

STUDY SET Section 5.4

VOCABULARY

1. algebraic

3. polynomial

5. monomial

7. binomial

9. P of x

CONCEPTS

11. Yes, because all of the exponents are positive integers.

13. Yes, because all of the exponents are positive integers.

15. Binomial, because there are two terms: $3x$ and 7.

17. Trinomial, because there are three terms: y^2, $4y$, and 3.

19. Monomial, because there is one term: $3x^2$.

21. Binomial, because there are two terms: $5t$ and 32.

23. Trinomial, because there are three terms: s^2, $23s$, and 31.

25. None of these, because there are four terms: $3x^5$, x^4, $3x^3$, and 7.

27. 4th degree, because the largest exponent is $3x^4$.

29. 2nd degree, because the largest exponent is $-2x^2$.

31. 1st degree, because the largest exponent is $3x$.

33. 4th degree, because the exponents of $-5r^2s^2$ add together to 4.

35. 12th degree, because the largest exponent is x^{12}.

37. 0th degree, because 38 can be written as $38x^0$.

NOTATION

39. $-2x^2+3x-1 = -2(2)^2+3(2)-1$
$= -2(4)+6-1$
$= -8+6-1$
$= -2-1$
$= -3$

PRACTICE

41. $P(2) = 5(2)-3$
$= 10-3$
$= 7$

43. $P(-1) = 5(-1)-3$
$= -5-3$
$= -8$

45. $P(\frac{1}{5}) = 5(\frac{1}{5})-3$
$= 1-3$
$= -2$

47. $P(-0.9) = 5(-0.9)-3$
$= -4.5-3$
$= -7.5$

49. $Q(0) = -(0)^2-4$
$= -0-4$
$= -4$

51. $Q(-1) = -(-1)^2-4$
$= -(1)-4$
$= -5$

53. $Q(1.3) = -(1.3)^2-4$
$= -(1.69)-4$
$= -5.69$

55. $Q(-13.6) = -(-13.6)^2 - 4$
$= -(184.96) - 4$
$= -188.96$

57. $R(0) = (0)^2 - 2(0) + 3$
$= (0) - 0 + 3$
$= 3$

59. $R(-2) = (-2)^2 - 2(-2) + 3$
$= (4) - (-4) + 3$
$= 4 + 4 + 3$
$= 11$

61. $R(0.9) = (0.9)^2 - 2(0.9) + 3$
$= (.81) - 1.8 + 3$
$= 2.01$

63. $R(-8.1) = (-8.1)^2 - 2(-8.1) + 3$
$= (65.61) - (-16.2) + 3$
$= 65.61 + 16.2 + 3$
$= 84.81$

65. Evaluate $x = -2$:
$x^2 - 3 = (-2)^2 - 3$
$= 4 - 3$
$= 1$
Evaluate $x = -1$:
$x^2 - 3 = (-1)^2 - 3$
$= 1 - 3$
$= -2$
Evaluate $x = 0$:
$x^2 - 3 = (0)^2 - 3$
$= 0 - 3$
$= -3$
Evaluate $x = 1$:
$x^2 - 3 = (1)^2 - 3$
$= 1 - 3$
$= -2$
Evaluate $x = 2$:
$x^2 - 3 = (2)^2 - 3$
$= 4 - 3$
$= 1$

67. Evaluate $x = -2$:
$x^3 + 2 = (-2)^3 + 2$
$= (-8) + 2$
$= -6$
Evaluate $x = -1$:
$x^3 + 2 = (-1)^3 + 2$
$= (-1) + 2$
$= 1$
Evaluate $x = 0$:
$x^3 + 2 = (0)^3 + 2$
$= (0) + 2$
$= 2$
Evaluate $x = 1$:
$x^3 + 2 = (1)^3 + 2$
$= (1) + 2$
$= 3$
Evaluate $x = 2$:
$x^3 + 2 = (2)^3 + 2$
$= (8) + 2$
$= 10$

69. $f(x) = x^2 - 3$

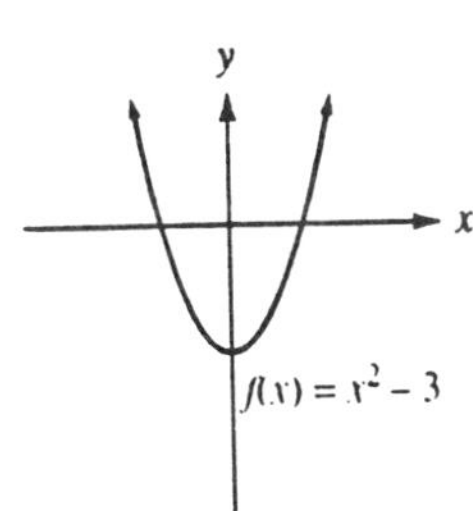

71. $f(x) = x^3 + 2$

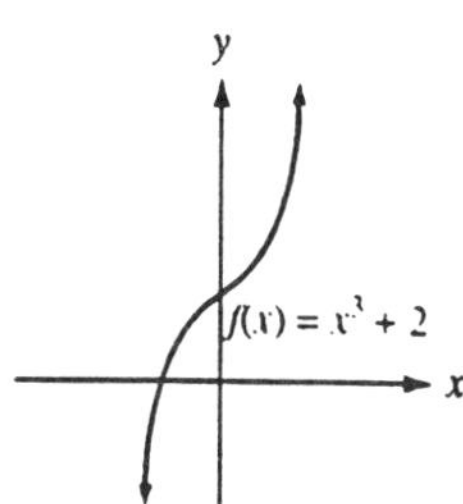

APPLICATIONS

73. $f(t) = -0.2t^2 + 11.12t$

$f(10.49) = -0.2(10.49)^2 + 11.12(10.49)$
$= -0.2(110.0401) + 11.12(10.49)$
$= -0.2(110.0401) + 116.6488$
$= -22.00802 + 116.6488$
$= 94.64078$

So, she finished
$100 - 94.64078 = 5.35922$ meters
or approximately 5.4 meters behind.

75. $f(t) = -16t^2 + 12t + 20$

$f(0.5) = -16(0.5)^2 + 12(0.5) + 20$
$= -16(0.25) + 12(0.5) + 20$
$= -4 + 6 + 20$
$= 22$ ft.

77. $f(v) = 0.04v^2 + 0.9v$

$f(30) = 0.04(30)^2 + 0.9(30)$
$= 0.04(900) + 0.9(30)$
$= 36 + 27$
$= 63$ ft.

79. $f(x) = -0.05x^2 + 2x$

$f(15) = -0.05(15)^2 + 2(15)$
$= -0.05(225) + 2(15)$
$= -11.25 + 30$
$= 18.75$ ft.

$f(20) = -0.05(20)^2 + 2(20)$
$= -0.05(400) + 2(20)$
$= -20 + 40$
$= 20$ ft.

$f(30) = -0.05(30)^2 + 2(30)$
$= -0.05(900) + 2(30)$
$= -45 + 60$
$= 15$ ft.

REVIEW

83. $0 = 4x - 3y \Rightarrow 0 = 4x - 3y$
$5x = 4y - 2 \Rightarrow 2 = -5x + 4y$

To eliminate the x's:

$5(0 = 4x - 3y)$
$4(2 = -5x + 4y)$

$5 \cdot 0 = 5 \cdot 4x - 5 \cdot 3y$
$4 \cdot 2 = 4 \cdot (-5)x + 4 \cdot 4y$

$0 = 20x - 15y$
$8 = -20x + 16y$

Add the two equations:

$8 = 0x + y$

Solve for y :

$8 = y$

Substitute 8 for y in second equation and solve for x:

$5x = 4(8) - 2$
$5x = 32 - 2$
$5x = 30$
$\frac{5x}{5} = \frac{30}{5}$
$x = 6.$

Check results in first equation:

$0 = 4(6) - 3(8)$
$0 = 24 - 24$
$0 = 0$

Answer: (6, 8)

85. $-4(3y + 2) \le 28$
$-12y - 8 \le 28$
$-12y - 8 + 8 \le 28 + 8$
$-12y \le 36$
$\frac{-12y}{-12} \ge \frac{36}{-12}$
$y \ge -3$

```
---------[======>
         -3
```

87. $(x^2x^4)^3 = (x^{2+4})^3 = (x^6)^3 = x^{6 \cdot 3} = x^{18}$

89. $\left(\frac{y^2y^5}{y^4}\right)^3 = \left(\frac{y^{2+5}}{y^4}\right)^3 = \left(\frac{y^7}{y^4}\right)^3$

$= (y^{7-4})^3 = (y^3)^3 = y^{3 \cdot 3} = y^9$

STUDY SET Section 5.5

VOCABULARY

1. polynomials

3. coefficients

CONCEPTS

5. terms

7. like terms

9. changing

11. $-(-2x^2 - 3x + 4) = 2x^2 + 3x - 4$

13. $(9x - 15) + (2x + 3)$
$= 9x - 15 + 2x + 3$
$= 9x + 2x - 15 + 3$
$= 11x - 12$ ft.

NOTATION

15. $(5x^2 + 3x) - (7x^2 - 2x)$
$= 5x^2 + 3x - 7x^2 + 2x$
$= 5x^2 - 7x^2 + 3x + 2x$
$= -2x^2 + 5x$

PRACTICE

17. $4y + 5y = 9y$

19. $8t^2 + 4t^2 = 12t^2$

21. $-32u^3 - 16^3 = -48u^3$

23. $1.8x - 1.9x = -0.1x$

25. $\frac{1}{2}s + \frac{3}{2}s = \frac{4}{2}s = 2s$

27. $3r - 4r + 7r = -r + 7r = 6r$

29. $-4ab + 4ab - ab = 0 - ab = -ab$

31. $(3x)^2 - 4x^2 + 10x^2$
$= 9x^2 - 4x^2 + 10x^2$
$= 5x^2 + 10x^2 = 15x^2$

33. $(3x + 7) + (4x - 3)$
$= 3x + 4x + 7 - 3$
$= 7x + 4$

35. $(4a + 3) - (2a - 4)$
$= 4a + 3 - 2a + 4$
$= 4a - 2a + 3 + 4$
$= 2a + 7$

37. $(2x + 3y) + (5x - 10y)$
$= 2x + 3y + 5x - 10y$
$= 2x + 5x + 3y - 10y$
$= 7x - 7y$

39. $(-8x - 3y) - (-11x + y)$
$= -8x - 3y + 11x - y$
$= -8x + 11x - 3y - y$
$= 3x - 4y$

41. $(3x^2 - 3x - 2) + (3x^2 + 4x - 3)$
$= 3x^2 - 3x - 2 + 3x^2 + 4x - 3$
$= 3x^2 + 3x^2 - 3x + 4x - 2 - 3$
$= 6x^2 + x - 5$

43. $(2b^2 + 3b - 5) - (2b^2 - 4b - 9)$
$= 2b^2 + 3b - 5 - 2b^2 + 4b + 9$
$= 2b^2 - 2b^2 + 3b + 4b - 5 + 9$
$= 0b^2 + 7b + 4$
$= 7b + 4$

45. $(2x^2 - 3x + 1) - (4x^2 - 3x + 2)$
$+ (2x^2 + 3x + 2)$
$= 2x^2 - 3x + 1 - 4x^2 + 3x - 2$
$+ 2x^2 + 3x + 2$
$= 2x^2 - 4x^2 + 2x^2 - 3x + 3x + 3x$
$+ 1 - 2 + 2$
$= 0x^2 + 3x + 1$
$= 3x + 1$

47.
$$\begin{array}{r} 3x^2 + 4x + 5 \\ \underline{2x^2 - 3x + 6} \\ 5x^2 + x + 11 \end{array}$$

49.
$$\begin{array}{r} 2x^3 - 3x^2 + 4x - 7 \\ \underline{-9x^3 - 4x^2 - 5x + 6} \\ -7x^3 - 7x^2 - x - 1 \end{array}$$

51. $\begin{array}{r} -3x^2 + 4x + 25 \\ \underline{5x^2 \qquad - 12} \\ 2x^2 + 4x + 13 \end{array}$

53. $\begin{array}{r} 3x^2 + 4x - 5 \\ \underline{-2x^2 - 2x + 3} \end{array}$

$\Rightarrow \begin{array}{r} 3x^2 + 4x - 5 \\ \underline{2x^2 + 2x - 3} \\ 5x^2 + 6x - 8 \end{array}$

55. $\begin{array}{r} 4x^3 + 4x^2 - 3x + 10 \\ \underline{5x^3 - 2x^2 - 4x - 4} \end{array}$

$\Rightarrow \begin{array}{r} 4x^3 + 4x^2 - 3x + 10 \\ \underline{-5x^3 + 2x^2 + 4x + 4} \\ -x^3 + 6x^2 + \ x + 14 \end{array}$

57. $\begin{array}{r} -2x^2y^2 \qquad\quad + 12y^2 \\ \underline{10x^2y^2 + 9xy - 24y^2} \end{array}$

$\Rightarrow \begin{array}{r} -2x^2y^2 \qquad\quad + 12y^2 \\ \underline{-10x^2y^2 - 9xy + 24y^2} \\ -12x^2y^2 - 9xy + 36y^2 \end{array}$

59. $[(3t^3 + t^2) + (-t^3 + 6t - 3)] - (t^3 - 2t^2 + 2)$

$= [3t^3 + t^2 - t^3 + 6t - 3] - (t^3 - 2t^2 + 2)$

$= [3t^3 - t^3 + t^2 + 6t - 3] - (t^3 - 2t^2 + 2)$

$= [2t^3 + t^2 + 6t - 3] - (t^3 - 2t^2 + 2)$

$= 2t^3 + t^2 + 6t - 3 - t^3 + 2t^2 - 2$

$= 2t^3 - t^3 + t^2 + 2t^2 + 6t - 3 - 2$

$= t^3 + 3t^2 + 6t - 5$

61. $[(-2x^2 - 7x + 1) + (-4x^2 + 8x - 1)] + (3x^2 + 4x - 7)$

$= [-2x^2 - 7x + 1 - 4x^2 + 8x - 1] + (3x^2 + 4x - 7)$

$= [-2x^2 - 4x^2 - 7x + 8x + 1 - 1] + (3x^2 + 4x - 7)$

$= [-6x^2 + x + 0] + (3x^2 + 4x - 7)$

$= -6x^2 + x + 3x^2 + 4x - 7$

$= -6x^2 + 3x^2 + x + 4x - 7$

$= -3x^2 + 5x - 7$

63. $2(x + 3) + 4(x - 2) = 2x + 6 + 4x - 8$

$= 2x + 4x + 6 - 8$

$= 6x - 2$

65. $-2(x^2 + 7x - 1) - 3(x^2 - 2x + 2)$

$= -2x^2 - 14x + 2 - 3x^2 + 6x - 6$

$= -2x^2 - 3x^2 - 14x + 6x + 2 - 6$

$= -5x^2 - 8x - 4$

67. $2(2y^2 - 2y + 2) - 4(3y^2 - 4y - 1) + 4y(y^2 - y - 1)$

$= 4y^2 - 4y + 4 - 12y^2 + 16y + 4 + 4y^3 - 4y^2 - 4y$

$= 4y^3 + 4y^2 - 12y^2 - 4y^2 - 4y + 16y - 4y + 4 + 4$

$= 4y^3 - 12y^2 + 8y + 8$

69. $2a(ab^2 - b) - 3b(a + 2ab)$
$+ b(b - a + a^2b)$

$= 2a^2b^2 - 2ab - 3ab - 6ab^2 + b^2$
$- ab + a^2b^2$

$= 2a^2b^2 + a^2b^2 - 2ab - 3ab - ab$
$- 6ab^2 + b^2$

$= 3a^2b^2 - 6ab - 6ab^2 + b^2$

71. $P = (x^2 + 3x + 1) + (x^2 + 3x + 1)$
$+ (x^2 - 4)$

$P = x^2 + 3x + 1 + x^2 + 3x + 1$
$+ x^2 - 4$

$P = x^2 + x^2 + x^2 + 3x + 3x$
$+ 1 + 1 - 4$

$P = 3x^2 + 6x - 2$ yd.

APPLICATIONS

73. Length $= (x^2 - 3x) + (1.5\pi) + (x^2 + 9x)$
$+ 2\pi + (x^2 + 5x) + \pi$

Length $= x^2 - 3x + 1.5\pi + x^2 + 9x$
$+ 2\pi + x^2 + 5x + \pi$

Length $= x^2 + x^2 + x^2 - 3x + 9x + 5x$
$+ 1.5\pi + 2\pi + \pi$

Length $= (3x^2 + 11x + 4.5\pi)$ in.

75. If $x = 10$,
$y = 900x + 105,000$
$y = 900(100) + 105,000$
$y = 90,000 + 105,000$
$y = \$114,000$

77. Value of two houses
$= (900x + 105,000) + (1000x + 120,000)$
$= 900x + 105,000 + 1000x + 120,000$
$= 900x + 1000x + 105,000 + 120,000$
$= 1900x + 225,000$

79. value = appreciation•time + purchase
$y = -1,100x + 6,600$

81. value of both computers
$= (\text{-}1,100x + 6,600) + (\text{-}1,700x + 9,200)$
$= \text{-}1,100x + 6,600 - 1,700x + 9,200$
$= \text{-}1,100x + 6,600 - 1,700x + 9,200$
$= \text{-}1,100x - 1,700x + 6,600 + 9,200$
$= -2,800x - 15,800$

REVIEW

85. The sum of the measures of the angles in a triangle is 180°.

87. $-4(3x - 3) \geq -12$
$\frac{-4(3x-3)}{-4} \leq \frac{-12}{-4}$
$3x - 3 \leq 3$
$3x - 3 + 3 \leq 3 + 3$
$3x \leq 6$
$\frac{3x}{3} \leq \frac{6}{3}$
$x \leq 2$

<=====]-----
2

STUDY SET Section 5.6
VOCABULARY

1. binomials

3. distributive

CONCEPTS

5. $(2x)(3x) = (2)(3)x \cdot x = 6x^2$

7. $(5)(3x) = (5)(3)x = 15x$

9. Perimeter:
$P = 2(3x - 3) - 2(2x) + 2(x + 3)$
$P = 6x - 6 - 4x + 2x + 6$
$P = 6x - 4x + 2x - 6 + 6$
$P = 4x$ cm.

Area:
$A = [(3x - 3) - 2x](x + 3)$
$A = [3x - 2 - 2x](x + 3)$
$A = [x - 2](x + 3)$
$A = x \cdot x + x \cdot 3 + (-2) \cdot x + (-2) \cdot 3$
$A = x^2 + 3x - 2x - 6$
$A = x^2 + x - 6\,\text{cm}^2$

NOTATION

11. $7x(3x^2 - 2x + 5)$
$= (7x) \cdot 3x^2 - (7x) \cdot 2x + (7x) \cdot 5$
$= 21x^3 - 14x + 35x$

PRACTICE

13. $(3x^2)(4x^3) = 3(4)x^2x^3 = 12x^5$

15. $(3b^2)(-2b)(4b^3) = 3(-2)(4)b^2bb^3$
$= -24b^{2+1+3} = -24b^6$

17. $(2x^2y^3)(3x^3y^2) = 2(3)x^2x^3y^3y^2$
$= 6x^5y^5$

19. $(x^2y^5)(x^2z^5) = x^2x^2y^5z^5 = x^4y^5z^5$

21. $3(x + 4) = 3x + 3 \cdot 4 = 3x + 12$

23. $-4(t + 7) = -4t + (-4)7$
$= -4t - 28$

25. $3x(x - 2) = 3x \cdot x - 3x \cdot 2$
$= 3x^2 - 6x$

27. $-2x^2(3x^2 - x)$
$= -2(3)x^2x^2 - (-2)x^2 \cdot x$
$= -6x^4 + 2x^3$

29. $3xy(x + y) = 3xy \cdot x + 3xy \cdot y$
$= 3x \cdot xy + 3xy \cdot y$
$= 3x^2y + 3xy^2$

31. $2x^2(3x^2 + 4x - 7)$
$= 2x^2 \cdot 3x^2 + 2x^2 \cdot 4x - 2x^2 \cdot 7$
$= 2(3)x^2 \cdot x^2 + 2(4)x^2 \cdot x - 2(7)x^2$
$= 6x^4 + 8x^3 - 14x^2$

33. $(3x)(-2x^2)(x + 4)$
$= 3(-2)x \cdot x^2(x + 4)$
$= -6x^3(x + 4)$
$= -6x^3 \cdot x + (-6x^3) \cdot 4$
$= -6x^4 + (-6)(4)x^3$
$= -6x^4 - 24x^3$

35. $(a + 4)(a + 5)$
$= a \cdot a + a \cdot 5 + 4 \cdot a + (4)5$
$= a^2 + 5a + 4a + 20$
$= a^2 + 9a + 20$

37. $(3x - 2)(x + 4)$
$= 3x \cdot x + 3x \cdot 4 + (-2)x + (-2)4$
$= 3x^2 + 12x - 2x - 8$
$= 3x^2 + 10x - 8$

39. $(2a + 4)(3a - 5)$
$= 2a \cdot 3a + 2a(-5) + 4(3)a + 4(-5)$
$= 2(3)a \cdot a + 2(-5)a + 12a - 20$
$= 6a^2 - 10a + 12a - 20$
$= 6a^2 + 2a - 20$

41. $(3x - 5)(2x + 1)$
$= 3x \cdot 2x + 3x \cdot 1 + (-5) \cdot 2x + (-5)(1)$
$= 3(2)x \cdot x + 3x - 10x - 5$
$= 6x^2 + 3x - 10x - 5$
$= 6x^2 - 7x - 5$

43. $(x+3)(2x-3)$
$= x \cdot 2x + x \cdot (-3) + 3 \cdot 2x + 3(-3)$
$= 2x \cdot x - 3x + 6x - 9$
$= 2x^2 - 3x + 6x - 9$
$= 2x^2 + 3x - 9$

45. $(2t+3s)(3t-s)$
$= 2t \cdot 3t + 2t \cdot (-s) + 3s \cdot 3t + 3s \cdot (-s)$
$= 2(3)t \cdot t - 2ts + 3(3)st + 3(-1)s \cdot s$
$= 6t^2 - 2st + 9st - 3s^2$
$= 6t^2 + 7st - 3s^2$

47. $(x+y)(x+z)$
$= x \cdot x + x \cdot z + y \cdot x + y \cdot z$
$= x^2 + xz + yx + yz$

49. $(4t-u)(-3t+u)$
$= 4t \cdot (-3t) + 4t \cdot u + (-u) \cdot (-3t) + (-u) \cdot u$
$= 4t \cdot (-3t) + 4t \cdot u + (-u) \cdot (-3t) + (-u) \cdot u$
$= 4(-3)t \cdot t + 4tu + (-1)(-3)u \cdot t - u^2$
$= -12t^2 + 4tu + 3tu - u^2$
$= -12t^2 + 7tu - u^2$

51. $4(2x+1)(x-2)$
$= 4(2x \cdot x - 2x \cdot 2 + x - 2)$
$= 4(2x^2 - 4x + x - 2)$
$= 4(2x^2 - 3x - 2)$
$= 4 \cdot 2x^2 - 4 \cdot 3x - 4 \cdot 2$
$= 8x^2 - 12x - 8$

53. $3a(a+b)(a-b)$
$= 3a(a \cdot a + a \cdot (-b) + b \cdot a + b \cdot (-b))$
$= 3a(a^2 - ab + ba - b^2)$
$= 3a(a^2 - b^2)$
$= 3a \cdot a^2 - 3a \cdot b^2$
$= 3a^3 - 3ab^2$

55. $2t(t+2) + 3t(t-5)$
$= 2t \cdot t + 2t \cdot 2 + 3t \cdot t - 3t \cdot 5$
$= 2t^2 + 4t + 3t^2 - 15t$
$= 2t^2 + 3t^2 + 4t - 15t$
$= 5t^2 - 11t$

57. $(x+y)(x-y) + x(x+y)$
$= x \cdot x + x \cdot y - x \cdot y - y \cdot y + x \cdot x + x \cdot y$
$= x^2 + xy - xy - y^2 + x^2 + xy$
$= x^2 + x^2 + xy - xy + xy - y^2$
$= 2x^2 + xy - y^2$

59. $(x+4)(x+4)$
$= x \cdot x + 4 \cdot x + 4 \cdot x + 4 \cdot 4$
$= x^2 + 4x + 4x + 16$
$= x^2 + 8x + 16$

61. $(t-3)(t-3)$
$= t \cdot t - 3 \cdot t - 3 \cdot t + (-3) \cdot (-3)$
$= t^2 - 3t - 3t + 9$
$= t^2 - 6t + 9$

63. $(r+4)(r-4)$
$= r \cdot r - 4 \cdot r + 4 \cdot r + (4)(-4)$
$= r^2 - 4r + 4r - 16$
$= r^2 - 16$

65. $(x+5)^2 = (x+5)(x+5)$
$= x \cdot x + 5 \cdot x + 5x + 5 \cdot 5$
$= x^2 + 5x + 5x + 25$
$= x^2 + 10x + 25$

67. $(2s+1)(2s+1)$
$= 2s \cdot 2s + s \cdot 1 + s \cdot 1 + 1 \cdot 1$
$= 2(2)s \cdot s + s + s + 1$
$= 4s^2 + 2s + 1$

69. $(4x+5)(4x-5)$
$= 4x \cdot 4x + 4x \cdot (-5) + 4x \cdot 5 + 5(-5)$
$= 16x^2 - 20x + 20x - 25$
$= 16x^2 - 25$

71. $(x-2y)^2 = (x-2y)(x-2y)$
$= x \cdot x + x \cdot (-2y) + x \cdot (-2y) + (-2y)(-2y)$
$= x^2 - 2xy - 2xy + 4y^2$
$= x^2 - 4xy + 4y^2$

73. $(2a-3b)^2 = (2a-3b)(2a-3b)$
$= 2a \cdot 2a + 2a \cdot (-3b) + 2a \cdot (-3b)$
$+ (-3b)(-3b)$
$= 4a^2 - 6ab - 6ab + 9b^2$
$= 4a^2 - 12ab + 9b^2$

75. $(4x + 5y)(4x - 5y)$
$= 4x \cdot 4x + 4x(-y) + 4x(-5y)$
$+ (-5y)(-5y)$
$= 16x^2 - 4xy + 4xy + 25y$
$= 16x^2 + 25y$

77. $A = \frac{1}{2}bh$
$= \frac{1}{2}(2x - 2)(4x - 2)$
$= \frac{1}{2}(2x \cdot 4x + (-2) \cdot 2x + (-2) \cdot 4x$
$+ (-2)(-2))$
$= \frac{1}{2}(8x^2 - 4x - 8x + 4)$
$= \frac{1}{2}(8x^2 - 12x + 4)$
$= \frac{1}{2} \cdot 8x^2 - \frac{1}{2} \cdot 12x + \frac{1}{2} \cdot 4$
$= 4x^2 - 6x + 2 \text{ cm}^2$

79. $A = \pi r^2$
$= \pi(x + 3)^2$
$= \pi(x + 3)(x + 3)$
$= \pi(x \cdot x + 3 \cdot x + 3 \cdot x + 3 \cdot 3)$
$= \pi(x^2 + 3x + 3x + 9)$
$= \pi(x^2 + 6x + 9) \text{ in}^2$

81. $(x + 2)(x^2 - 2x + 3)$
$= x \cdot x^2 - x \cdot 2x + 3 \cdot x + 2 \cdot x^2$
$- 2 \cdot 2x + 2 \cdot 3$
$= x^3 - 2x^2 + 3x + 2x^2 - 4x + 6$
$= x^3 - 2x^2 + 2x^2 + 3x - 4x + 6$
$= x^3 - x + 6$

83. $(4t + 3)(t^2 + 2t + 3)$
$= 4t \cdot t^2 + 4t \cdot 2t + 4t \cdot 3$
$+ 3 \cdot t^2 + 3 \cdot 2t + 3 \cdot 3$
$= 4t^3 + 8t^2 + 12t + 3t^2 + 6t + 9$
$= 4t^3 + 8t^2 + 3t^2 + 12t + 6t + 9$
$= 4t^3 + 11t^2 + 18 + 9$

85. $(-3x + y)(x^2 - 8xy + 16y^2)$
$= (-3x + y) \cdot x^2$
$+ (-3x + y) \cdot (-8xy)$
$+ (-3x + y) \cdot 16y^2$

$= -3x \cdot x^2 + y \cdot x^2$
$- 3x \cdot (-8xy) + y \cdot (-8xy)$
$- 3x \cdot 16y^2 + y \cdot 16y^2$

$= -3x^3 + x^2y + (-3)(-8)x^2y$
$- 8xy^2 - 3(16)xy^2 + 16y^3$

$= -3x^3 + x^2y + 24x^2y$
$- 8xy^2 - 48xy^2 + 16y^3$

$= -3x^2 + 25x^2y - 56xy^2 + 16y^3$

87. $(x - 2)(x^2 + 2x + 4)$
$= (x - 2) \cdot x^2 + (x - 2) \cdot 2x + (x - 2) \cdot 4$

$= x \cdot x^2 - 2x^2 + x \cdot 2x - 2 \cdot 2x$
$+ 4x - 2 \cdot 4$

$= x^3 - 2x^2 + 2x^2 - 4x + 4x - 8$

$= x^3 - 8$

89.
$$\begin{array}{r} x^2 - 2x + 1 \\ x + 2 \\ \hline x^3 - 2x^2 + x \\ 2x^2 - 4x + 2 \\ \hline x^3 + 0x^2 - 3x + 2 \end{array}$$
$= x^3 - 3x + 2$

91.
$$\begin{array}{r} 4x^2 + 3x - 4 \\ 3x + 2 \\ \hline 12x^3 + 9x^2 - 12x \\ 8x^2 + 69x - 8 \\ \hline 12x^3 + 17x^2 - 6x - 8 \end{array}$$

93. $(s - 4)(s + 1) = s^2 + 5$
$s \cdot s + s \cdot 1 - 4 \cdot s - 4 \cdot 1 = s^2 + 5$
$s^2 + s - 4s - 4 = s^2 + 5$
$s^2 - 3s - 4 = s^2 + 5$
$s^2 - s^2 - 3s - 4 = s^2 - s^2 + 5$
$-3s - 4 = 5$
$-3s - 4 + 4 = 5 + 4$
$-3s = 9$
$\frac{-3s}{-3} = \frac{9}{-3}$
$s = -3$

95. $z(z+2) = (z+4)(z-4)$
$z \cdot z + 2 \cdot z = z \cdot z + (-4) \cdot z + 4 \cdot z + 4(-4)$
$z^2 + 2z = z^2 - 4z + 4z - 16$
$z^2 + 2z = z^2 - 16$
$z^2 - z^2 + 2z = z^2 - z^2 - 16$
$2z = -16$
$\frac{2z}{2} = \frac{-16}{2}$
$z = -8$

97. $(x+4)(x-4) = (x-2)(x+6)$

$x \cdot x - 4 \cdot x + 4 \cdot x + 4(-4)$
$= x \cdot x + 6 \cdot x - 2 \cdot x + 6(-2)$

$x^2 - 4x + 4x - 16$
$= x^2 + 6x - 2x - 12$

$x^2 - 16 = x^2 + 4x - 12$
$x^2 - x^2 - 16 = x^2 - x^2 + 4x - 12$
$-16 = 4x - 12$
$-16 + 12 = 4x - 12 + 12$
$-4 = 4x$
$\frac{-4}{4} = \frac{4x}{4}$
$-1 = x$

99. $(a-3)^2 = (a+3)^2$
$(a-3)(a-3) = (a+3)(a+3)$

$a \cdot a - 3 \cdot a - 3 \cdot a - 3(-3)$
$= a \cdot a + 3 \cdot a + 3 \cdot a + 3(3)$

$a^2 - 3a - 3a + 9 = a^2 + 3a + 3a + 9$
$a^2 - 6a + 9 = a^2 + 6a + 9$
$a^2 - a^2 - 6a + 9 = a^2 - a^2 + 6a + 9$
$-6a + 9 = 6a + 9$
$-6a - 6a + 9 = 6a - 6a + 9$
$-12a + 9 = 9$
$-12a + 9 - 9 = 9 - 9$
$-12a = 0$
$\frac{-12a}{-12} = \frac{0}{-12}$
$a = 0$

APPLICATIONS

101. a) corn: $A = x \cdot x = x^2$ ft.
tomatoes: $A = 6 \cdot x = 6x$ ft.
beans: $A = 5 \cdot x = 5x$ ft.
carrots: $A = 5 \cdot 6 = 30$ ft.
Total area:
$x^2 + 6x + 5x + 30$
$= x^2 + 11x + 30$ ft.
b) length: $x + 6$ ft
width: $x + 5$ ft
Area $= l \cdot w$
$= (x+6)(x+5)$
$= x \cdot x + 5 \cdot x + 6 \cdot x + 6(5)$
$= x^2 + 5x + 6x + 30$
$= x^2 + 11x + 30\ \text{ft}^2$
c) The areas are the same even though found in different ways.

103. Analyze: The difference between the squares of two consecutive integers is 11. We need to find two numbers. Because they are consecutive numbers, the second one is in terms of the first.
Form: Let x = first integer.
$x + 1$ = second integer.
Key word: difference
Translation: subtract
Key word: squares
Translation: power two.
square of 2nd minus square of 1st is 11.
$(x+1)^2 - x^2 = 11$
Solve: $(x+1)^2 - x^2 = 11$
$(x+1)(x+1) - x^2 = 11$
$(x \cdot x + 1 \cdot x + 1 \cdot x + 1 \cdot 1) - x^2 = 11$
$x^2 + x + x + 1 - x^2 = 11$
$x^2 - x^2 + x + x + 1 = 11$
$2x + 1 = 11$
$2x + 1 - 1 = 11 - 1$
$2x = 10$
$\frac{2x}{2} = \frac{10}{2}$
$x = 5$
State: The first integer is 5 and the second is 6.
Check: $(5+1)^2 - (5)^2 = 11$
$(6)^2 - (5)^2 = 11$
$36 - 25 = 11$
$11 = 11$

105. Analyze: The radius of the large millstone is 3 meters greater than the small one. Their areas differ by 15π m^2. Area of a circle is πr^2.

Form: Let r = radius of the small millstone.

$r + 3$ = radius of large millstone.

Key word: differ

Translation: subtract

area of large minus area of small is 15π.

$\pi(r+3)^2 - \pi r^2 = 15\pi$

Solve: $\pi(r+3)^2 - \pi r^2 = 15\pi$

$\pi(r+3)(r+3) - \pi r^2 = 15\pi$

$\pi(r \cdot r + 3 \cdot r + 3 \cdot r + 3(3)) - \pi r^2 = 15\pi$

$\pi(r^2 + 3r + 3r + 9) - \pi r^2 = 15\pi$

$\pi(r^2 + 6r + 9) - \pi r^2 = 15\pi$

$\pi \cdot r^2 + \pi \cdot 6r + \pi \cdot 9 - \pi r^2 = 15\pi$

$\pi r^2 + 6\pi r + 9\pi - \pi r^2 = 15\pi$

$\pi r^2 - \pi r^2 + 6\pi r + 9\pi = 15\pi$

$6\pi r + 9\pi = 15\pi$

$6\pi r + 9\pi - 9\pi = 15\pi - 9\pi$

$6\pi r = 6\pi$

$\frac{6\pi r}{6\pi} = \frac{6\pi}{6\pi}$

$r = 1$

State: The radius of the small millstone is 1 meter and the radius of the large millstone is 4 meters.

Check: $\pi(1+3)^2 - \pi(1)^2 = 15\pi$

$\pi(4)^2 - \pi(1)^2 = 15\pi$

$\pi 16 - \pi 1 = 15\pi$

$16\pi - \pi = 15\pi$

$15\pi = 15\pi$

107. Analyze: The distance between bases for baseball is 30 ft greater than for softball. The area of the diamond for baseball is 4,500 ft^2 greater than the area for (no suggestions).

The area for baseball and softball diamonds is $A = s^2$.

Form: Let d = distance between bases for softball.

$d + 30$ = distance between bases for baseball.

Key word: greater

Translation: add

area of baseball is 4,500 plus area of softball.

$(d+30)^2 = 4,500 + d^2$

Solve: $(d+30)^2 = 4,500 + d^2$

$(d+30)(d+30) = 4,500 + d^2$

$d \cdot d + 30 \cdot d + 30 \cdot d + 30(30) = 4,500 + d^2$

$d^2 + 30d + 30d + 900 = 4,500 + d^2$

$d^2 + 60d + 900 = 4,500 + d^2$

$d^2 - d^2 + 60d + 900 = 4,500 + d^2 - d^2$

$60d + 900 = 4,500$

$60d + 900 - 900 = 4,500 - 900$

$60d = 3,600$

$\frac{60d}{60} = \frac{3600}{60}$

$d = 60$

State: The distance between bases for softball is 60 ft. and for baseball is 90 ft.

Check: $(60+30)^2 = 4,500 + 60^2$

$(90)^2 = 4,500 + 60^2$

$8{,}100 = 4,500 + 3{,}600$

$8{,}100 = 8{,}100$

111. slope $= \frac{rise}{run} = \frac{1}{1} = 1$

113. slope $= \frac{rise}{run} = \frac{-2}{3} = -\frac{2}{3}$

115. y-intercept is $(-2, 0)$

STUDY SET Section 5.7

VOCABULARY

1. polynomial

3. two

5. reciprocal

CONCEPTS

7. $\frac{1}{b} \cdot a = \frac{a}{b}$

9. In the numerator and denominator, a common factor of 2 was divided out.

11. a) $d = rt$

$\frac{d}{r} = \frac{rt}{r}$

$\frac{d}{r} = t$

b) $t = \frac{6x^3}{2x} = \frac{\overset{3}{\cancel{6}}x \cdot x \cdot \cancel{x}}{\underset{1}{\cancel{2}}\cancel{x}} = 3x^2$

13. $\frac{10x+35}{5} = \frac{10x}{5} + \frac{35}{5} = 2x + 7$

NOTATION

15. $\frac{a^2b^3}{a^3b^2} = \frac{a \cdot a \cdot b \cdot b \cdot b}{a \cdot a \cdot a \cdot b \cdot b} = \frac{\cancel{a} \cdot \cancel{a} \cdot b \cdot \cancel{b} \cdot \cancel{b}}{\cancel{a} \cdot \cancel{a} \cdot a \cdot \cancel{b} \cdot \cancel{b}} = \frac{b}{a}$

PRACTICE

17. $\frac{5}{15} = \frac{5}{3 \cdot 5} = \frac{1}{3}$

19. $\frac{-125}{75} = \frac{-5 \cdot 25}{3 \cdot 25} = \frac{-5 \cdot 25}{3 \cdot 25} = -\frac{5}{3}$

21. $\frac{120}{160} = \frac{40 \cdot 3}{40 \cdot 4} = \frac{3}{4}$

23. $\frac{-3{,}612}{-3{,}612} = 1$

25. $\frac{x^5}{x^2} = x^{5-2} = x^3$

27. $\frac{r^3s^2}{rs^3} = r^{3-1}s^{2-3} = r^2s^{-1} = \frac{r^2}{s}$

29. $\frac{8x^3y^2}{4xy^3} = \frac{8}{4}x^{3-1}y^{2-3} = 2x^2y^{-1} = \frac{2x^2}{y}$

31. $\frac{12u^5v}{-4u^2v^3} = -\frac{12}{4}u^{5-2}v^{1-3}$

$= -3u^3v^{-2} = -\frac{3u^3}{v^2}$

33. $\frac{-16r^3y^2}{-4r^2y^4} = \frac{-16}{-4}r^{3-2}y^{2-4} = 4r^1y^{-2} = \frac{4r}{y^2}$

35. $\frac{-65rs^2t}{15r^2s^3t} = \frac{-65}{15}r^{1-2}s^{2-3}t^{1-1}$

$= \frac{-13}{3}r^{-1}s^{-1}t^0 = -\frac{13}{3rs}$

37. $\frac{x^2x^3}{xy^6} = \frac{x^{2+3}}{xy^6} = \frac{x^5}{xy^6} = \frac{x^{5-1}}{y^6} = \frac{x^4}{y^6}$

39. $\frac{(a^3b^4)^3}{ab^4} = \frac{a^{3 \cdot 3}b^{4 \cdot 3}}{ab^4} = \frac{a^9b^{12}}{ab^4}$

$= a^{9-1}b^{12-4} = a^8b^8$

41. $\frac{15(r^2s^3)^2}{-5(rs^5)^3} = \frac{15}{-5}\frac{r^{2 \cdot 2}s^{3 \cdot 2}}{r^3s^{5 \cdot 3}}$

$= -3\frac{r^4s^6}{r^3s^{15}} = -3r^{4-3}s^{6-15}$

$= -3r^1s^{-9} = -\frac{3r}{s^9}$

43. $\frac{-32(x^3y)^3}{128(x^2y^2)^3} = \frac{-32x^{3 \cdot 3}y^3}{128x^{2 \cdot 3}y^{2 \cdot 3}} = \frac{-32x^9y^3}{128x^6y^6}$

$= \frac{-32}{128}x^{9-6}y^{3-6} = \frac{1}{4}x^3y^{-3} = \frac{x^3}{4y^3}$

45. $\frac{-(4x^3y^3)^2}{(x^2y^4)^3} = \frac{-4^2x^{3 \cdot 2}y^{3 \cdot 2}}{x^{2 \cdot 3}y^{4 \cdot 3}} = \frac{-4x^6y^6}{x^6y^{12}}$

$= -4x^{6-6}y^{6-12} = -4x^0y^{-6}$

$= \frac{-4}{y^6}$

47. $\frac{(a^2b^4)^3}{(a^4)^3} = \frac{a^{2 \cdot 3}b^{4 \cdot 3}}{a^{4 \cdot 3}} = \frac{a^6b^{12}}{a^{12}}$

$= a^{6-12}b^{12} = a^{-6}b^{12} = \frac{b^{12}}{a^6}$

49. $\frac{6x+9}{3} = \frac{1}{3}(6x+9)$

$= \frac{1}{3}\bullet 6x + \frac{1}{3}\bullet 9 = 2x + 3$

51. $\frac{5x-10y}{25xy} = \frac{1}{25xy}(5x - 10y)$

$= \frac{5x}{25xy} - \frac{10y}{25xy} = \frac{1}{5y} - \frac{2}{5x}$

53. $\frac{3x^2+6y^3}{3x^2y^2} = \frac{1}{3x^2y^2}(3x^2 + 6y^3)$

$= \frac{3x^2}{3x^2y^2} + \frac{6y^3}{3x^2y^2} = \frac{1}{y^2} + \frac{2y}{x^2}$

55. $\frac{15a^3b^2-10a^2b^3}{5a^2b^2} = \frac{1}{5a^2b^2}(15a^3b^2 - 10a^2b^3)$

$= \frac{15a^3b^2}{5a^2b^2} - \frac{10a^2b^3}{5a^2b^2} = 3a - 2b$

57. $\frac{4x-2y+8z}{4xy} = \frac{1}{4xy}(4x - 2y + 8z)$

$= \frac{4x}{4xy} - \frac{2y}{4xy} + \frac{8z}{4xy}$

$= \frac{1}{y} - \frac{1}{2x} + \frac{2z}{xy}$

59. $\frac{12x^3y^2-8x^2y-4x}{4xy}$

$= \frac{1}{4xy}(12x^3y^2 - 8x^2y - 4x)$

$= \frac{12x^3y^2}{4xy} - \frac{8x^2y}{4xy} - \frac{4x}{4xy}$

$= \frac{12}{4}x^{3-1}y^{2-1} - \frac{8}{4}x^{2-1}y^{1-1}$
$- \frac{4}{4}x^{1-1}y^{-1}$

$= 3x^2y^1 - 2x^1y^0 - x^0y^{-1}$

$= 3x^2y - 2x - \frac{1}{y}$

61. $\frac{-25x^2y+30xy^2-5xy}{-5xy}$

$= \frac{1}{-5xy}(-25x^2y + 30xy^2 - 5xy)$

$= \frac{-25x^2y}{-5xy} + \frac{30xy^2}{-5xy} - \frac{5xy}{-5xy}$

$= \frac{-25}{-5}x^{2-1}y^{1-1} + \frac{30}{-5}x^{1-1}y^{2-1}$
$- \frac{5}{-5}x^{1-1}y^{1-1}$

$= 5x^1y^0 - 6x^0y^1 + 1x^0y^0$

$= 5x - 6y - 1$

63. $\frac{5x(4x-2y)}{2y} = \frac{20x^2-10xy}{2y}$

$= \frac{1}{2y}(20x^2 - 10xy)$

$= \frac{20x^2}{2y} - \frac{10xy}{2y} = \frac{10x^2}{y} - 5x$

65. $\frac{(-2x)^3+(3x^2)^2}{6x^2} = \frac{-8x^3+9x^4}{6x^2}$

$= \frac{1}{6x^2}(-8x^3 + 9x^4)$

$= \frac{-8x^3}{6x^2} - \frac{9x^4}{6x^2} = -\frac{4x}{3} - \frac{3x^2}{2}$

67. $\frac{4x^2y^2-2(x^2y^2+xy)}{2xy} = \frac{4x^2y^2-2x^2y^2-2xy}{2xy}$

$= \frac{1}{2xy}(4x^2y^2 - 2x^2y^2 - 2xy)$

$= \frac{4x^2y^2}{2xy} - \frac{2x^2y^2}{2xy} - \frac{2xy}{2xy}$

$= 2xy - xy - 1 = xy - 1$

69. $\frac{(3x-y)(2x-3y)}{6xy} = \frac{6x^2-9xy-2xy+3y^2}{6xy}$

$= \frac{6x^2-11xy+3y^2}{6xy}$

$= \frac{1}{6xy}(6x^2 - 11xy + 3y^2)$

$= \frac{6x^2}{6xy} - \frac{11xy}{6xy} - \frac{3y^2}{6xy}$

$= \frac{x}{y} - \frac{11}{6} + \frac{y}{2x}$

71. $\frac{(a+b)^2-(a-b)^2}{2ab}$

$= \frac{(a^2+ab+ab+b^2)-(a^2-ab-ab+b^2)}{2ab}$

$= \frac{(a^2+2ab+b^2)-(a^2-2ab+b^2)}{2ab}$

$= \frac{a^2+2ab+b^2-a^2+2ab-b^2}{2ab}$

$= \frac{a^2-a^2+2ab+2ab+b^2-b^2}{2ab}$

$= \frac{4ab}{2ab} = 2$

APPLICATIONS

73. $P = 6x^2 - 3x + 9$ and there are 3 sides. So, one side is perimeter divided by 3.

length of side $= \frac{6x^2-3x+9}{3}$

$= \frac{6x^2}{3} - \frac{3x}{3} + \frac{9}{3}$

$= 2x^2 - x + 3$ in.

75. Volume is calculated by $V = l \cdot w \cdot h$. We know the volume, length, and width. We want to find the height.

$V = l \cdot w \cdot h$

$36x^3 - 24x^2 = (4x)(3x)h$

$36x^3 - 24x^2 = 12x^2 h$

$\frac{36x^3-24x^2}{12x^2} = \frac{12x^2 h}{12x^2}$

$\frac{36x^3}{12x^2} - \frac{24x^2}{12x^2} = h$

$3x - 2 = h$

The height is $3x - 2$ ft.

77. Yes. $l = \frac{P-2w}{2}$

$l = \frac{P}{2} - \frac{2w}{2}$

$l = \frac{P}{2} - w$

79. No. $\frac{0.08x+5}{x} = \frac{0.08x}{x} + \frac{5}{x}$

$= 0.08 + \frac{5}{x}$

REVIEW

83. Binomial, because it has two terms.

85. None of the above, because it has more than three terms.

87. Degree two, because the largest exponent for a term is two.

STUDY SET Section 5.8

VOCABULARY

1. divisor; dividend

3. remainder

CONCEPTS

5. $4x^3 - 2x^2 + 7x + 6$

7. $6x^4 - x^3 + 2x^2 + 9x$

9. $0x^3$ and $0x$

11. $x^3 + 3x^2 + 9x + 27$ is the only reasonable quotient because $x^3 \bullet x = x^4$.

13. a) $d = rt$

$$\frac{d}{t} = \frac{rt}{t}$$
$$r = \frac{d}{t}$$

b) $r = \frac{d}{t}$

$$r = \frac{x^2+x-12}{x+4}$$

$$\begin{array}{r} x-3 \\ x+4\overline{)x^2+\ x-12} \\ \underline{x^2+4x} \\ -3x-12 \\ \underline{-3x-12} \\ 0 \end{array}$$

$r = x - 3$

15. It is correct.

$$(3x+2)(x+2) = 3x^2 + 6x + 2x + 4$$
$$= 3x^2 + 8x + 4$$

NOTATION

17.

$$\begin{array}{r} x+2 \\ x+2\overline{)x^2+4x+4} \\ \underline{x^2+2x} \\ 2x+4 \\ \underline{2x+4} \\ 0 \end{array}$$

answer : $x + 2$

PRACTICE

19.

$$\begin{array}{r} x+6 \\ x-2\overline{)x^2+4x-12} \\ \underline{x^2-2x} \\ 6x-12 \\ \underline{6x-12} \\ 0 \end{array}$$

answer : $x + 6$

21.

$$\begin{array}{r} y+12 \\ y+1\overline{)y^2+13y+12} \\ \underline{y^2+y} \\ 12y+12 \\ \underline{12y+12} \\ 0 \end{array}$$

answer : $y + 12$

23.

$$\begin{array}{r} 3a-2 \\ 2a+3\overline{)6a^2+5a-6} \\ \underline{6a^2+9a} \\ -4a-6 \\ \underline{-4a-6} \\ 0 \end{array}$$

answer : $3z - 2$

25.

$$\begin{array}{r} b+3 \\ 3b+2\overline{)3b^2+11b+6} \\ \underline{3b^2+\ 2b} \\ 9b+6 \\ \underline{9b+6} \\ 0 \end{array}$$

answer : $b + 3$

27. Write as $5x+3\overline{)10x^2+11x+3}$

$$\begin{array}{r} 2x+1 \\ 5x+3\overline{)10x^2+11x+3} \\ \underline{10x^2+\ 6x} \\ 5x+3 \\ \underline{5x+3} \\ 0 \end{array}$$

answer : $2x + 1$

29. Write as $2x+4\overline{)2x^2-10x-28}$

$$\begin{array}{r} x-7 \\ 2x+4\overline{)2x^2-10x-28} \\ \underline{2x^2+4x} \\ -14x-28 \\ \underline{-14x-28} \\ 0 \end{array}$$

answer : $x-7$

31. Write as $2x-1\overline{)6x^2+x-2}$

$$\begin{array}{r} 3x+2 \\ 2x-1\overline{)6x^2+x-2} \\ \underline{6x^2-3x} \\ 4x-2 \\ \underline{4x-2} \\ 0 \end{array}$$

answer : $3x+2$

33. Write as $x+3\overline{)2x^2+5x-3}$

$$\begin{array}{r} 2x-1 \\ x+3\overline{)2x^2+5x-3} \\ \underline{2x^2+6x} \\ -x-3 \\ \underline{-x-3} \\ 0 \end{array}$$

answer : $2x-1$

35.

$$\begin{array}{r} x^2+2x-1 \\ 2x+3\overline{)2x^3+7x^2+4x-3} \\ \underline{2x^3+3x^2} \\ 4x^2+4x \\ \underline{4x^2+6x} \\ -2x-3 \\ \underline{-2x-3} \\ 0 \end{array}$$

answer : x^2+2x-1

37.

$$\begin{array}{r} 2x^2+2x+1 \\ 3x+2\overline{)6x^3+10x^2+7x+2} \\ \underline{6x^3+4x^2} \\ 6x^2+7x \\ \underline{6x^2+4x} \\ 3x+2 \\ \underline{3x+2} \\ 0 \end{array}$$

answer : $2x^2+2x+1$

39.

$$\begin{array}{r} x^2+x+1 \\ 2x+1\overline{)2x^3+3x^2+3x+1} \\ \underline{2x^3+x^2} \\ 2x^2+3x \\ \underline{2x^2+x} \\ 2x+1 \\ \underline{2x+1} \\ 0 \end{array}$$

answer : x^2+x+1

41.

$$\begin{array}{r} x+1 \\ 2x+3\overline{)2x^2+5x+2} \\ \underline{2x^2+3x} \\ 2x+2 \\ \underline{2x+3} \\ -1 \end{array}$$

answer: $x+1+\frac{-1}{2x+3}$

43.

$$\begin{array}{r} 2x+2 \\ 2x+1\overline{)4x^2+6x-1} \\ \underline{4x^2+2x} \\ 4x-1 \\ \underline{4x+2} \\ -3 \end{array}$$

answer: $2x+2+\frac{-3}{2x+1}$

45.

$$\begin{array}{r} x^2+2x+1 \\ x+1\overline{)\,x^3+3x^2+3x+1} \\ \underline{x^3+x^2}\qquad\qquad\quad \\ 2x^2+3x\qquad \\ \underline{2x^2+2x}\qquad \\ x+1 \\ \underline{x+1} \\ 0 \end{array}$$

answer : x^2+2x+1

47.

$$\begin{array}{r} x^2+2x-1 \\ 2x+3\overline{)\,2x^3+7x^2+4x+3} \\ \underline{2x^3+3x^2}\qquad\qquad\quad \\ 4x^2+4x\qquad \\ \underline{4x^2+6x}\qquad \\ -2x+3 \\ \underline{-2x-3} \\ 6 \end{array}$$

answer : $x^2+2x-1+\frac{6}{2x+3}$

49.

$$\begin{array}{r} 2x^2+8x+14 \\ x-2\overline{)\,2x^3+4x^2-2x+3} \\ \underline{2x^3-4x^2}\qquad\qquad\quad \\ 8x^2-2x\qquad \\ \underline{8x^2-16x}\qquad \\ 14x+3 \\ \underline{14x-28} \\ 31 \end{array}$$

answer : $2x^2+8x+14+\frac{31}{x-2}$

51.

$$\begin{array}{r} x+1 \\ x-1\overline{)\,x^2+0x-1} \\ \underline{x^2-x}\qquad \\ x-1 \\ \underline{x-1} \\ 0 \end{array}$$

answer : $x+1$

53.

$$\begin{array}{r} 2x-3 \\ 2x+3\overline{)\,4x^2+0x-9} \\ \underline{4x^2+6x}\qquad \\ -6x-9 \\ \underline{-6x-9} \\ 0 \end{array}$$

answer : $2x-3$

55.

$$\begin{array}{r} x^2-x+1 \\ x+1\overline{)\,x^3+0x^2+0x+1} \\ \underline{x^3+x^2}\qquad\qquad\quad \\ -x^2+0x\qquad \\ \underline{-x^2-x}\qquad \\ x+1 \\ \underline{x+1} \\ 0 \end{array}$$

answer : x^2-x+1

57.

$$\begin{array}{r} a^2-3a+10 \\ a+3\overline{)\,a^3+0a^2+a+0} \\ \underline{a^3+3a^2}\qquad\qquad\quad \\ -3a^2+a\qquad \\ \underline{-3a^2-9a}\qquad \\ 10a+0 \\ \underline{10a+30} \\ -30 \end{array}$$

answer : $x^2+2x-1+\frac{-30}{a+3}$

59.

$$\begin{array}{r} 5x^2-x+4 \\ 3x-4\overline{)\,15x^3-23x^2+16x+0} \\ \underline{15x^3-20x^2}\qquad\qquad\quad \\ -3x^2+16x\qquad \\ \underline{-3x^2+4x}\qquad \\ 12x+0 \\ \underline{12x-16} \\ 16 \end{array}$$

answer : $5\,x^2-x+4+\frac{16}{3x-4}$

APPLICATIONS

61. $A = x^2 - 2x - 24$

a) We know $A = l \cdot w$ and $w = x + 4$.
To find l, solve for l.

$$A = l \cdot w$$
$$\frac{A}{w} = \frac{lw}{w}$$
$$\frac{A}{w} = l$$
$$l = \frac{x^2-2x-24}{x+4}$$

$$\begin{array}{r} x - 6 \\ x + 4 \overline{) x^2 - 2x - 24} \\ \underline{x^2 + 4x} \\ -6x - 24 \\ \underline{-6x - 24} \\ 0 \end{array}$$

So, the length is $x - 6$ in.

b) $P = 2l + 2w$
$$P = 2(x - 6) + 2(x + 4)$$
$$P = 2x - 12 + 2x + 8$$
$$P = 2x + 2x - 12 + 8$$
$$P = 4x - 4$$

So, the perimeter is $4x - 4$ in.

63. The number of poles is the total length divided by interval between poles.

$$\begin{array}{r} 4x^2 + 3x + 7 \\ 2x - 3 \overline{) 8x^3 - 6x^2 + 5x - 21} \\ \underline{8x^3 - 12x^2} \\ 6x^2 + 5x \\ \underline{6x^2 - 9x} \\ 14x - 21 \\ \underline{14x - 21} \\ 0 \end{array}$$

So, the number of poles is $4x^2 + 3x + 7$.

REVIEW

67. 20, 21, 22, 24, 25, 26, 27, 28, 30

69. $3(2x^2 - 4x + 5) + 2(x^2 + 3x - 7)$

$$= 6x^2 - 12x + 15 + 2x^2 + 6x - 14$$
$$= 6x^2 + 2x^2 - 12x + 6x + 15 - 14$$
$$= 8x^2 - 6x + 1$$

71. The slopes of parallel lines are the same.

CHAPTER 5 REVIEW

1. a) $-3x^4 = -3 \cdot x \cdot x \cdot x \cdot x$

b) $\left(\frac{1}{2}pq\right)^3 = \left(\frac{1}{2}pq\right)\left(\frac{1}{2}pq\right)\left(\frac{1}{2}pq\right)$

3. a) $x^3x^2 = x^{3+2} = x^5$

b) $xx^8 = x^{1+8} = x^9$

c) $(y^7)^3 = y^{7 \cdot 3} = y^{21}$

d) $(x^{21})^2 = x^{21 \cdot 2} = x^{42}$

e) $(ab)^3 = a^3b^3$

f) $(3x)^4 = 3^4x^4 = 81x^4$

g) $b^3b^4b^5 = b^{3+4+5}b^{12}$

h) $-z^2(z^3y^2) = -z^2z^3y^2 = -z^{2+3}y^2$

$= -z^5y^2$

i) $(-16s)^2s = (-16)^2s^2s = 256s^{2+1}$

$= 256s^3$

j) $-3y(y^5) = -3yy^5 = -3y^{1+5}$

$= -3y^6$

k) $(x^2x^3)^3 = (x^{2+3})^3 = (x^5)^3 = x^{5 \cdot 3}$
$= x^{15}$

l) $(2x^2y)^2 = 2^2x^{2 \cdot 2}y^2 = 4x^4y^2$

m) $\frac{x^7}{x^3} = x^{7-3} = x^4$

n) $\left(\frac{x^2y}{xy^2}\right)^2 = (x^{2-1}y^{1-2})^2 = (x^1y^{-1})^2$

$= x^{1 \cdot 2}y^{-1 \cdot 2} = x^2y^{-2} = \frac{x^2}{y^2}$

o) $\frac{8(y^2x)^2}{4(yx^2)^2} = \frac{8}{4}\frac{y^{2 \cdot 2}x^{1 \cdot 2}}{y^{1 \cdot 2}x^{2 \cdot 2}} = 2\frac{y^4x^2}{y^2x^4}$

$= 2y^{4-2}x^{2-4} = 2y^2x^{-2} = \frac{2y^2}{x^2}$

p) $\frac{(5y^2z^3)^3}{25(yz)^5} = \frac{5^3y^{2 \cdot 3}z^{3 \cdot 3}}{25y^5z^5} = \frac{125y^6z^9}{25y^5z^5}$

$= \frac{125}{25}y^{6-5}z^{9-5} = 5yz^4$

5. a) $x^0 = 1$

b) $(3x^2y^2)^0 = 1$

c) $(3x^0)^2 = (3 \cdot 1)^2 = 3^2 = 9$

d) $10^{-3} = \frac{1}{10^3} = \frac{1}{100}$

e) $\left(\frac{3}{4}\right)^{-1} = \frac{1}{\frac{3}{4}} = 1 \div \frac{3}{4} = 1 \times \frac{4}{3} = \frac{4}{3}$

f) $-5^{-2} = -\frac{1}{5^2} = -\frac{1}{25}$

g) $x^{-5} = \frac{1}{x^5}$

h) $a^2b^{-3} = \frac{a^2}{b^3}$

i) $x^{-2}x^3 = x^{-2+3} = x^1 = x$

j) $-6y^4y^{-5} = -6y^{4-5} = -6y^{-1}$

$= -\frac{6}{y}$

k) $\frac{x^{-3}}{x^7} = x^{-3-7} = x^{-10} = \frac{1}{x^{10}}$

l) $(x^{-3}x^{-4})^{-2} = \left(x^{-3+(-4)}\right)^{-2}$

$= (x^{-7})^{-2} = x^{(-7) \cdot (-2)} = x^{14}$

m) $\left(\frac{x^2}{x}\right)^{-5} = (x^{2-1})^{-5} = (x^1)^{-5}$

$= x^{1 \cdot (-5)} = x^{-5} = \frac{1}{x^5}$

5. n) $\left(\frac{15z^4}{5z^3}\right)^{-2} = \left(\frac{15}{5}z^{4-3}\right)^{-2} = (3z^1)^{-2}$

$= \frac{1}{(3z)^2} = \frac{1}{3^2z^2} = \frac{1}{9z^2}$

7. a) $728 = 7.28 \times 10^2$
b) $9,370,000 = 9.37 \times 10^6$
c) $0.0136 = 1.36 \times 10^{-2}$
d) $0.00942 = 9.42 \times 10^{-3}$
e) $0.018 \times 10^{-2} = (1.8 \times 10^{-2}) \times 10^{-2}$
$= 1.8 \times 10^{-2+(-2)} = 1.8 \times 10^{-4}$
f) $753 \times 10^3 = (7.53 \times 10^2) \times 10^3$
$= 7.53 \times 10^{2+3} = 7.53 \times 10^5$

9. a) $\frac{(0.00012)(0.00004)}{0.00000016} = \frac{(1.2\times10^{-4})(4\times10^{-5})}{1.6\times10^{-7}}$

$= \frac{(1.2)(4)\times10^{-4}10^{-5}}{1.6\times10^{-7}} = \frac{4.8\times10^{-9}}{1.6\times10^{-7}}$

$= \frac{4.8}{1.6} \times 10^{-9-(-7)} = 3 \times 10^{-2}$

$= 0.03$

b) $\frac{(4,800)(20,000)}{600,000} = \frac{(4.8\times10^3)(2\times10^4)}{6\times10^5}$

$= \frac{(4.8)(2)\times10^3 10^4}{6\times10^5} = \frac{9.6\times10^7}{6\times10^5}$

$= \frac{9.6}{6} \times 10^{7-5} = 1.6 \times 10^2$

$= 160$

11. a) Yes.
b) No, because it has a negative exponent.

13. $P(x) = 3x^2 + 2x + 1$
a) $P(3) = 3(3)^2 + 2(3) + 1$
$= 3(9) + 2(3) + 1$
$= 27 + 6 + 1$
$= 34$

b) $P(0) = 3(0)^2 + 2(0) + 1$
$= 3(0) + 2(0) + 1$
$= 0 + 0 + 1$
$= 1$

c) $P(-2) = 3(-2)^2 + 2(-2) + 1$
$= 3(4) + 2(-2) + 1$
$= 12 - 4 + 1$
$= 9$

d) $P(-0.2) = 3(-0.2)^2 + 2(-0.2) + 1$
$= 3(0.04) + 2(-0.2) + 1$
$= 0.12 - 0.4 + 1$
$= 0.72$

15. a) $3x^6 + 5x^5 - x^6 = 3x^6 - x^6 + 5x^5$
$= 2x^6 + 5x^5$

b) $x^2y^2 - 3x^2y^2 = -2x^2y^2$

c) $-2x^2yz + 3yx^2z = -2x^2yz + 3x^2yz$
$= x^2yz$

d) $(3x^2 + 2x) + (5x^2 - 8x)$
$= 3x^2 + 2x + 5x^2 - 8x$
$= 3x^2 + 5x^2 + 2x - 8x$
$= 8x^2 - 6x$

e) $(7a^2 + 2a - 5) - (3a^2 - 2a + 1)$
$= 7a^2 + 2a - 5 - 3a^2 + 2a - 1$
$= 7a^2 - 3a^2 + 2a + 2a - 5 - 1$
$= 4a^2 + 4a - 6$

f) $3(9x^2 + 3x + 7) - 2(11x^2 - 5x + 9)$
$= 3 \cdot 9x^2 + 3 \cdot 3x + 3 \cdot 7$
$- 2 \cdot 11x^2 + 2 \cdot 5x - 2 \cdot 9$
$= 27x^2 + 9x + 21 - 22x^2 + 10x - 18$
$= 27x^2 - 22x^2 + 9x + 10x + 21 - 18$
$= 5x^2 + 19x + 3$

17. a) $(2x^2)(5x) = 2(5)x^2x = 10x^3$

b) $(-6x^4z^3)(x^6z^2) = -6x^4x^6z^3z^2$
$= -6x^{10}z^5$

c) $(2rst)(\text{-}3r^2s^3t^4) = 2(\text{-}3)rr^2ss^3tt^4$
$= -6r^3s^4t^5$

d) $5b^3 \bullet 6b^2 \bullet 4b^6 = 5 \bullet 6 \bullet 4b^3b^2b^6 = 120b^{11}$

19. a) $(x+3)(x+2)$
$= x \bullet x + 2x + 3x + 3(2)$
$= x^2 + 2x + 3x + 6$
$= x^2 + 5x + 6$

b) $(2x+1)(x-1)$
$= 2x \bullet x - 2x + x + 1(-1)$
$= 2x^2 - 2x + x - 1$
$= 2x^2 - x - 1$

c) $(3a-3)(2a+2)$
$= 3a \bullet 2a + 3a \bullet 2 - 3 \bullet 2a - 3 \bullet 2$
$= 6a^2 + 6a - 6a - 6$
$= 6a^2 - 6$

d) $6(a-1)(a+1)$
$= 6(a \bullet a + a - a - 1(1))$
$= 6(a^2 + a - a - 1)$
$= 6(a^2 - 1)$
$= 6a^2 - 6$

e) $(a-b)(2a+b)$
$= a \bullet 2a + a \bullet b - 2a \bullet b - b \bullet b$
$= 2a^2 + ab - 2ab - b^2$
$= 2a^2 - ab - b^2$

f) $(-3x-y)(2x+y)$
$= \text{-}3x \bullet 2x - 3x \bullet y - 2x \bullet y - y \bullet y$
$= -6x^2 - 3xy - 2xy - y^2$
$= -6x^2 - 5xy - y^2$

21. a) $(3x+1)(x^2+2x+1)$
$= (3x+1) \bullet x^2 + (3x+1) \bullet 2x$
$+ (3x+1) \bullet 1$
$= 3x \bullet x^2 + 1 \bullet x^2 + 3x \bullet 2x + 1 \bullet 2x$
$+ 3x \bullet 1 + 1 \bullet 1$
$= 3x^3 + x^2 + 6x^2 + 2x + 3x + 1$
$= 3x^3 + 7x^2 + 5x + 1$

b) $(2a-3)(4a^2+6a+9)$
$= (2a-3) \bullet 4a^2 + (2a-3) \bullet 6a$
$+ (2a-3) \bullet 9$
$= 2a \bullet 4a^2 - 3 \bullet 4a^2 + 2a \bullet 6a - 3 \bullet 6a$
$+ 2a \bullet 9 - 3 \bullet 9$
$= 8a^3 - 12a^2 + 12a^2 - 18a$
$+ 18a - 27$
$= 8a^3 - 27$

23. Perimeter: $P = 2l + 2w$
$P = 2(x+6) + 2(2x-1)$
$P = 2x + 12 + 4x - 2$
$P = 2x + 4x + 12 - 2$
$P = 6x + 10$ in.

Area: $A = lw$
$A = (x+6)(2x-1)$
$A = x \bullet 2x - x \bullet 1 + 6 \bullet 2x - 6 \bullet 1$
$A = 2x^2 - x + 12x - 6$
$A = 2x^2 + 11x - 6$ in^2

Volume: $V = l \bullet w \bullet h$
$V = (x+6)(2x-1)(3x)$
$V = (2x^2 + 11x - 6)(3x)$ [got it from the work on area]
$V = 2x^2 \bullet 3x + 11x \bullet 3x - 6 \bullet 3x$
$V = 6x^3 + 33x^2 - 18x$ in^3

25. a) $\frac{8x+6}{2} = \frac{1}{2}(8x+6)$

$= \frac{1}{2} \bullet 8x + \frac{1}{2} \bullet 6$

$= 4x + 3$

b) $\frac{14xy-21x}{7xy} = \frac{1}{7xy}(14xy - 21x)$

$= \frac{14xy}{7xy} - \frac{21x}{7xy} = 2 - \frac{3}{y}$

25. c) $\frac{15a^2b+20ab^2-25ab}{5ab}$

$= \frac{1}{5ab}(15a^2b + 20ab^2 - 25ab)$

$= \frac{15a^2b}{5ab} + \frac{20ab^2}{5ab} - \frac{25ab}{5ab}$

$= 3a + 4b - 5$

d) $\frac{(x+y)^2+(x-y)^2}{-2xy}$

$= \frac{(x+y)(x+y)+(x-y)(x-y)}{-2xy}$

$= \frac{(x^2+xy+xy+y^2)+(x^2-xy-xy+y^2)}{-2xy}$

$= \frac{(x^2+2xy+y^2)+(x^2-2xy+y^2)}{-2xy}$

$= \frac{x^2+2xy+y^2+x^2-2xy+y^2}{-2xy}$

$= \frac{x^2+x^2+2xy-2xy+y^2+y^2}{-2xy}$

$= \frac{2x^2+2y^2}{-2xy}$

$= \frac{1}{-2xy}(2x^2 + 2y^2)$

$= \frac{2x^2}{-2xy} + \frac{2y2}{-2xy}$

$= -\frac{x}{y} - \frac{y}{x}$

27. a)

$$\begin{array}{r} x + 1 \\ x + 2\,\overline{)\, x^2 + 3x + 5} \\ \underline{x^2 + 2x} \\ x + 5 \\ \underline{x + 2} \\ 3 \end{array}$$

answer : $x + 1 + \frac{3}{x+2}$

b)

$$\begin{array}{r} x - 5 \\ x - 1\,\overline{)\, x^2 - 6x + 5} \\ \underline{x^2 - x} \\ -5x + 5 \\ \underline{-5x + 5} \\ 0 \end{array}$$

answer : $x - 5$

c) Write as $x + 3\,\overline{)\, 2x^2 + 7x + 3}$

$$\begin{array}{r} 2x + 1 \\ x + 3\,\overline{)\, 2x^2 + 7x + 3} \\ \underline{2x^2 + 6x} \\ x + 3 \\ \underline{x + 3} \\ 0 \end{array}$$

answer : $2x + 1$

d) Write as $3x - 1\,\overline{)\, 3x^2 + 14x - 2}$

$$\begin{array}{r} x + 5 \\ 3x - 1\,\overline{)\, 3x^2 + 14x - 2} \\ \underline{3x^2 - x} \\ 15x - 2 \\ \underline{15x - 5} \\ 3 \end{array}$$

answer : $x + 5 + \frac{3}{3x-1}$

e) Write as $2x - 1\,\overline{)\, 6x^3 + x^2 + 0x + 1}$

$$\begin{array}{r} 3x^2 + 2x + 1 \\ 2x - 1\,\overline{)\, 6x^3 + x^2 + 0x + 1} \\ \underline{6x^3 - 3x^2} \\ 4x^2 + 0x \\ \underline{4x^2 - 2x} \\ 2x + 1 \\ \underline{2x - 1} \\ 2 \end{array}$$

answer : $3x^2 + 2x + 1 + \frac{2}{2x-1}$

27. f) Write as

$$3x+1\overline{)\,9x^3+0x^2-13x-4}$$

$$\begin{array}{r} 3x^2-x-4 \\ 3x+1\overline{)\,9x^3+0x^2-13x-4} \\ \underline{9x^3+3x^2} \\ -3x^2-13x \\ \underline{-3x^2-\ \ x} \\ -12x-4 \\ \underline{-12x-4} \\ 0 \end{array}$$

answer : $3x^2-x-4$

29. We know the distance, $8x^2+2x-3$, and the time, $2x-1$. We can use the formula, $d=rt$, to solve for r.

$d=rt$

$\frac{d}{t}=\frac{rt}{t}$

$\frac{d}{t}=r$

So, $r=\frac{8x^2+2x-3}{2x-1}$.

Write as $2x-1\overline{)\,8x^2+2x-3}$

$$\begin{array}{r} 2x+1 \\ 2x-1\overline{)\,8x^2+2x-3} \\ \underline{8x^2-4x} \\ 6x-3 \\ \underline{6x-3} \\ 0 \end{array}$$

Thus, the rate of travel is $4x+3$ inches per minute.

STUDY SET Section 6.1

VOCABULARY

1. square root

3. positive

5. right

CONCEPTS

7. The number 25 has two square roots. They are 5 and -5.

9. $c^2 = a^2 + b^2$

11. If a and b are positive numbers and $a = b$ then $\sqrt{a} = \sqrt{b}$.

13. a) Divide both sides by 2.
b) Take the square root of both sides.

15. If $x = 0$, $\sqrt{0} = 0$.
If $x = \frac{1}{81}$, $\sqrt{\frac{1}{81}} = \frac{1}{9}$.
If $x = 0.4$, $\sqrt{0.4} = 0.4$.
If $x = 36$, $\sqrt{36} = 6$.
If $x = 400$, $\sqrt{400} = 20$.

NOTATION

17.
$$\begin{aligned} c^2 &= a^2 + b^2 \\ &= (5)^2 + (12)^2 \\ &= 25 + 144 \\ &= 169 \\ \sqrt{c^2} &= \sqrt{169} \\ c &= 13 \end{aligned}$$

19. multiplication. $3\sqrt{2} = 3 \times \sqrt{2}$

PRACTICE

21. $\sqrt{25} = 5$, because $5^2 = 25$.

23. $-\sqrt{81} = -9$, because $9^2 = 81$.

25. $\sqrt{1.21} = 1.1$, because $1.1^2 = 1.21$.

27. $\sqrt{196} = 14$, because $14^2 = 196$.

29. $\sqrt{\frac{9}{256}} = \frac{3}{16}$, because $\left(\frac{3}{16}\right)^2 = \frac{9}{256}$.

31. $-\sqrt{289} = -17$, because $17^2 = 289$.

33. $-\sqrt{2500} = -50$, because $50^2 = 2500$.

35. $\sqrt{3600} = 60$, because $60^2 = 3600$.

37. $\sqrt{2} \approx 1.414$

39. $\sqrt{11} \approx 3.317$

41. $\sqrt{95} \approx 9.747$

43. $\sqrt{428} \approx 20.688$

45. $-\sqrt{9,876} \approx -99.378$

47. $\sqrt{21.35} \approx 4.621$

49. $\sqrt{0.3588} \approx 0.599$

51. $-\sqrt{0.8372} \approx -0.915$

53. $2\sqrt{3} \approx 2(1.732) \approx 3.464$

55. $\frac{2+\sqrt{3}}{2} \approx \frac{2+(1.732)}{2} \approx \frac{3.732}{2} \approx 1.866$

57. $\sqrt{9} = 3$, rational
$\sqrt{17} \approx 4.123$, irrational
$\sqrt{49} = 7$, rational
$\sqrt{-49}$, imaginary

59.

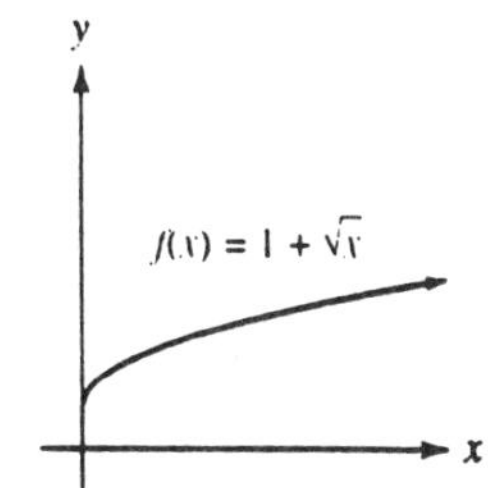

61.

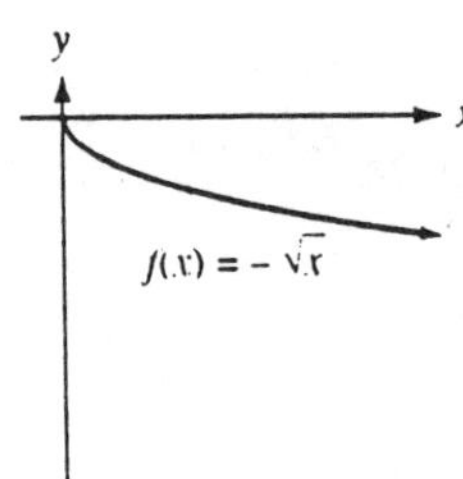

63. $c^2 = a^2 + b^2$
$c^2 = (4)^2 + (3)^2$
$c^2 = 16 + 9$
$c^2 = 25$
$\sqrt{c^2} = \sqrt{25}$
$c = 5$

65. $c^2 = a^2 + b^2$
$(17)^2 = (15)^2 + b^2$
$289 = 225 + b^2$
$289 - 225 = 225 - 225 + b^2$
$64 = b^2$
$\sqrt{64} = \sqrt{b^2}$
$8 = b$

67. $c^2 = a^2 + b^2$
$(34)^2 = a^2 + (16)^2$
$1156 = a^2 + 256$
$1156 - 256 = a^2 + 256 - 256$
$900 = a^2$
$\sqrt{900} = \sqrt{b^2}$
$30 = b$

APPLICATIONS

69. **Analyze:** The wall, ground, and ladder form a right triangle with the ladder as the hypotenuse. The ladder is 20ft. long. The window is 16 ft. high. Let b be the distance the base of the ladder is from the wall.

Form:

Hypotenuse squared	is	height squared	plus	base squared

$20^2 = 16^2 + b^2$

Solve: $20^2 = 16^2 + b^2$
$400 = 256 + b^2$
$400 - 256 = 256 - 256 + b^2$
$144 = b^2$
$\sqrt{144} = \sqrt{b^2}$
$12 = b$

State: The base of the ladder is 12 ft. from the wall.

71. **Analyze:** The wire, pole, and ground form a right triangle with the wire as the hypotenuse. The wire is 34ft. long and the point on the ground is 16 ft. from the pole. Let h represent the height of the pole.

Form:

Wire squared	is	height squared	plus	ground squared

$34^2 = h^2 + 16^2$

Solve: $34^2 = h^2 + 16^2$
$1156 = h^2 + 256$
$1156 - 256 = h^2 + 256 - 256$
$900 = h^2$
$\sqrt{900} = \sqrt{h^2}$
$30 = h$

State: The height of the pole is 30 ft.

73. **Analyze:** The paths from home to first, from first to second, and from second to home form a right triangle with the path from second to home as the hypotenuse. The base paths, which are the legs of the triangle, are 90 ft. Let d represent the distance between second and home.

Form:

2nd-home squared	is	home-1st squared	plus	1st-2nd squared

$d^2 = 90^2 + 90^2$

Solve: $d^2 = 90^2 + 90^2$
$d^2 = 900 + 900$
$d^2 = 1800$
$\sqrt{d^2} = \sqrt{1800}$
$d \approx 127.3$ ft.

State: The distance between second and home is approximately 127.3 ft.

75. **Analyze:** The path traveled east and north combined with the "as the crow flies" from a right triangle with this side as the hypotenuse. The legs of the triangle have distance of 4.2 and 4.0 miles. Let d represent the path of "as the crow flies".

Form:

"as the crow flies" squared is east squared plus west squared

$$d^2 = 4.2^2 + 4.0^2$$

Solve: $d^2 = 4.2^2 + 4.0^2$

$d^2 = 17.64 + 16$

$d^2 = 33.64$

$\sqrt{d^2} = \sqrt{33.64}$

$d \approx 5.8$ ft.

State: The distance from camp is approximately 5.8 miles.

77. **Analyze:** The tight end had gained 6 yds. We need to find the length of BC. Because the angle is 45°, AB and BC have the same length. The hypotenuse is 5 yds. Let b represent the length of BC. We have that AB is also b.

Form:

$c^2 = a^2 + b^2$

$5^2 = b^2 + b^2$

Solve: $5^2 = b^2 + b^2$

$5^2 = 2b^2$

$25 = 2b^2$

$\frac{25}{2} = \frac{2b^2}{2}$

$12.5 = b^2$

$\sqrt{12.5} = \sqrt{b^2}$

$\sqrt{12.5} = b$

$b \approx 3.53$

State: The tight end gained a total of $5 + 3.53 = 8.53$ yds, which was short of the first down.

79. **Analyze:** The two sides and the diagonal of the wrestling ring form a right triangle. Two legs are 18 ft. long. Let d represent the diagonal of the ring.

Form:

the diagonal squared is side squared plus side squared

$$d^2 = 18^2 + 18^2$$

Solve: $d^2 = 18^2 + 18^2$

$d^2 = 324 + 324$

$d^2 = 648$

$\sqrt{d^2} = \sqrt{648}$

$d = \sqrt{648}$

$b \approx 25.5$

State: The diagonal of the wrestling ring is 25.5 ft. long.

81. **Analyze:** Area of triangle: $A = \frac{1}{2}bh$. The height of the isosceles triangle can be determined by using the right triangle formed by the height, half the base, and the side of the isosceles triangle. The side is 26 in. and half the base is 10 in. Let h represent the height of the triangle.

Form: $c^2 = a^2 + b^2$

$26^2 = h^2 + 10^2$

Solve: $26^2 = h^2 + 10^2$

$676 = h^2 + 100$

$676 - 100 = h^2 + 100 - 100$

$576 = h^2$

$\sqrt{576} = \sqrt{h^2}$

$\sqrt{648} = h$

$h \approx 75.3$

The area of the isosceles triangle is

$A = \frac{1}{2}(20)(75.3)$

$A = 10(75.3)$

$A = 753$

State: The area of the isosceles triangle is 753 in^2.

83. a) The hypotenuse is the leg times $\sqrt{2}$.

$h = (6)\sqrt{2}$

$h \approx 8.5$ in.

b) The hypotenuse is the leg times $\frac{\sqrt{3}}{2}$.

$h = (9)\frac{\sqrt{3}}{2}$

$h \approx 7.8$ in.

REVIEW

87. $(3s^2 - 3s - 2) + (3s^2 + 4s - 3)$

$= 3s^2 - 3s - 2 + 3s^2 + 4s - 3$

$= 3s^2 + 3s^2 - 3s + 4s - 2 - 3$

$= 6s^2 + s - 5$

89. $(3x - 2)(x + 4)$

$= 3x \cdot x + 3x \cdot 4 + (-2) \cdot x + (-2)(4)$

$= 3x^2 + 12x - 2x - 8$

$= 3x^2 + 10x - 8$

STUDY SET Section 6.2

VOCABULARY

1. cube

3. function

CONCEPTS

5. cube; root

7. $(-6)^3$

9. a) Divide both sides by 3.
b) Take the cube root of both sides.

11. a) $f(1) = \sqrt[3]{1} = 1$
b) $f(-\frac{1}{27}) = \sqrt[3]{-\frac{1}{27}} = -\frac{1}{3}$
c) $f(125) = \sqrt[3]{125} = 5$
d) $f(0.008) = \sqrt[3]{0.008} = 0.2$
e) $f(1,000) = \sqrt[3]{1,000} = 10$

NOTATION

13. index; radicand

15. 2

PRACTICE

17. $\sqrt[3]{8} = 2$, because $2^3 = 8$.

19. $\sqrt[3]{0} = 0$, because $0^3 = 0$.

21. $\sqrt[3]{-8} = -2$, because $(-2)^3 = -8$.

23. $\sqrt[3]{-64} = -4$, because $(-4)^3 = -64$.

25. $\sqrt[3]{\frac{1}{125}} = \frac{1}{5}$, because $(\frac{1}{5})^3 = \frac{1}{125}$.

27. $\sqrt[3]{-1} = -1$, because $(-1)^3 = -1$.

29. $-\sqrt[3]{64} = -4$, because $(4)^3 = 64$.

31. $\sqrt[3]{729} = 9$, because $(9)^3 = 729$.

33. $\sqrt[3]{32,100} \approx 31.78$

35. $\sqrt[3]{-0.11324} \approx -0.48$

37.

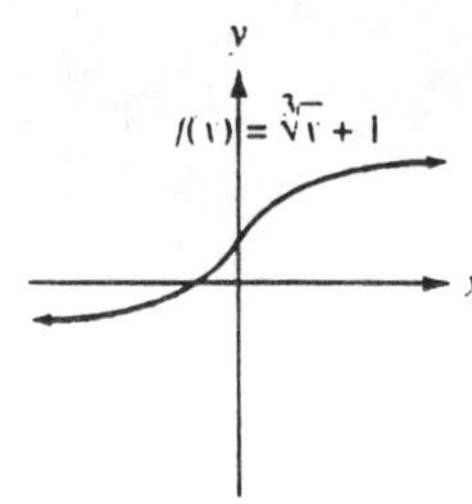

39.

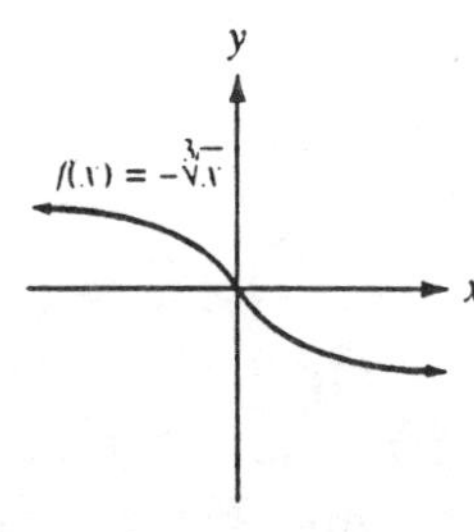

41. $\sqrt[4]{16} = 2$, because $2^4 = 16$.

43. $-\sqrt[5]{32} = -(2) = -2$,
because $2^5 = 32$.

45. $\sqrt[6]{1} = 1$, because $1^6 = 1$.

47. $\sqrt[5]{-32} = -2$, because $(-2)^5 = -8$.

49. $\sqrt[4]{125} \approx 3.34$

51. $\sqrt[5]{-6,000} \approx -5.70$

53. $\sqrt{x^2} = x$, because $(x)^2 = x^2$.

55. $\sqrt{x^6} = x^3$, because $(x^3)^2 = x^6$.

57. $\sqrt{x^{10}} = x^5$, because $(x^5)^2 = x^{10}$.

59. $\sqrt[4]{x^4} = x$, because $(x)^4 = x^4$.

61. $\sqrt{4z^2} = 2z$, because $(2z)^2 = 4z^2$.

63. $-\sqrt{x^4y^2} = -(x^2y) = -x^2y$,
because $(x^2y)^2 = x^4y^2$.

65. $-\sqrt{0.04y^2} = -(0.2y) = -0.2y$,
because $(0.2y)^2 = 0.04y^2$.

67. $-\sqrt{25x^4z^{12}} = -(5x^2y^6) = -5x^2y^6$,
because $(5x^2y^6)^2 = 25x^4y^{12}$.

69. $\sqrt{36z^{36}} = 6z^{18}$,
because $(6z^{18})^2 = 36z^{36}$.

71. $-\sqrt{625z^2} = -25z$,
because $(25z)^2 = 625z^2$.

73. $\sqrt[3]{y^6} = y^2$, because $(y^2)^3 = y^6$.

75. $\sqrt[5]{f^5} = f$, because $(f)^5 = f^5$.

77. $\sqrt[3]{27y^3} = 3y$, because $(3y)^3 = 27y^3$.

79. $\sqrt[3]{-p^6q^3} = -p^2q$,
because $(p^2q)^3 = p^6q^3$.

APPLICATIONS

81. $V = s^3$
$(2) = s^3$
$\sqrt[3]{2} = \sqrt[3]{s^3}$
$\sqrt[3]{2} = s$
$s \approx 1.26$ ft.

83. $P = 400$
$S = \sqrt[3]{\frac{P}{0.02}}$
$= \sqrt[3]{\frac{(400)}{0.02}}$
$= \sqrt[3]{20,000}$
≈ 27.14 mph

85. From the information:
$C = 27,000$
$n = 4$
$S = 8,000$

$r = 1 - \sqrt[n]{\frac{S}{C}}$

$r = 1 - \sqrt[4]{\frac{8,000}{27,000}}$

$r = 1 - \sqrt[4]{\frac{8}{27}}$

$r \approx 1 - .7378$
$r \approx 0.2622$
$0.2622 = 26.22\%$

REVIEW

89. $m^5m^2 = m^{5+2} = m^7$

91. $(3^2)^4 = 3^{2\cdot 4} = 3^8$

93. $(x^2x^3)^5 = (x^{2+3})^5$
$= (x^5)^5$
$= x^{5\cdot 5}$
$= x^{25}$

95. $4x^3(6x^5) = 4(6)x^3\cdot x^5$
$= 24x^{3+5}$
$= 24x^8$

STUDY SET Section 6.3

VOCABULARY

1. radical

3. extraneous

CONCEPTS

5. If $a = b$, then $a^2 = b^2$.

7. a) Take the square root of both sides.
b) Square both sides.

9. Both sides of the equation were not squared--only the left side was squared.

11. a) trapezoid
b) $A = (-2, 5)$ $B = (2, 5)$
$C = (8, -3)$ $D = (-8, -3)$
c) $AB:$ Let $x_1 = -2$, $y_1 = 5$,
$x_2 = 2$, & $y_2 = 5$.

$$d = \sqrt{(x_2 - x_1)^2 + (y_2 - y_1)^2}$$
$$d = \sqrt{((2) - (-2))^2 + ((5) - (5))^2}$$
$$d = \sqrt{(2+2)^2 + (5-5)^2}$$
$$d = \sqrt{(4)^2 + (0)^2}$$
$$d = \sqrt{16+0} = \sqrt{16} = 4$$

$BC:$ Let $x_1 = 2$, $y_1 = 5$,
$x_2 = 8$, & $y_2 = -3$.

$$d = \sqrt{((8) - (2))^2 + ((-3) - (5))^2}$$
$$d = \sqrt{(8-2)^2 + (-3-5)^2}$$
$$d = \sqrt{(6)^2 + (-8)^2}$$
$$d = \sqrt{36+64} = \sqrt{100} = 10$$

$CD:$ Let $x_1 = 8$, $y_1 = -3$,
$x_2 = -8$, & $y_2 = -3$.

$$d = \sqrt{((-8) - (8))^2 + ((-3) - (-3))^2}$$
$$d = \sqrt{(-8-8)^2 + (-3+3)^2}$$
$$d = \sqrt{(-16)^2 + (0)^2}$$
$$d = \sqrt{256+0} = \sqrt{256} = 16$$

$DA:$ Let $x_1 = -8$, $y_1 = -3$,
$x_2 = -2$, & $y_2 = 5$.

$$d = \sqrt{((-2) - (-8))^2 + ((5) - (-3))^2}$$
$$d = \sqrt{(-2+8)^2 + (5+3)^2}$$
$$d = \sqrt{(6)^2 + (8)^2}$$
$$d = \sqrt{36+64} = \sqrt{100} = 10$$

NOTATION

13. $\sqrt{x-3} = 5$

$$\left(\sqrt{x-3}\right)^2 = (5)^2$$
$$x - 3 = 25$$
$$x - 3 + 3 = 25 + 3$$
$$x = 28$$

PRACTICE

15. $\sqrt{x} = 3$

$$(\sqrt{x})^2 = (3)^2$$
$$x = 9$$

check: $\sqrt{(9)} = 3$
$3 = 3$

17. $\sqrt{2a} = 4$

$$\left(\sqrt{2a}\right)^2 = (4)^2$$
$$2a = 16$$
$$\frac{2a}{2} = \frac{16}{2}$$
$$a = 8$$

check: $\sqrt{2(8)} = 4$
$\sqrt{16} = 4$
$4 = 4$

19. $\sqrt{r} + 4 = 0$

$$\sqrt{r} + 4 - 4 = 0 - 4$$
$$\sqrt{r} = -4$$
$$(\sqrt{r})^2 = (-4)^2$$
$$r = 16$$

check: $\sqrt{2(8)} = 4$
$\sqrt{16} = 4$
$4 = 4$

21. $-\sqrt{x} = -5$
$-1(-\sqrt{x}) = -1(-5)$
$\sqrt{x} = 5$
$\left(\sqrt{x}\right)^2 = (5)^2$
$x = 25$

check: $-\sqrt{(25)} = -5$
$-(5) = -5$
$-5 = -5$

23. $10 - \sqrt{s} = 7$
$10 - 10 - \sqrt{s} = 7 - 10$
$-\sqrt{s} = -3$
$-1(-\sqrt{s}) = -1(-3)$
$\sqrt{s} = 3$
$\left(\sqrt{s}\right)^2 = (3)^2$
$s = 9$

check: $10 - \sqrt{(9)} = 7$
$10 - 3 = 7$
$7 = 7$

25. $\sqrt{x+3} = 2$
$\left(\sqrt{x+3}\right)^2 = (2)^2$
$x + 3 = 4$
$x + 3 - 3 = 4 - 3$
$x = 1$

check: $\sqrt{(1)+3} = 2$
$\sqrt{4} = 2$
$2 = 2$

27. $\sqrt{3-T} = -2$
$\left(\sqrt{3-T}\right)^2 = (-2)^2$
$3 - T = 4$
$3 - 3 - T = 4 - 3$
$-T = 1$
$-1(-T) = -1(1)$
$T = -1$

check: $\sqrt{3-(-1)} = -2$
$\sqrt{3+1} = -2$
$\sqrt{4} = -2$
$2 = -2$ No solution.

29. $\sqrt{6+2x} = 4$
$\left(\sqrt{6+2x}\right)^2 = (4)^2$
$6 + 2x = 16$
$6 - 6 + 2x = 16 - 6$
$2x = 10$
$\frac{2x}{2} = \frac{10}{2}$
$x = 5$

check: $\sqrt{6+2(5)} = 4$
$\sqrt{6+10} = 4$
$\sqrt{16} = 4$
$4 = 4$

31. $\sqrt{5x-5} - 5 = 0$
$\sqrt{5x-5} - 5 + 5 = 0 + 5$
$\sqrt{5x-5} = 5$
$\left(\sqrt{5x-5}\right)^2 = (5)^2$
$5x - 5 = 25$
$5x - 5 + 5 = 25 + 5$
$5x = 30$
$\frac{5x}{5} = \frac{30}{5}$
$x = 6$

check: $\sqrt{5(6)-5} - 5 = 0$
$\sqrt{30-5} - 5 = 0$
$\sqrt{25} - 5 = 0$
$5 - 5 = 0$
$0 = 0$

33. $\sqrt{x+3} + 5 = 12$
$\sqrt{x+3} + 5 - 5 = 12 - 5$
$\sqrt{x+3} = 7$
$\left(\sqrt{x+3}\right)^2 = (7)^2$
$x + 3 = 49$
$x + 3 - 3 = 49 - 3$
$x = 46$

check: $\sqrt{(46)+3} + 5 = 12$
$\sqrt{30-5} - 5 = 0$
$\sqrt{49} + 5 = 12$
$7 + 5 = 12$
$12 = 12$

35. $x - 3 = \sqrt{x^2 - 15}$

$(x-3)^2 = \left(\sqrt{x^2 - 15}\right)^2$

$(x-3)(x-3) = x^2 - 15$

$x^2 - 6x + 9 = x^2 - 15$

$x^2 - x^2 - 6x + 9 = x^2 - x^2 - 15$

$-6x + 9 = -15$

$-6x + 9 - 9 = -15 - 9$

$-6x = -24$

$\frac{-6x}{-6} = \frac{-24}{-6}$

$x = 4$

check: $(4) - 3 = \sqrt{(4)^2 - 15}$

$1 = \sqrt{16 - 15}$

$1 = \sqrt{1}$

$1 = 1$

37. $\sqrt{3t - 9} = \sqrt{t+1}$

$\left(\sqrt{3t-9}\right)^2 = \left(\sqrt{t+1}\right)^2$

$3t - 9 = t + 1$

$3t - t - 9 + 9 = t - t + 1 + 9$

$2t = 10$

$\frac{2t}{2} = \frac{10}{2}$

$t = 5$

check: $\sqrt{3(5) - 9} = \sqrt{(5) + 1}$

$\sqrt{15 - 9} = \sqrt{6}$

$\sqrt{6} = \sqrt{6}$

39. $\sqrt{10 - 3x} = \sqrt{2x + 20}$

$\left(\sqrt{10 - 3x}\right)^2 = \left(\sqrt{2x+20}\right)^2$

$10 - 3x = 2x + 20$

$10 - 3x - 2x = 2x - 2x + 20$

$10 - 5x = 20$

$10 - 10 - 5x = 20 - 10$

$-5x = 10$

$\frac{-5x}{-5} = \frac{10}{-5}$

$x = -2$

check: $\sqrt{10 - 3(-2)} = \sqrt{2(-2) + 20}$

$\sqrt{10 + 6} = \sqrt{-4 + 20}$

$\sqrt{16} = \sqrt{16}$

$4 = 4$

41. $\sqrt{3c - 8} - \sqrt{c} = 0$

$\sqrt{3c - 8} - \sqrt{c} + \sqrt{c} = 0 + \sqrt{c}$

$\sqrt{3c - 8} = \sqrt{c}$

$\left(\sqrt{3c-8}\right)^2 = \left(\sqrt{c}\right)^2$

$3c - 8 = c$

$3c - c - 8 = c - c$

$2c - 8 = 0$

$2c - 8 + 8 = 0 + 8$

$2c = 8$

$\frac{2c}{2} = \frac{8}{2}$

$c = 4$

check: $\sqrt{3(4) - 8} - \sqrt{(4)} = 0$

$\sqrt{12 - 8} - \sqrt{4} = 0$

$\sqrt{4} - \sqrt{4} = 0$

$2 - 2 = 0$

$0 = 0$

43. $x - 1 = \sqrt{x^2 - 4x + 9}$

$(x-1)^2 = \left(\sqrt{x^2 - 4x + 9}\right)^2$

$(x-1)(x-1) = x^2 - 4x + 9$

$x^2 - 2x + 1 = x^2 - 4x + 9$

$x^2 - x^2 - 2x + 1 = x^2 - x^2 - 4x + 9$

$-2x + 1 = -4x + 9$

$-2x + 4x + 1 = -4x + 4x + 9$

$2x + 1 = 9$

$2x + 1 - 1 = 9 - 1$

$2x = 8$

$\frac{2x}{2} = \frac{8}{2}$

$x = 4$

check: $(4) - 1 = \sqrt{(4)^2 - 4(4) + 9}$

$3 = \sqrt{16 - 16 + 9}$

$3 = \sqrt{9}$

$3 = 3$

45. $\sqrt{4m^2+6m+6} = -2m$

$\left(\sqrt{4m^2+6m+6}\right)^2 = (-2m)^2$

$4m^2+6m+6 = 4m^2$

$4m^2-4m^2+6m+6 = 4m^2-4m^2$

$6m+6=0$

$6m+6-6=0-6$

$6m = -6$

$\frac{6m}{6} = \frac{-6}{6}$

$x = -1$

check: $\sqrt{4(-1)^2+6(-1)+6} = -2(-1)$

$\sqrt{4(1)-6+6}=2$

$\sqrt{4-6+6}=2$

$\sqrt{4}=2$

$2=2$

47. $\sqrt{3x+3} = 3\sqrt{x-1}$

$\left(\sqrt{3x+3}\right)^2 = \left(3\sqrt{x-1}\right)^2$

$3x+3 = 3^2(x-1)$

$3x+3 = 9(x-1)$

$3x+3 = 9x-9$

$3x-9x+3 = 9x-9x-9$

$-6x+3 = -9$

$-6x+3-3 = -9-3$

$-6x = -12$

$\frac{-6x}{-6} = \frac{-12}{-6}$

$x=2$

check: $\sqrt{3(2)+3} = 3\sqrt{(2)-1}$

$\sqrt{6+3} = 3\sqrt{1}$

$\sqrt{9} = 3(1)$

$3=3$

49. $2\sqrt{3x+4} = \sqrt{5x+9}$

$\left(2\sqrt{3x+4}\right)^2 = \left(\sqrt{5x+9}\right)^2$

$2^2(3x+4) = 5x+9$

$4(3x+4) = 5x+9$

$12x+16 = 5x+9$

$12x-5x+16 = 5x-5x+9$

$7x+16=9$

$7x+16-16 = 9-16$

$7x = -7$

$\frac{7x}{7} = \frac{-7}{7}$

$x = -1$

check: $2\sqrt{3(-1)+4} = \sqrt{5(-1)+9}$

$2\sqrt{-3+4} = 3\sqrt{-5+9}$

$2\sqrt{1} = 3\sqrt{4}$

$2(1) = 3(2)$

$2=6$

51. $\sqrt[3]{x-1} = 4$

$\left(\sqrt[3]{x-1}\right)^3 = (4)^3$

$x-1=64$

$x-1+1 = 64+1$

$x=65$

check: $\sqrt[3]{(65)-1} = 4$

$\sqrt[3]{64}=4$

$4=4$

53. $\sqrt[3]{\frac{1}{2}x-3} = 2$

$\left(\sqrt[3]{\frac{1}{2}x-3}\right)^3 = (2)^3$

$\frac{1}{2}x-3 = 8$

$\frac{1}{2}x-3+3 = 8+3$

$\frac{1}{2}x = 11$

$2\left(\frac{1}{2}x\right) = 2(11)$

$x=22$

check: $\sqrt[3]{\frac{1}{2}(22)-3} = 2$

$\sqrt[3]{11-3} = 2$

$\sqrt[3]{8}=2$

$2=2$

55. Let $x_1 = 3$, $y_1 = -4$,
$x_2 = 0$, & $y_2 = 0$.

$d = \sqrt{(x_2-x_1)^2+(y_2-y_1)^2}$

$d = \sqrt{((0)-(3))^2+((0)-(-4))^2}$

$d = \sqrt{(0-3)^2+(0+4)^2}$

$d = \sqrt{(-3)^2+(4)^2}$

$d = \sqrt{9+16} = \sqrt{25} = 5$

57. Let $x_1 = 2,\ y_1 = 4,$
$x_2 = 5,\ \&\ y_2 = 9.$

$$d = \sqrt{(x_2 - x_1)^2 + (y_2 - y_1)^2}$$
$$d = \sqrt{((5) - (2))^2 + ((9) - (4))^2}$$
$$d = \sqrt{(5 - 2)^2 + (9 - 4)^2}$$
$$d = \sqrt{(3)^2 + (5)^2}$$
$$d = \sqrt{9 + 25} = \sqrt{34} \approx 5.83$$

59. Let $x_1 = -2,\ y_1 = -8,$
$x_2 = 3,\ \&\ y_2 = 4.$

$$d = \sqrt{(x_2 - x_1)^2 + (y_2 - y_1)^2}$$
$$d = \sqrt{((3) - (-2))^2 + ((4) - (-8))^2}$$
$$d = \sqrt{(3 + 2)^2 + (4 + 8)^2}$$
$$d = \sqrt{(5)^2 + (12)^2}$$
$$d = \sqrt{25 + 144} = \sqrt{169} = 13$$

61. Let $x_1 = 6,\ y_1 = 8,$
$x_2 = 12,\ \&\ y_2 = 16.$

$$d = \sqrt{(x_2 - x_1)^2 + (y_2 - y_1)^2}$$
$$d = \sqrt{((12) - (6))^2 + ((16) - (8))^2}$$
$$d = \sqrt{(12 - 6)^2 + (16 - 8)^2}$$
$$d = \sqrt{(6)^2 + (8)^2}$$
$$d = \sqrt{36 + 64} = \sqrt{100} = 10$$

63. Let $t = 3.25$

$$t = \frac{\sqrt{s}}{4}$$
$$3.25 = \frac{\sqrt{s}}{4}$$
$$4(3.25) = 4\left(\frac{\sqrt{s}}{4}\right)$$
$$13 = \sqrt{s}$$
$$(13)^2 = \left(\sqrt{s}\right)^2$$
$$169 = s$$

The height of the waterfall is 169 ft.

65. Let $t = 8.91.$

$$t = 1.11\sqrt{L}$$
$$8.91 = 1.11\sqrt{L}$$
$$\frac{8.91}{1.11} = \frac{1.11\sqrt{L}}{1.11}$$
$$\frac{8.91}{1.11} = \sqrt{L}$$
$$\left(\frac{8.91}{1.11}\right)^2 = \left(\sqrt{L}\right)^2$$
$$\left(\frac{8.91}{1.11}\right)^2 = L$$
$$64.4 \approx L$$

The pendulum is about 64.4 ft.

67. Let $k = 3.24\ \&\ s = 55$

$$s = k\sqrt{d}$$
$$55 = 3.24\sqrt{d}$$
$$\frac{55}{3.24} = \frac{3.24\sqrt{d}}{3.24}$$
$$\frac{55}{3.24} = \sqrt{d}$$
$$\left(\frac{55}{3.24}\right)^2 = \left(\sqrt{d}\right)^2$$
$$\left(\frac{55}{3.24}\right)^2 = d$$
$$288 \approx L$$

The car will skid about 288 ft.

69. **Analyze:** From the illustration, the distance from the earth's center is the sum of the distance from the center to the surface and the altitude. The distance from the center of the earth to the surface is 6.4×10^6 m. The speed of the satellite is 7×10^3 meters per second. We want to find the altitude. In order to find the altitude, we first need to find the distance of the satellite from the center of the earth.

Form: distance from center is 6.4×10^6 + alt.

$$r = 6.4 \times 10^6 + a$$

Also, we have

$$\sqrt{r} = \frac{2.029 \times 10^7}{s}$$
$$\sqrt{r} = \frac{2.029 \times 10^7}{7 \times 10^3}$$

Solve:

$\sqrt{r} = \frac{2.029 \times 10^7}{7 \times 10^3}$

$\sqrt{r} = \frac{2.029}{7} \times \frac{10^7}{10^3}$

$\sqrt{r} \approx 2.90 \times 10^3$

$\left(\sqrt{r}\right)^2 \approx (2.90 \times 10^3)^2$

$r \approx 8.41 \times 10^6$

Now, to find a :

$r = 6.4 \times 10^6 + a$

$8.41 \times 10^6 = 6.4 \times 10^6 + a$

$8.41 \times 10^6 - 6.4 \times 10^6$

$= 6.4 \times 10^6 - 6.4 \times 10^6 + a$

$2.01 \times 10^6 = a$

State: The altitude of the satellite is about 2×10^6 meters.

71. $r = \sqrt{\frac{3V}{\pi h}}$

$(r)^2 = \left(\sqrt{\frac{3V}{\pi h}}\right)^2$

$r^2 = \frac{3V}{\pi h}$

$\pi h(r^2) = \pi h\left(\frac{3V}{\pi h}\right)$

$\pi h r^2 = 3V$

$\frac{\pi h r^2}{3} = \frac{3V}{3}$

$\frac{\pi h r^2}{3} = V$

REVIEW

77. $x + y = 5$

$x + 1 = y$

Substitute the second equation into the first equation.

$x + y = 5$

$x + (x + 1) = 5$

$2x + 1 = 5$

$2x + 1 - 1 = 5 - 1$

$2x = 4$

$\frac{2x}{2} = \frac{4}{2}$

$x = 2$

Find y :

$x + 1 = y$

$(2) + 1 = y$

$3 = y$

The solution is $(2,\ 3)$.

79. $2x + 3y = 0$

$3x - 2y = 13$

Multiply the first equation by -3 and the second equation by 2.

$-3(2x + 3y) = -3(0)$

$-6x - 9y = 0$

$2(3x - 2y) = 2(13)$

$6x - 4y = 26$

Now, we have

$-6x - 9y = 0$

$6x - 4y = 26$

$-13y = 26$

$\frac{-13y}{-13} = \frac{26}{-13}$

$y = -2$

Find x :

$-6x - 9(-2) = 0$

$-6x + 18 = 0$

$-6x + 18 - 18 = 0 - 18$

$-6x = -18$

$\frac{-6x}{-6} = \frac{-18}{-6}$

$x = 3$

The solution is $(3,\ -2)$.

STUDY SET Section 6.4

VOCABULARY

1. perfect

3. simplify

CONCEPTS

5. product; $\sqrt{ab} = \sqrt{a}\sqrt{b}$

7. Line 2 is not true.
$\sqrt{9+4} \neq \sqrt{9} + \sqrt{4}$
There is no addition property of radicals.

9. a) Area of square $= s^2$.

$$28 = s^2$$
$$\sqrt{28} = \sqrt{s^2}$$
$$\sqrt{4 \cdot 7} = s$$
$$\sqrt{4}\sqrt{7} = s$$
$$2\sqrt{7}\text{ in.} = s$$

b) $2\sqrt{7}$ in. ≈ 5.3 in.

11. $a = 5,\ b = 10,\ \&\ c = 3$

$$\sqrt{b^2 - 4ac} = \sqrt{(10)^2 - 4(5)(3)}$$
$$= \sqrt{100 - 60}$$
$$= \sqrt{40}$$
$$= \sqrt{4 \cdot 10}$$
$$= \sqrt{4}\sqrt{10}$$
$$= 2\sqrt{10}$$

13. $a = -1,\ b = 6,\ \&\ c = 9$

$$\sqrt{b^2 - 4ac} = \sqrt{(6)^2 - 4(-1)(9)}$$
$$= \sqrt{36 + 36}$$
$$= \sqrt{72}$$
$$= \sqrt{36 \cdot 2}$$
$$= \sqrt{36}\sqrt{2}$$
$$= 6\sqrt{2}$$

NOTATION

15. $\sqrt{80a^3b^2} = \sqrt{16 \cdot 5 \cdot a^2 \cdot a \cdot b^2}$

$$= \sqrt{16a^2b^2 \cdot 5a}$$
$$= \sqrt{16a^2b^2}\sqrt{5a}$$
$$= 4ab\sqrt{5a}$$

PRACTICE

17. multiplication

19. $\sqrt{20} = \sqrt{4 \cdot 5} = \sqrt{4}\sqrt{5} = 2\sqrt{5}$

21. $\sqrt{50} = \sqrt{25 \cdot 2} = \sqrt{25}\sqrt{2} = 5\sqrt{2}$

23. $\sqrt{45} = \sqrt{9 \cdot 5} = \sqrt{9}\sqrt{5} = 3\sqrt{5}$

25. $\sqrt{98} = \sqrt{49 \cdot 2} = \sqrt{49}\sqrt{2} = 7\sqrt{2}$

27. $\sqrt{48} = \sqrt{16 \cdot 3} = \sqrt{16}\sqrt{3} = 4\sqrt{3}$

29. $-\sqrt{200} = -\sqrt{100 \cdot 2} = -\sqrt{100}\sqrt{2}$
$= -10\sqrt{2}$

31. $\sqrt{192} = \sqrt{64 \cdot 3} = \sqrt{64}\sqrt{3} = 8\sqrt{3}$

33. $\sqrt{250} = \sqrt{25 \cdot 10} = \sqrt{25}\sqrt{10} = 5\sqrt{10}$

35. $2\sqrt{24} = 2\sqrt{4 \cdot 6} = 2\sqrt{4}\sqrt{6} = 2(2)\sqrt{6}$
$= 4\sqrt{6}$

37. $-2\sqrt{28} = -2\sqrt{4 \cdot 7} = -2\sqrt{4}\sqrt{7}$
$= -2(2)\sqrt{7} = -4\sqrt{7}$

39. $\sqrt{25x} = \sqrt{25}\sqrt{x} = 5\sqrt{x}$

41. $\sqrt{a^2b} = \sqrt{a^2}\sqrt{b} = a\sqrt{b}$

43. $\sqrt{9x^2y} = \sqrt{9x^2}\sqrt{y} = 3x\sqrt{y}$

45. $\frac{1}{5}x^2y\sqrt{50x^2y^2} = \frac{1}{5}x^2y\sqrt{25 \cdot 2x^2y^2}$
$= \frac{1}{5}x^2y\sqrt{25x^2y^2}\sqrt{2}$
$= \frac{1}{5}x^2y(5xy)\sqrt{2}$
$= \frac{1}{5}(5)x^2xyy\sqrt{2}$
$= x^3y^2\sqrt{2}$

47. $-12x\sqrt{16x^2y^3} = -12x\sqrt{16x^2y^2y}$
$= -12x\sqrt{16x^2y^2}\sqrt{y}$
$= -12x(4xy)\sqrt{y}$
$= -12(4)xxy\sqrt{y}$
$= -48x^2y\sqrt{y}$

49. $-\frac{2}{5}\sqrt{80mn^2} = -\frac{2}{5}\sqrt{16 \cdot 5mn^2}$
$= -\frac{2}{5}\sqrt{16n^2}\sqrt{5m}$
$= -\frac{2}{5}(4n)\sqrt{5m}$
$= -\frac{8}{5}n\sqrt{5m}$
$= \frac{-8n\sqrt{5m}}{5}$

51. $\sqrt{\frac{25}{9}} - \frac{\sqrt{25}}{\sqrt{9}} - \frac{5}{3}$

53. $\sqrt{\frac{81}{64}} = \frac{\sqrt{81}}{\sqrt{64}} = \frac{9}{8}$

55. $\sqrt{\frac{26}{25}} = \frac{\sqrt{26}}{\sqrt{25}} = \frac{\sqrt{26}}{5}$

57. $-\sqrt{\frac{20}{49}} = -\frac{\sqrt{20}}{\sqrt{49}}$
$= -\frac{\sqrt{4 \cdot 5}}{7}$
$= -\frac{\sqrt{4}\sqrt{5}}{7}$
$= -\frac{2\sqrt{5}}{7}$

59. $\sqrt{\frac{48}{81}} = \frac{\sqrt{48}}{\sqrt{81}}$
$= \frac{\sqrt{16 \cdot 3}}{9}$
$= \frac{\sqrt{16}\sqrt{3}}{9}$
$= \frac{4\sqrt{3}}{9}$

61. $\sqrt{\frac{32}{25}} = \frac{\sqrt{32}}{\sqrt{25}} = \frac{\sqrt{16 \cdot 2}}{5} = \frac{\sqrt{16}\sqrt{2}}{5}$
$= \frac{4\sqrt{2}}{5}$

63. $\sqrt{\frac{72x^3}{y^2}} = \frac{\sqrt{72x^3}}{\sqrt{y^2}}$
$= \frac{\sqrt{36 \cdot 2x^2x}}{y}$
$= \frac{\sqrt{36x^2}\sqrt{2x}}{y}$
$= \frac{6x\sqrt{2x}}{y}$

65. $\sqrt{\frac{125n^5}{64n}} = \sqrt{\frac{125n^4}{64}}$
$= \frac{\sqrt{125n^4}}{\sqrt{64}}$
$= \frac{\sqrt{25 \cdot 5n^4}}{8}$
$= \frac{\sqrt{25n^4}\sqrt{5}}{8}$
$= \frac{5n^2\sqrt{5}}{8}$

67. $\sqrt{\frac{128m^3n^5}{36mn^7}} = \sqrt{\frac{32m^2}{9n^2}}$
$= \frac{\sqrt{32m^2}}{\sqrt{9n^2}}$
$= \frac{\sqrt{16 \cdot 2m^2}}{3n}$
$= \frac{\sqrt{16m^2}\sqrt{2}}{3n}$
$= \frac{4m\sqrt{2}}{3n}$

69. $\sqrt{\frac{12r^7s^7}{81r^5s^2}} = \sqrt{\frac{4r^2s^5}{27}}$
$= \frac{\sqrt{4r^2s^5}}{\sqrt{27}}$
$= \frac{\sqrt{4r^2s^4s}}{\sqrt{9 \cdot 3}}$
$= \frac{\sqrt{4r^2s^4}\sqrt{s}}{\sqrt{9}\sqrt{3}}$
$= \frac{2rs^2\sqrt{s}}{3\sqrt{3}}$

or

$$\sqrt{\frac{12r^7s^7}{81r^5s^2}} = \sqrt{\frac{12r^2s^5}{81}}$$
$$= \frac{\sqrt{12r^2s^5}}{\sqrt{81}}$$
$$= \frac{\sqrt{4\cdot 3r^2s^4s}}{9}$$
$$= \frac{\sqrt{4r^2s^4}\sqrt{3s}}{9}$$
$$= \frac{2rs^2\sqrt{3s}}{3}$$

71. $\sqrt[3]{24} = \sqrt[3]{8\cdot 3} = \sqrt[3]{8}\sqrt[3]{3} = 2\sqrt[3]{3}$

73. $\sqrt[3]{-128} = \sqrt[3]{-64\cdot 2} = \sqrt[3]{-64}\sqrt[3]{2}$
$= -4\sqrt[3]{3}$

75. $\sqrt[3]{8x^3} = 2x$

77. $\sqrt[3]{-64x^5} = \sqrt[3]{-64x^3}\sqrt[3]{x^2}$
$= -4x\sqrt[3]{x^2}$

79. $\sqrt[3]{54x^3z^6} = \sqrt[3]{27\cdot 2x^3z^6} = \sqrt[3]{27x^3z^6}\sqrt[3]{2}$
$= 3xz^2\sqrt[3]{2}$

81. $\sqrt[3]{-81x^2y^3} = \sqrt[3]{-27\cdot 3x^2y^3}$
$= \sqrt[3]{-27y^3}\sqrt[3]{3x^2} = -3y\sqrt[3]{3x^2}$

83. $\sqrt[3]{\frac{27m^3}{8n^6}} = \frac{\sqrt[3]{27m^3}}{\sqrt[3]{8n^6}} = \frac{3m}{2n^2}$

85. $\sqrt[3]{\frac{16r^4s^5}{1{,}000t^3}} = \frac{\sqrt[3]{16r^4s^5}}{\sqrt[3]{1{,}000t^3}}$
$$= \frac{\sqrt{8\cdot 2r^3rs^3s^2}}{10t}$$
$$= \frac{\sqrt{8r^3s^3}\sqrt{2rs^2}}{10t}$$
$$= \frac{2rs\sqrt{2rs^2}}{10t}$$
$$= \frac{rs\sqrt{2rs^2}}{5t}$$

APPLICATIONS

87. a) $L = 54$
$$t = \pi\sqrt{\frac{L}{32}}$$
$$= \pi\sqrt{\frac{(54)}{32}}$$
$$= \pi\sqrt{\frac{27}{16}}$$
$$= \pi\frac{\sqrt{27}}{\sqrt{16}}$$
$$= \pi\frac{\sqrt{9\cdot 3}}{\sqrt{16}}$$
$$= \pi\frac{\sqrt{9}\sqrt{3}}{\sqrt{16}}$$
$$= \pi\frac{3\sqrt{3}}{4}$$

The time it takes to swing is $\frac{3\pi}{4}\sqrt{3}$ sec.

b) $\frac{3\pi}{4}\sqrt{3}$ secs ≈ 4.1 secs

89. a) $R_1 = 9,\ R_2 = 5,\ \&\ h = 8.$
$S = \pi(R_1 + R_2)\sqrt{(R_1 - R_2)^2 + h^2}$
$S = \pi(9 + 5)\sqrt{(9 - 5)^2 + 8^2}$
$S = \pi(14)\sqrt{(4)^2 + 8^2}$
$S = \pi(14)\sqrt{16 + 64}$
$S = 14\pi\sqrt{80}$
$S = 14\pi\sqrt{16\cdot 5}$
$S = 14\pi\sqrt{16}\sqrt{5}$
$S = 14\pi(4)\sqrt{5}$
$S = 56\pi\sqrt{5}\text{ in}^2$

b) $56\pi\sqrt{5}\text{ in}^2 \approx 393.4\text{ in}^2$

REVIEW

93. $(-2a^3)(3a^2) = -2(3)a^3a^2$
$= -6a^{3+2} = -6a^5$

95. $(0,\ 3)$ is the y-intercept. So, $b = 3$ and
$m = -2.$
$y = mx + b$
$y = -2x + 3$

97. $-x > -5$
$-1(-x) < -1(-5)$
$x < 5$

STUDY SET Section 6.5

VOCABULARY

1. radicals

3. simplified

CONCEPTS

5. No, because $\sqrt{2} \neq \sqrt{3}$.

7. Yes, because $\sqrt[3]{13a} = \sqrt[3]{13a}$.

9. The radicals don't have the same radicand. So, they can't be combined.

11. The two terms are not like terms. So, they can't be combined.

13. If $x = 3$;

$\sqrt{x} + \sqrt{3} = \sqrt{(3)} + \sqrt{3}$

$= \sqrt{3} + \sqrt{3} = 2\sqrt{3}$

If $x = 12$;

$\sqrt{x} + \sqrt{3} = \sqrt{(12)} + \sqrt{3}$

$= \sqrt{4 \cdot 3} + \sqrt{3} = \sqrt{4}\sqrt{3} + \sqrt{3}$

$= 2\sqrt{3} + \sqrt{3}$

$= 3\sqrt{3}$

If $x = 27$;

$\sqrt{x} + \sqrt{3} = \sqrt{(27)} + \sqrt{3}$

$- \sqrt{9 \cdot 3} + \sqrt{3} = \sqrt{9}\sqrt{3} + \sqrt{3}$

$= 3\sqrt{3} + \sqrt{3}$

$= 4\sqrt{3}$

If $x = 48$;

$\sqrt{x} + \sqrt{3} = \sqrt{(48)} + \sqrt{3}$

$= \sqrt{16 \cdot 3} + \sqrt{3} = \sqrt{16}\sqrt{3} + \sqrt{3}$

$= 4\sqrt{3} + \sqrt{3}$

$= 5\sqrt{3}$

NOTATION

15. $3\sqrt{80} + 4\sqrt{125} = 3\sqrt{16 \cdot 5} + 4\sqrt{25 \cdot 5}$

$= 3\sqrt{16}\sqrt{5} + 4\sqrt{25}\sqrt{5}$

$= 3(4)\sqrt{5} + 4(5)\sqrt{5}$

$= 12\sqrt{5} + 20\sqrt{5}$

$= 32\sqrt{5}$

PRACTICE

17. $5\sqrt{7} + 4\sqrt{7} = 9\sqrt{7}$

19. $\sqrt{2} - 4\sqrt{2} = -3\sqrt{2}$

21. $5 + 3\sqrt{3} + 3\sqrt{3} = 5 + 6\sqrt{3}$

23. $-1 + 2\sqrt{6} - 3\sqrt{6} = -1 - \sqrt{6}$

25. $\sqrt{12} + \sqrt{27} = \sqrt{4 \cdot 3} + \sqrt{9 \cdot 3}$

$= \sqrt{4}\sqrt{3} + \sqrt{9}\sqrt{3}$

$= 2\sqrt{3} + 3\sqrt{3}$

$= 5\sqrt{3}$

27. $\sqrt{18} - \sqrt{8} = \sqrt{9 \cdot 2} - \sqrt{4 \cdot 2}$

$= \sqrt{9}\sqrt{2} - \sqrt{4}\sqrt{2}$

$= 3\sqrt{2} + 2\sqrt{2}$

$= 5\sqrt{2}$

29. $2\sqrt{45} + 2\sqrt{80} = 2\sqrt{9 \cdot 5} + 2\sqrt{16 \cdot 5}$

$= 2\sqrt{9}\sqrt{5} + 2\sqrt{16}\sqrt{5}$

$= 2(3)\sqrt{5} + 2(4)\sqrt{5}$

$= 6\sqrt{5} + 8\sqrt{5}$

$= 14\sqrt{5}$

31. $2\sqrt{80} - 3\sqrt{125} = 2\sqrt{16 \cdot 5} - 3\sqrt{25 \cdot 5}$

$= 2\sqrt{16}\sqrt{5} - 3\sqrt{25}\sqrt{5}$

$= 2(4)\sqrt{5} - 3(5)\sqrt{5}$

$= 8\sqrt{5} - 15\sqrt{5}$

$= -7\sqrt{5}$

33. $\sqrt{20}+\sqrt{180}=\sqrt{4\cdot5}+\sqrt{36\cdot5}$
$=\sqrt{4}\sqrt{5}+\sqrt{36}\sqrt{5}$
$=2\sqrt{5}+6\sqrt{5}$
$=8\sqrt{5}$

35. $\sqrt{12}-\sqrt{48}=\sqrt{4\cdot3}-\sqrt{16\cdot3}$
$=\sqrt{4}\sqrt{3}-\sqrt{16}\sqrt{3}$
$=2\sqrt{3}-4\sqrt{3}$
$=-2\sqrt{3}$

37. $\sqrt{288}-3\sqrt{200}=\sqrt{144\cdot2}-3\sqrt{100\cdot2}$
$=\sqrt{144}\sqrt{2}-3\sqrt{100}\sqrt{2}$
$=12\sqrt{2}-3(10)\sqrt{2}$
$=12\sqrt{2}-30\sqrt{2}$
$=-18\sqrt{2}$

39. $2\sqrt{28}+2\sqrt{112}=2\sqrt{4\cdot7}+2\sqrt{16\cdot7}$
$=2\sqrt{4}\sqrt{7}+2\sqrt{16}\sqrt{7}$
$=2(2)\sqrt{7}+2(4)\sqrt{7}$
$=4\sqrt{7}+8\sqrt{7}$
$=12\sqrt{7}$

41. $\sqrt{20}+\sqrt{45}+\sqrt{80}$
$=\sqrt{4\cdot5}+\sqrt{9\cdot5}+\sqrt{16\cdot5}$
$=\sqrt{4}\sqrt{5}+\sqrt{9}\sqrt{5}+\sqrt{16}\sqrt{5}$
$=2\sqrt{5}+3\sqrt{5}+4\sqrt{5}$
$=9\sqrt{5}$

43. $\sqrt{200}-\sqrt{75}+\sqrt{48}$
$=\sqrt{100\cdot2}-\sqrt{25\cdot3}+\sqrt{16\cdot3}$
$=\sqrt{100}\sqrt{2}-\sqrt{25}\sqrt{3}+\sqrt{16}\sqrt{3}$
$=10\sqrt{2}-5\sqrt{3}+4\sqrt{3}$
$=10\sqrt{2}-\sqrt{3}$

45. $8\sqrt{6}-5\sqrt{2}-3\sqrt{6}$
$=8\sqrt{6}-3\sqrt{6}-5\sqrt{2}$
$=5\sqrt{6}-5\sqrt{2}$

47. $\sqrt{24}+\sqrt{150}+\sqrt{240}$
$=\sqrt{4\cdot6}+\sqrt{25\cdot6}+\sqrt{16\cdot15}$
$=\sqrt{4}\sqrt{6}+\sqrt{25}\sqrt{6}+\sqrt{16}\sqrt{15}$
$=2\sqrt{6}+5\sqrt{6}+4\sqrt{15}$
$=7\sqrt{6}+4\sqrt{15}$

49. $\sqrt{48}-\sqrt{8}+\sqrt{27}-\sqrt{32}$
$=\sqrt{16\cdot3}-\sqrt{4\cdot2}+\sqrt{9\cdot3}-\sqrt{16\cdot2}$
$=\sqrt{16}\sqrt{3}-\sqrt{4}\sqrt{2}+\sqrt{9}\sqrt{3}-\sqrt{16}\sqrt{2}$
$=4\sqrt{3}-2\sqrt{2}+3\sqrt{3}-4\sqrt{2}$
$=4\sqrt{3}+3\sqrt{3}-2\sqrt{2}-4\sqrt{2}$
$=7\sqrt{3}-6\sqrt{2}$

51. $\sqrt{2x^2}+\sqrt{8x^2}=\sqrt{2\cdot x^2}+\sqrt{4\cdot2\cdot x^2}$
$=\sqrt{x^2}\sqrt{2}+\sqrt{4}\sqrt{x^2}\sqrt{2}$
$=x\sqrt{2}+2x\sqrt{2}$
$=3x\sqrt{2}$

53. $\sqrt{2x^3}+\sqrt{8x^3}$
$=\sqrt{2\cdot x^2\cdot x}+\sqrt{4\cdot2\cdot x^2\cdot x}$
$=\sqrt{x^2}\sqrt{2x}+\sqrt{4}\sqrt{x^2}\sqrt{2x}$
$=x\sqrt{2x}+2x\sqrt{2x}$
$=3x\sqrt{2x}$

55. $\sqrt{18x^2y} - \sqrt{27x^2y}$

$= \sqrt{9 \bullet 2 \bullet x^2 \bullet y} - \sqrt{9 \bullet 3 \bullet x^2 \bullet y}$

$= \sqrt{9}\sqrt{x^2}\sqrt{2y} - \sqrt{9}\sqrt{x^2}\sqrt{3y}$

$= 3x\sqrt{2y} - 3x\sqrt{3y}$

57. $\sqrt{32x^5} - \sqrt{18x^5}$

$= \sqrt{16 \bullet 2 \bullet x^4 \bullet x} - \sqrt{9 \bullet 2 \bullet x^4 \bullet x}$

$= \sqrt{16}\sqrt{x^4}\sqrt{2x} - \sqrt{9}\sqrt{x^4}\sqrt{2x}$

$= 4x^2\sqrt{2x} - 3x^2\sqrt{2x}$

$= x^2\sqrt{2x}$

59. $3\sqrt{54x^2} + 5\sqrt{24x^2}$

$= 3\sqrt{9 \bullet 6 \bullet x^2} + 5\sqrt{4 \bullet 6 \bullet x^2}$

$= 3\sqrt{9}\sqrt{x^2}\sqrt{6} + 5\sqrt{4}\sqrt{x^2}\sqrt{6}$

$= 3(3)(x)\sqrt{6} + 5(2)(x)\sqrt{6}$

$= 9x\sqrt{6} + 10x\sqrt{6}$

$= 19x\sqrt{6}$

61. $y\sqrt{490y} - 2\sqrt{360y^2}$

$= y\sqrt{49 \bullet 10 \bullet y} - 2\sqrt{36 \bullet 10 \bullet y^2 \bullet y}$

$= y\sqrt{49}\sqrt{10y} - 2\sqrt{36}\sqrt{y^2}\sqrt{10y}$

$= y(7)\sqrt{10y} - 2(6)(y)\sqrt{10y}$

$= 7y\sqrt{10y} - 12y\sqrt{10y}$

$= -5y\sqrt{10y}$

63. $\sqrt{20x^3y} + \sqrt{45x^5y^3} - \sqrt{80x^7y^5}$

$= \sqrt{4 \bullet 5 \bullet x^2 \bullet x \bullet y} + \sqrt{9 \bullet 5 \bullet x^4 \bullet x \bullet y^2 \bullet y}$

$- \sqrt{16 \bullet 5 \bullet x^6 \bullet x \bullet y^4 \bullet y}$

$= \sqrt{4}\sqrt{x^2}\sqrt{5xy} + \sqrt{9}\sqrt{x^2}\sqrt{y^2}\sqrt{5xy}$

$- \sqrt{16}\sqrt{x^6}\sqrt{y^4}\sqrt{5xy}$

$= 2x\sqrt{5xy} + 3xy\sqrt{5xy} - 4x^3y^2\sqrt{5xy}$

65. $\sqrt[3]{3} + \sqrt[3]{3} = 2\sqrt[3]{3}$

67. $2\sqrt[3]{15} - 3\sqrt[3]{15} = -\sqrt[3]{15}$

69. $\sqrt[3]{16} + \sqrt[3]{54} = \sqrt[3]{8 \bullet 2} + \sqrt[3]{27 \bullet 2}$

$= \sqrt[3]{8}\sqrt[3]{2} + \sqrt[3]{27}\sqrt[3]{2}$

$= 2\sqrt[3]{2} + 3\sqrt[3]{2}$

$= 5\sqrt[3]{2}$

71. $\sqrt[3]{81} - \sqrt[3]{24} = \sqrt[3]{27 \bullet 3} - \sqrt[3]{8 \bullet 3}$

$= \sqrt[3]{27}\sqrt[3]{3} - \sqrt[3]{8}\sqrt[3]{3}$

$= 3\sqrt[3]{3} - 2\sqrt[3]{3}$

$= \sqrt[3]{3}$

73. $\sqrt[3]{40} + \sqrt[3]{125} = \sqrt[3]{8 \bullet 5} + \sqrt[3]{125}$

$= \sqrt[3]{8}\sqrt[3]{5} + \sqrt[3]{125}$

$= 2\sqrt[3]{5} + 5$

75. $\sqrt[3]{x^4} - \sqrt[3]{x^7} = \sqrt[3]{x^3 \bullet x} - \sqrt[3]{x^6 \bullet x}$

$= \sqrt[3]{x^3}\sqrt[3]{x} - \sqrt[3]{x^6}\sqrt[3]{x}$

$= x\sqrt[3]{x} - x^2\sqrt[3]{x}$

77. $\sqrt[3]{192x^4y^5} - \sqrt[3]{24x^4y^5}$

$= \sqrt[3]{64 \bullet 3 \bullet x^3 \bullet x \bullet y^3 \bullet y^2}$

$- \sqrt[3]{8 \bullet 3 \bullet x^3 \bullet x \bullet y^3 \bullet y^2}$

$= \sqrt[3]{64}\sqrt[3]{x^3}\sqrt[3]{y^3}\sqrt[3]{3xy^2}$

$- \sqrt[3]{8}\sqrt[3]{x^3}\sqrt[3]{y^3}\sqrt[3]{3xy^2}$

$= 4xy\sqrt[3]{3xy^2} - 2xy\sqrt[3]{3xy^2}$

$= 2xy\sqrt[3]{3xy^2}$

79. $\sqrt[3]{135x^7y^4} - \sqrt[3]{40x^7y^4}$

$= \sqrt[3]{27 \cdot 5 \cdot x^6 \cdot x \cdot y^3 \cdot y}$

$- \sqrt[3]{8 \cdot 5 \cdot x^6 \cdot x \cdot y^3 \cdot y}$

$= \sqrt[3]{8}\sqrt[3]{x^6}\sqrt[3]{y^3}\sqrt[3]{5xy}$

$- \sqrt[3]{8}\sqrt[3]{x^6}\sqrt[3]{y^3}\sqrt[3]{5xy}$

$= 3x^2y\sqrt[3]{5xy} - 2x^2y\sqrt[3]{5xy}$

$= x^2y\sqrt[3]{5xy}$

APPLICATIONS

81. length of arm $= 5\sqrt{12} + 5\sqrt{48}$

$= 5\sqrt{4 \cdot 3} + 5\sqrt{16 \cdot 3}$

$= 5\sqrt{4}\sqrt{3} + 5\sqrt{16}\sqrt{3}$

$= 5(2)\sqrt{3} + 5(4)\sqrt{3}$

$= 10\sqrt{3} + 20\sqrt{3}$

$= 30\sqrt{3}$

The length of the arm when straightened is $30\sqrt{3}$ in. long.

83. Length of motor

$= 10\sqrt{50} - \left(\sqrt{128} + 5\sqrt{18}\right).$

$= 10\sqrt{25 \cdot 2} - \left(\sqrt{64 \cdot 2} + 5\sqrt{9 \cdot 2}\right)$

$= 10\sqrt{25}\sqrt{2} - \left(\sqrt{64}\sqrt{2} + 5\sqrt{9}\sqrt{2}\right)$

$= 10(5)\sqrt{2} - \left(8\sqrt{2} + 5(3)\sqrt{2}\right)$

$= 50\sqrt{2} - \left(8\sqrt{2} + 15\sqrt{2}\right)$

$= 50\sqrt{2} - 23\sqrt{2}$

$= 27\sqrt{2}$

The length of the motor is $27\sqrt{2}$ cm.

REVIEW

87. $3^{-2} = \frac{1}{3^2} = \frac{1}{9}$

89. $-3^2 = -(3)^2 = -9$

91. $x^{-3} = \frac{1}{x^3}$

93. $3^0 = 1$

STUDY SET Section 6.6

VOCABULARY

1. rationalizing

3. conjugate

5. coefficient

CONCEPTS

7. $\sqrt{11}$

9. $\sqrt{7}$

11. 3 and 5.

13. Rational: $\frac{\sqrt{5}}{3}, \frac{-\sqrt{2}}{8}, \frac{1+\sqrt{3}}{4}$
Irrational: $\frac{2}{\sqrt{6}}, \frac{9}{7-\sqrt{10}}$

15. a) $\sqrt{2}+\sqrt{3}$ not possible to add.
b) $\sqrt{2}-\sqrt{3}$ not possible to subtract.
c) $\sqrt{2}\sqrt{3}=\sqrt{2 \cdot 3}=\sqrt{6}$
d) $\frac{\sqrt{2}}{\sqrt{3}}=\frac{\sqrt{2}}{\sqrt{3}}\frac{\sqrt{3}}{\sqrt{3}}=\frac{\sqrt{2 \cdot 3}}{\sqrt{3 \cdot 3}}=\frac{\sqrt{6}}{\sqrt{9}}=\frac{\sqrt{6}}{3}$

NOTATION

17. $\left(\sqrt{x}+\sqrt{2}\right)\left(\sqrt{x}-3\sqrt{2}\right)$
$=\sqrt{x}\sqrt{x}-\sqrt{x}\left(3\sqrt{2}\right)$
$+\sqrt{2}\sqrt{x}-\sqrt{2}\left(3\sqrt{2}\right)$
$=x-3\sqrt{2x}+\sqrt{2x}-3\sqrt{2}\sqrt{2}$
$=x-2\sqrt{2x}-3(4)$
$=x-2\sqrt{2x}-12$

PRACTICE

19. $\left(\sqrt{5}\right)^2=\sqrt{5}\sqrt{5}=\sqrt{5 \cdot 5}=\sqrt{25}=5$

21. $\left(3\sqrt{6}\right)^2=3\sqrt{6} \cdot 3\sqrt{6}=3(3)\sqrt{6}\sqrt{6}$
$=9\sqrt{6 \cdot 6}=9\sqrt{36}=9(6)=54$

23. $\sqrt{2}\sqrt{8}=\sqrt{2 \cdot 8}=\sqrt{16}=4$

25. $\sqrt{7}\sqrt{3}=\sqrt{7 \cdot 3}=\sqrt{21}$

27. $\sqrt{8}\sqrt{7}=\sqrt{8 \cdot 7}=\sqrt{56}=\sqrt{4 \cdot 14}$
$=\sqrt{4}\sqrt{14}=2\sqrt{14}$

29. $\sqrt{6}\sqrt{3}=\sqrt{6 \cdot 3}=\sqrt{18}=\sqrt{9 \cdot 2}$
$=\sqrt{9}\sqrt{2}=3\sqrt{2}$

31. $\sqrt{2}\sqrt{x}=\sqrt{2x}$

33. $\left(-\sqrt[3]{9}\right)^3=\left(-\sqrt[3]{9}\right)\left(-\sqrt[3]{9}\right)\left(-\sqrt[3]{9}\right)$
$=-\sqrt[3]{9 \cdot 9 \cdot 9}=-\sqrt[3]{279}=-9$

35. $\sqrt{x^3}\sqrt{x^5}=\sqrt{x^3x^5}=\sqrt{x^8}=x^4$

37. $\left(2\sqrt{5}\right)\left(2\sqrt{3}\right)=2(2)\sqrt{5}\sqrt{3}$
$=4\sqrt{5 \cdot 3}=4\sqrt{15}$

39. $\left(-5\sqrt{6}\right)\left(4\sqrt{3}\right)=-5(4)\sqrt{6}\sqrt{3}$
$=-20\sqrt{6 \cdot 3}=-20\sqrt{18}$
$=-20\sqrt{9 \cdot 2}=-20\sqrt{9}\sqrt{2}$
$=-20(3)\sqrt{2}=-60\sqrt{2}$

41. $\left(2\sqrt[3]{4}\right)\left(3\sqrt[3]{3}\right)=2(3)\sqrt[3]{4}\sqrt[3]{3}$
$=6\sqrt[3]{4 \cdot 3}=6\sqrt[3]{12}$

43. $(4\sqrt{x})(-2\sqrt{x})=4(-2)\sqrt{x}\sqrt{x}$
$=-8\sqrt{x \cdot x}=-8\sqrt{x^2}=-8x$

45. $\sqrt{8x}\sqrt{2x^3}=\sqrt{8x \cdot 2x^3}=\sqrt{8(2)xx^3}$
$=\sqrt{16x^4}=\sqrt{16}\sqrt{x^4}=4x^2$

47. $\sqrt{2}\left(\sqrt{2}+1\right)=\sqrt{2}\sqrt{2}+\sqrt{2}$
$=\sqrt{2 \cdot 2}+\sqrt{2}=\sqrt{4}+\sqrt{2}$
$=2+\sqrt{2}$

49. $\sqrt{3}\left(\sqrt{27}-1\right) = \sqrt{3}\sqrt{27} - \sqrt{3}$

$= \sqrt{3 \cdot 27} - \sqrt{3} = \sqrt{81} - \sqrt{3}$

$= 9 - \sqrt{3}$

51. $\sqrt{3}\left(\sqrt{6}+1\right) = \sqrt{3}\sqrt{6} + \sqrt{3}$

$= \sqrt{3 \cdot 6} + \sqrt{3} = \sqrt{18} + \sqrt{3}$

$= \sqrt{9 \cdot 2} + \sqrt{3} = \sqrt{9}\sqrt{2} + \sqrt{3}$

$= 3\sqrt{2} + \sqrt{3}$

53. $\sqrt{x}\left(\sqrt{3x}-2\right) = \sqrt{x}\sqrt{3x} - 2\sqrt{x}$

$= \sqrt{x \cdot 3x} - 2\sqrt{x}$

$= \sqrt{3x^2} - 2\sqrt{x}$

$= \sqrt{x^2 \cdot 3} - 2\sqrt{x}$

$= \sqrt{x^2}\sqrt{3} - 2\sqrt{x}$

$= x\sqrt{3} - 2\sqrt{x}$

55. $2\sqrt{x}\left(\sqrt{9x}+3\right)$

$= 2\sqrt{x}\sqrt{9x} + 2(3)\sqrt{x}$

$= 2\sqrt{x \cdot 9x} + 6\sqrt{x}$

$= 2\sqrt{9x^2} + 6\sqrt{x}$

$= 2\sqrt{9 \cdot x^2} + 6\sqrt{x}$

$= 2\sqrt{9}\sqrt{x^2} + 6\sqrt{x}$

$= 2(3)x + 6\sqrt{x}$

$= 6x + 6\sqrt{x}$

57. $\sqrt[3]{7}\left(\sqrt[3]{49}-2\right) = \sqrt[3]{7}\sqrt[3]{49} - 2\sqrt[3]{7}$

$= \sqrt[3]{7 \cdot 49} - 2\sqrt[3]{7}$

$= \sqrt[3]{343} - 2\sqrt[3]{7}$

$= 7 - 2\sqrt[3]{7}$

59. $\left(\sqrt{2}+1\right)\left(\sqrt{2}-1\right)$

$= \sqrt{2}\sqrt{2} - \sqrt{2} + \sqrt{2} - 1$

$= \sqrt{4} - 1$

$= 2 - 1 = 1$

61. $\left(\sqrt{7}-x\right)\left(\sqrt{2}+x\right)$

$= \sqrt{7}\sqrt{2} + \sqrt{7}x - \sqrt{2}x - x^2$

$= \sqrt{7 \cdot 2} + \sqrt{7}x - \sqrt{2}x - x^2$

$= \sqrt{14} + \sqrt{7}x - \sqrt{2}x - x^2$

63. $(\sqrt{x}+2)(\sqrt{x}+3)$

$= \sqrt{x}\sqrt{x} + 3\sqrt{x} + 2\sqrt{x} + 6$

$= \sqrt{x \cdot x} + 5\sqrt{x} + 6$

$= \sqrt{x^2} + 5\sqrt{x} + 6$

$= x + 5\sqrt{x} + 6$

65. $\left(\sqrt{6}+1\right)^2 = \left(\sqrt{6}+1\right)\left(\sqrt{6}+1\right)$

$= \sqrt{6}\sqrt{6} + \sqrt{6} + \sqrt{6} + 1$

$= \sqrt{6 \cdot 6} + 2\sqrt{6} + 1$

$= \sqrt{36} + 2\sqrt{6} + 1$

$= 6 + 2\sqrt{6} + 1$

$= 7 + 2\sqrt{6}$

67. $\left(\sqrt{2x}+3\right)\left(\sqrt{8x}-6\right)$

$= \sqrt{2x}\sqrt{8x} - 6\sqrt{2x} + 3\sqrt{8x} - 18$

$= \sqrt{2x \cdot 8x} - 6\sqrt{2x} + 3\sqrt{4 \cdot 2x} - 18$

$= \sqrt{16x^2} - 6\sqrt{2x} + 3\sqrt{4}\sqrt{2x} - 18$

$= \sqrt{16}\sqrt{x^2} - 6\sqrt{2x} + 3(2)\sqrt{2x} - 18$

$= 4x - 6\sqrt{2x} + 6\sqrt{2x} - 18$

$= 4x - 18$

69. $\left(\sqrt[3]{2}+1\right)\left(\sqrt[3]{2}+3\right)$

$= \sqrt[3]{2}\sqrt[3]{2}+3\sqrt[3]{2}+\sqrt[3]{2}+3$

$= \sqrt[3]{2 \cdot 2}+4\sqrt[3]{2}+3$

$= \sqrt[3]{4}+4\sqrt[3]{2}+3$

71. $\frac{\sqrt{12x^3}}{\sqrt{27x}} = \sqrt{\frac{12x^3}{27x}} = \sqrt{\frac{4x^2}{9}} = \frac{\sqrt{4x^2}}{\sqrt{9}}$
$= \frac{2x}{3}$

73. $\frac{\sqrt{18x}}{\sqrt{25x}} = \sqrt{\frac{18x}{25x}} = \sqrt{\frac{18}{25}} = \frac{\sqrt{18}}{\sqrt{25}}$

$= \frac{\sqrt{9 \cdot 2}}{\sqrt{25}} = \frac{\sqrt{9}\sqrt{2}}{\sqrt{25}} = \frac{3\sqrt{2}}{5}$

75. $\frac{\sqrt{196x}}{\sqrt{49x^3}} = \sqrt{\frac{196x}{49x^3}} = \sqrt{\frac{4}{x^2}} = \frac{\sqrt{4}}{\sqrt{x^2}}$
$= \frac{2}{x}$

77. $\frac{\sqrt[3]{16x^6}}{\sqrt[3]{54x^3}} = \sqrt[3]{\frac{16x^6}{54x^3}} = \sqrt[3]{\frac{8x^3}{27}} = \frac{\sqrt[3]{8x^3}}{\sqrt[3]{27}}$

$= \frac{\sqrt[3]{8}\sqrt[3]{x^3}}{\sqrt[3]{27}} = \frac{2x}{3}$

79. $\frac{1}{\sqrt{3}} = \frac{1}{\sqrt{3}}\frac{\sqrt{3}}{\sqrt{3}} = \frac{\sqrt{3}}{\sqrt{9}} = \frac{\sqrt{3}}{3}$

81. $\frac{2}{\sqrt{7}} = \frac{2}{\sqrt{7}}\frac{\sqrt{7}}{\sqrt{7}} = \frac{2\sqrt{7}}{\sqrt{49}} = \frac{2\sqrt{7}}{7}$

83. $\frac{9}{\sqrt{27}} = \frac{9}{\sqrt{27}}\frac{\sqrt{27}}{\sqrt{27}} = \frac{9\sqrt{27}}{\sqrt{729}}$

$= \frac{9\sqrt{9 \cdot 3}}{\sqrt{729}} = \frac{9\sqrt{9}\sqrt{3}}{\sqrt{729}} = \frac{9(3)\sqrt{3}}{27}$

$= \frac{27\sqrt{3}}{27} = \sqrt{3}$

85. $\frac{5}{\sqrt[3]{5}} = \frac{5}{\sqrt[3]{5}}\frac{\sqrt[3]{25}}{\sqrt[3]{25}} = \frac{5\sqrt[3]{25}}{\sqrt[3]{125}}$

$= \frac{5\sqrt[3]{25}}{5} = \sqrt[3]{25}$

87. $\frac{3}{\sqrt{32}} = \frac{3}{\sqrt{32}}\frac{\sqrt{32}}{\sqrt{32}} = \frac{3\sqrt{32}}{\sqrt{1024}}$

$= \frac{3\sqrt{16 \cdot 2}}{\sqrt{1024}} = \frac{3\sqrt{16}\sqrt{2}}{\sqrt{1024}} = \frac{3(4)\sqrt{2}}{32}$

$= \frac{12\sqrt{2}}{32} = \frac{3\sqrt{2}}{8}\sqrt{3}$

89. $\frac{4}{\sqrt[3]{4}} = \frac{4}{\sqrt[3]{4}}\frac{\sqrt[3]{16}}{\sqrt[3]{16}} = \frac{4\sqrt[3]{16}}{\sqrt[3]{64}}$

$= \frac{4\sqrt[3]{16}}{4} = \sqrt[3]{16}$

91. $\frac{\sqrt{5}}{\sqrt{3}} = \frac{\sqrt{5}}{\sqrt{3}}\frac{\sqrt{3}}{\sqrt{3}} = \frac{\sqrt{15}}{\sqrt{9}} = \frac{\sqrt{15}}{3}$

93. $\frac{10}{\sqrt{x}} = \frac{10}{\sqrt{x}}\frac{\sqrt{x}}{\sqrt{x}} = \frac{10\sqrt{x}}{\sqrt{x^2}} = \frac{10\sqrt{x}}{x}$

95. $\frac{\sqrt{9}}{\sqrt{2x}} = \frac{\sqrt{9}}{\sqrt{2x}}\frac{\sqrt{2x}}{\sqrt{2x}} = \frac{\sqrt{18x}}{\sqrt{4x^2}} = \frac{\sqrt{9 \cdot 2x}}{2x}$

$= \frac{\sqrt{9}\sqrt{2x}}{2x} = \frac{3\sqrt{2x}}{2x}$

97. $\frac{\sqrt[3]{5}}{\sqrt[3]{2}} = \frac{\sqrt[3]{5}}{\sqrt[3]{2}}\frac{\sqrt[3]{4}}{\sqrt[3]{4}} = \frac{\sqrt[3]{20}}{\sqrt[3]{8}} = \frac{\sqrt[3]{20}}{2}$

99. $\frac{3}{\sqrt{3}-1} = \frac{3}{\sqrt{3}-1}\frac{\sqrt{3}+1}{\sqrt{3}+1}$

$= \frac{3(\sqrt{3}+1)}{\sqrt{3}\sqrt{3}+\sqrt{3}-\sqrt{3}-1}$

$= \frac{3(\sqrt{3}+1)}{\sqrt{9}-1}$

$= \frac{3(\sqrt{3}+1)}{3-1}$

$= \frac{3(\sqrt{3}+1)}{2}$

101. $\frac{3}{\sqrt{7}+2} = \frac{3}{\sqrt{7}+2}\frac{\sqrt{7}-2}{\sqrt{7}-2}$

$= \frac{3(\sqrt{7}-2)}{\sqrt{7}\sqrt{7}-2\sqrt{7}+2\sqrt{7}-4}$

$= \frac{3(\sqrt{7}-2)}{\sqrt{49}-4} = \frac{3(\sqrt{7}-2)}{7-4}$

$= \frac{3(\sqrt{7}-2)}{3} = \sqrt{7}-2$

103. $\frac{12}{3-\sqrt{3}} = \frac{12}{3-\sqrt{3}}\frac{3+\sqrt{3}}{3+\sqrt{3}}$

$= \frac{12(3+\sqrt{3})}{9+3\sqrt{3}-3\sqrt{3}-\sqrt{3}\sqrt{3}}$

$= \frac{12(3+\sqrt{3})}{9-\sqrt{9}} = \frac{12(3+\sqrt{3})}{9-3}$

$= \frac{12(3+\sqrt{3})}{6} = 2(3+\sqrt{3})$

$= 6+2\sqrt{3}$

105. $\frac{-\sqrt{3}}{\sqrt{3}+1} = \frac{-\sqrt{3}}{\sqrt{3}+1}\frac{\sqrt{3}-1}{\sqrt{3}-1}$

$= \frac{-\sqrt{3}\sqrt{3}+\sqrt{3}}{\sqrt{3}\sqrt{3}-\sqrt{3}+\sqrt{3}-1}$

$= \frac{-\sqrt{9}+\sqrt{3}}{\sqrt{9}-1} = \frac{-3+\sqrt{3}}{3-1}$

$= \frac{-3+\sqrt{3}}{2} = \frac{\sqrt{3}-3}{2}$

107. $\frac{5}{\sqrt{3}+\sqrt{2}} = \frac{5}{\sqrt{3}+\sqrt{2}}\frac{\sqrt{3}-\sqrt{2}}{\sqrt{3}-\sqrt{2}}$

$= \frac{5(\sqrt{3}-\sqrt{2})}{\sqrt{3}\sqrt{3}-\sqrt{3}\sqrt{2}+\sqrt{3}\sqrt{2}-\sqrt{2}\sqrt{2}}$

$= \frac{5(\sqrt{3}-\sqrt{2})}{\sqrt{9}-\sqrt{4}} = \frac{5(\sqrt{3}-\sqrt{2})}{3-2}$

$= \frac{5(\sqrt{3}-\sqrt{2})}{1} = 5(\sqrt{3}-\sqrt{2})$

109. $\frac{\sqrt{x}+2}{\sqrt{x}-2} = \frac{\sqrt{x}+2}{\sqrt{x}-2}\frac{\sqrt{x}+2}{\sqrt{x}+2}$

$= \frac{\sqrt{x}\sqrt{x}+2\sqrt{x}+2\sqrt{x}+4}{\sqrt{x}\sqrt{x}+2\sqrt{x}-2\sqrt{x}-4}$

$= \frac{\sqrt{x^2}+4\sqrt{x}+4}{\sqrt{x^2}-4} = \frac{x+4\sqrt{x}+4}{x-4}$

111. Let $r = 6\sqrt{3}$

$A = \pi r^2$

$= \pi\left(6\sqrt{3}\right)^2 = \pi\left(6\sqrt{3}\right)\left(6\sqrt{3}\right)$

$= \pi(6)(6)\left(\sqrt{3}\right)\left(\sqrt{3}\right)$

$= \pi 36\sqrt{9} = \pi 36(3) = \pi 108$

The area of the lawn covered by one rotation is 108π in^2.

113. $w = 20\sqrt{3}$ & $l = 30\sqrt{6}$

$A = l \bullet w$

$= \left(20\sqrt{3}\right)\left(30\sqrt{6}\right)$

$= (20)(30)(\sqrt{3})(\sqrt{6})$

$= 600\sqrt{18} = 600\sqrt{9 \bullet 2}$

$= 600\sqrt{9}\sqrt{2} = 600(3)\sqrt{2}$

$= 1800\sqrt{2}$

The playing surface is $1800\sqrt{2}$ in^2.

REVIEW

117. If $x = -2$

$3x - 7 = 5x + 1$

$3(-2) - 7 = 5(-2) + 1$

$-6 - 7 = -10 + 1$

$-13 = -9$

No, -2 is not a solution.

119. Addition is performed first because it is in the parentheses.

121. $(x-4)(x+4)$

$= x \bullet x + 4 \bullet x - 4 \bullet x - 4 \bullet 4$

$= x^2 + x - 4x - 16$

$= x^2 - 16$

STUDY SET Section 6.7

VOCABULARY

1. rational

CONCEPTS

3. $x^m x^n = x^{m+n}$

5. $\left(\frac{x}{y}\right)^n = \frac{x^n}{y^n}$

7. $x^{-1} = \frac{1}{x}$

9. $x^{1/n} = \sqrt[n]{x}$

11. $\sqrt{5} = 5^{1/2}$

13. $8^{4/3} = \left(\sqrt[3]{8}\right)^4$

15. If $x = 0$,
$x^{1/2} = (0)^{1/2} = \sqrt{0} = 0.$
If $x = 1$,
$x^{1/2} = (1)^{1/2} = \sqrt{1} = 1.$
If $x = 4$,
$x^{1/2} = (4)^{1/2} = \sqrt{4} = 2.$
If $x = 9$,
$x^{1/2} = (9)^{1/2} = \sqrt{9} = 3.$

17.

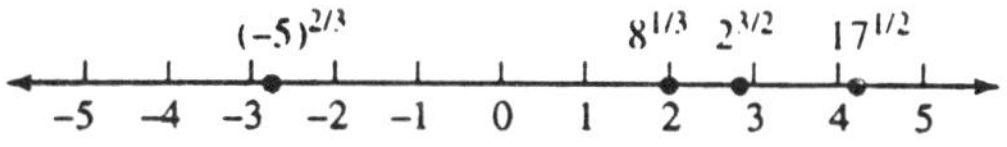

NOTATION

19. $(-216)^{4/3} = \left(\sqrt[3]{-216}\right)^4 = (-6)^4$
$= 1,296$

PRACTICE

21. $81^{1/2} = \sqrt{81} = 9$

23. $-144^{1/2} = -\sqrt{144} = -12$

25. $\left(\frac{1}{4}\right)^{1/2} = \sqrt{\frac{1}{4}} = \frac{\sqrt{1}}{\sqrt{4}} = \frac{1}{2}$

27. $\left(\frac{4}{49}\right)^{1/2} = \sqrt{\frac{4}{49}} = \frac{\sqrt{4}}{\sqrt{49}} = \frac{2}{7}$

29. $27^{1/3} = \sqrt[3]{27} = 3$

31. $-125^{1/3} = -\sqrt[3]{125} = -5$

33. $(-8)^{1/3} = \sqrt[3]{-8} = -2$

35. $\left(\frac{27}{64}\right)^{1/3} = \sqrt[3]{\frac{27}{64}} = \frac{\sqrt[3]{27}}{\sqrt[3]{64}} = \frac{3}{4}$

37. $81^{3/2} = \left(\sqrt{81}\right)^3 = (9)^3 = 729$

39. $25^{3/2} = \left(\sqrt{25}\right)^3 = (5)^3 = 125$

41. $125^{2/3} = \left(\sqrt[3]{125}\right)^2 = (5)^2 = 25$

43. $1,000^{2/3} = \left(\sqrt[3]{1,000}\right)^2 = (10)^2 = 100$

45. $(-8)^{2/3} = \left(\sqrt[3]{-8}\right)^2 = (-2)^2 = 4$

47. $\left(\frac{8}{27}\right)^{2/3} = \left(\sqrt[3]{\frac{8}{27}}\right)^2 = \left(\frac{\sqrt[3]{8}}{\sqrt[3]{27}}\right)^2$
$= \left(\frac{2}{3}\right)^2 = \frac{2^2}{3^2} = \frac{4}{9}$

49. $6^{3/5}6^{2/5} = 6^{3/5+2/5} = 6^{5/5} = 6^1 = 6$

51. $5^{2/3}5^{4/3} = 5^{2/3+4/3} = 5^{6/3} = 5^2 = 25$

53. $(7^{2/5})^{5/2} = 7^{(2/5)\cdot(5/2)} = 7^{10/10}$
$= 7^1 = 7$

55. $(5^{2/7})^7 = 5^{(2/7)\cdot(7)} = 5^2 = 25$

57. $\frac{8^{3/2}}{8^{1/2}} = 8^{3/2-1/2} = 8^{2/2} = 8^1 = 8$

59. $\frac{5^{11/3}}{5^{2/3}} = 5^{11/3-2/3} = 5^{9/3} = 5^3 = 125$

61. $4^{-1/2} = \frac{1}{4^{1/2}} = \frac{1}{\sqrt{4}} = \frac{1}{2}$

63. $27^{-2/3} = \frac{1}{27^{2/3}} = \frac{1}{\left(\sqrt[3]{27}\right)^2} = \frac{1}{3^2} = \frac{1}{9}$

65. $16^{-3/2} = \frac{1}{16^{3/2}} = \frac{1}{\left(\sqrt{16}\right)^3} = \frac{1}{4^3} = \frac{1}{64}$

67. $(-27)^{-4/3} = \frac{1}{(-27)^{4/3}} = \frac{1}{\left(\sqrt[3]{-27}\right)^4}$

$= \frac{1}{(-3)^4} = \frac{1}{-81} = -\frac{1}{81}$

69. $\left(x^{1/2}\right)^2 = x^{(1/2)\cdot 2} = x^1 = x$

71. $\left(x^{12}\right)^{1/6} = x^{12\cdot(1/6)} = x^{12/6} = x^2$

73. $x^{5/6}x^{7/6} = x^{5/6+7/6} = x^{12/6} = x^2$

75. $y^{4/7}y^{10/7} = y^{4/7+10/7} = y^{14/7} = y^2$

77. $\frac{x^{3/5}}{x^{1/5}} = x^{3/5-1/5} = x^{2/5}$

79. $\frac{x^{1/7}x^{3/7}}{x^{2/7}} = \frac{x^{1/7+3/7}}{x^{2/7}} = \frac{x^{4/7}}{x^{2/7}}$

$= x^{4/7-2/7} = x^{2/7}$

81. $x^{2/3}x^{3/4} = x^{2/3+3/4} = x^{8/12+9/12}$

$= x^{17/12}$

83. $\left(b^{1/2}\right)^{3/5} = b^{(1/2)\cdot(3/5)} = b^{3/10}$

85. $\frac{t^{2/3}}{t^{2/5}} = t^{2/3-2/5} = t^{10/15-6/15} = t^{4/15}$

87. $\left(\frac{x^{4/5}}{x^{2/15}}\right)^3 = \left(x^{4/5-2/15}\right)^3$

$= \left(x^{12/15-2/15}\right)^3 = \left(x^{10/15}\right)^3$

$= \left(x^{2/3}\right)^3 = x^{(2/3)\cdot 3} = x^2$

APPLICATIONS

89. $V = 2744$

$A = V^{2/3} = (2744)^{2/3}$

$= \left(\sqrt[3]{2744}\right)^2 = (14)^2 = 196$

Each speaker takes up 196 in^2. So, two speakers take up $2\cdot 196 = 392$ in^2.

91. $h = 24$ & $r = 10$

$s = (r^2 + h^2)^{1/2}$

$= ((10)^2 + (24)^2)^{1/2}$

$= (100 + 576)^{1/2} = (676)^{1/2}$

$= \sqrt{676} = 26$

The length of the string of lights is 26 ft.

93. $V = 2\pi$

$r = \left(\frac{3V}{4\pi}\right)^{1/3} = \left(\frac{3(2\pi)}{4\pi}\right)^{1/3}$

$= \left(\frac{6\pi}{4\pi}\right)^{1/3} = \left(\frac{3}{2}\right)^{1/3}$

$= \sqrt[3]{\frac{3}{2}} \approx 1.1$

The radius is about 1.1 in.

REVIEW

97. $x = 3$ is a vertical line with a x-intercept of (3, 0).

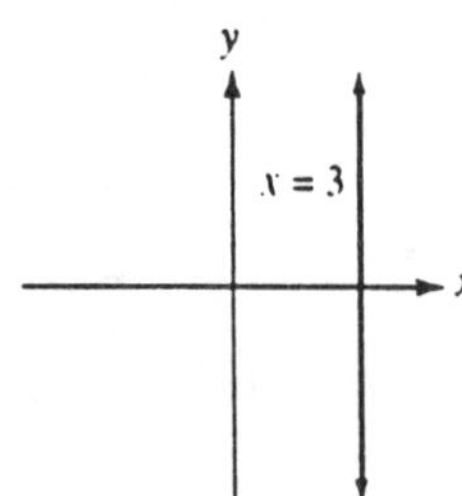

99. $-2x + y = 4$

Solve for y.

$-2x + 2x + y = 2x + 4$

$y = 2x + 4$

Graph.

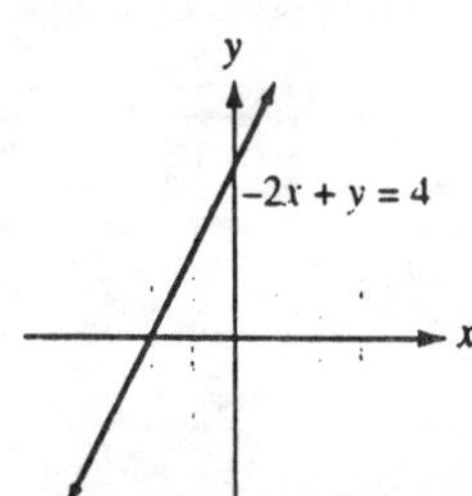

CHAPTER 6 REVIEW

1. square; square

3. a) $\sqrt{21} \approx 4.583$
   b) $-\sqrt{15} \approx -3.873$
   c) $2\sqrt{7} \approx 5.292$
   d) $\sqrt{751.9} \approx 27.421$

5. a)

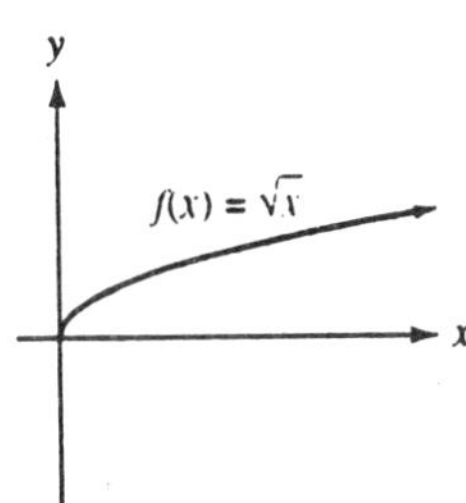

b)

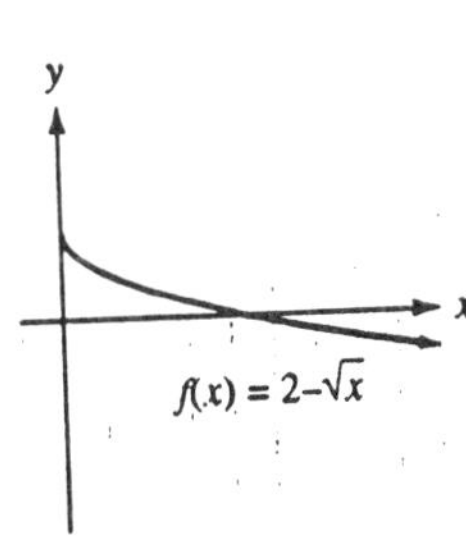

7. **Analyze:** The rope, mast, and base form a right triangle, where the rope is the hypotenuse.
Form: Let $c = 53$ and $a = 28$. We need to find the second leg.
$c^2 = a^2 + b^2$
$53^2 = 28^2 + b^2$
Solve: $53^2 = 28^2 + b^2$
$2809 = 784 + b^2$
$2809 - 784 = 784 - 784 + b^2$
$2025 = b^2$
$\sqrt{2025} = \sqrt{b^2}$
$45 = b$
State: The mast is 45 ft. tall.

9. $5^3 = 125$

11. a) $\sqrt[3]{16} \approx 2.520$
    b) $\sqrt[3]{-102.35} \approx -4.678$
    c) $\sqrt[4]{6} \approx 1.565$
    d) $\sqrt[5]{34,500} \approx 8.083$

13. The volume of a cube can be determined by $V = s^3$. We have $V = 1,728$. We want to find s.
$V = s^3$
$1728 = s^3$
$\sqrt[3]{1728} = \sqrt[3]{s^3}$
$12 = s$
The length of the edge of one of the dice is 12 mm.

15. a) $\sqrt{x} = 9$
$\left(\sqrt{x}\right)^2 = (9)^2$
$x = 81$
check: $\sqrt{(81)} = 9$
$9 = 9$

b) $\sqrt{5b} = 10$
$\left(\sqrt{5b}\right)^2 = (10)^2$
$5b = 100$
$\frac{5b}{5} = \frac{100}{5}$
$b = 20$
check: $\sqrt{5(20)} = 10$
$\sqrt{100} = 10$
$10 = 10$

c) $\sqrt{h+3} = 3$
$\left(\sqrt{h+3}\right)^2 = (3)^2$
$h + 3 = 9$
$h + 3 - 3 = 9 - 3$
$h = 6$
check: $\sqrt{(6)+3} = 3$
$\sqrt{9} = 3$
$3 = 3$

d) $\sqrt{2x+10} = 2$

$\left(\sqrt{2x+10}\right)^2 = (2)^2$
$2x + 10 = 4$
$2x + 10 - 10 = 4 - 10$
$2x = -6$
$\frac{2x}{2} = \frac{-6}{2}$
$b = -3$

check: $\sqrt{2(-3)+10} = 2$

$\sqrt{-6+10} = 2$
$2 = 2$

e) $\sqrt{3x+4} + 5 = 3$

$\sqrt{3x+4} + 5 - 5 = 3 - 5$
$\sqrt{3x+4} = -2$
$\left(\sqrt{3x+4}\right)^2 = (-2)^2$
$3x + 4 = 4$
$3x + 4 - 4 = 4 - 4$
$3x = 0$
$\frac{3x}{3} = \frac{0}{3}$
$x = 0$

check: $\sqrt{3(0)+4} + 5 = 3$

$\sqrt{0+4} + 5 = 3$
$\sqrt{4} + 5 = 3$
$2 + 5 = 3$
$7 \neq 3$
No solution.

f) $\sqrt{2(r+4)} = 2\sqrt{r}$

$\left(\sqrt{2(r+4)}\right)^2 = \left(2\sqrt{r}\right)^2$
$2(r+4) = 2^2(r)$
$2r + 8 = 4r$
$2r - 4r + 8 - 8 = 4r - 4r - 8$
$-2r = -8$
$\frac{-2r}{-2} = \frac{-8}{-2}$
$x = 4$

check: $\sqrt{2((4)+4)} = 2\sqrt{(4)}$

$\sqrt{2(8)} = 2(2)$
$\sqrt{16} = 4$
$4 = 4$

g) $\sqrt{p^2-3} = p + 3$

$\left(\sqrt{p^2-3}\right)^2 = (p+3)^2$
$p^2 - 3 = (p+3)(p+3)$
$p^2 - 3 = p^2 + 3p + 3p + 9$
$p^2 - 3 = p^2 + 6p + 9$
$p^2 - p^2 - 3 = p^2 - p^2 + 6p + 9$
$-3 = 6p + 9$
$-3 - 9 = 6p + 9 - 9$
$-12 = 6p$
$\frac{-12}{6} = \frac{6p}{6}$
$-2 = p$

check: $\sqrt{(-2)^2 - 3} = (-2) + 3$

$\sqrt{4-3} = 1$
$\sqrt{1} = 1$
$1 = 1$

h) $\sqrt[3]{x-1} = 3$

$\left(\sqrt{x-1}\right)^3 = (3)^3$
$x - 1 = 27$
$x - 1 + 1 = 27 + 1$
$x = 28$

check: $\sqrt[3]{(28)-1} = 3$

$\sqrt[3]{27} = 3$
$3 = 3$

17. a) Let $x_1 = -7, y_1 = 12$
$x_2 = -4,\ y_2 = 8$

$$d = \sqrt{(x_2 - x_1)^2 + (y_2 - y_1)^2}$$

$$= \sqrt{((-4) - (-7))^2 + ((8) - (12))^2}$$

$$= \sqrt{(-4+7)^2 + (8-12)^2}$$

$$= \sqrt{(3)^2 + (-4)^2}$$

$$= \sqrt{9+16} = \sqrt{25} = 5$$

b) Let $x_1 = -15, y_1 = -3$
$x_2 = -10,\ y_2 = -16$

$$d = \sqrt{(x_2 - x_1)^2 + (y_2 - y_1)^2}$$

$$= \sqrt{((\text{-}10) - (\text{-}15))^2 + ((\text{-}16) - (\text{-}3))^2}$$

$$= \sqrt{(-10 + 15)^2 + (-16 + 3)^2}$$

$$= \sqrt{(5)^2 + (-13)^2}$$

$$= \sqrt{25 + 169} = \sqrt{194} \approx 13.93$$

19. a) $\sqrt{\frac{16}{25}} = \frac{\sqrt{16}}{\sqrt{25}} = \frac{4}{5}$

b) $\sqrt{\frac{60}{49}} = \frac{\sqrt{60}}{\sqrt{49}} = \frac{\sqrt{4 \cdot 15}}{\sqrt{49}} = \frac{\sqrt{4}\sqrt{15}}{\sqrt{49}}$

$$= \frac{2\sqrt{15}}{7}$$

c) $\sqrt[3]{\frac{1{,}000}{27}} = \frac{\sqrt[3]{1{,}000}}{\sqrt[3]{27}} = \frac{10}{3}$

d) $\sqrt{\frac{242x^4}{169x^2}} = \sqrt{\frac{242x^2}{169}} = \frac{\sqrt{242x^2}}{\sqrt{169}}$

$$= \frac{\sqrt{121 \cdot 2x^2}}{\sqrt{169}} = \frac{\sqrt{121}\sqrt{x^2}\sqrt{2}}{\sqrt{169}}$$

$$= \frac{11x\sqrt{2}}{13}$$

21. a) $\sqrt{2} + \sqrt{8} - \sqrt{18}$

$$= \sqrt{2} + \sqrt{4 \cdot 2} - \sqrt{9 \cdot 2}$$

$$= \sqrt{2} + \sqrt{4}\sqrt{2} - \sqrt{9}\sqrt{2}$$

$$= \sqrt{2} + 2\sqrt{2} - 3\sqrt{2}$$

$$= 0$$

b) $\sqrt{3} + 4 + \sqrt{27} - 7$

$$= 4 - 7 + \sqrt{3} + \sqrt{27}$$

$$= -3 + \sqrt{3} + \sqrt{9 \cdot 3}$$

$$= -3 + \sqrt{3} + \sqrt{9}\sqrt{3}$$

$$= -3 + \sqrt{3} + 3\sqrt{3}$$

$$= -3 + 4\sqrt{3}$$

c) $3\sqrt{5} + 5\sqrt{45}$

$$= 3\sqrt{5} + 5\sqrt{9 \cdot 5}$$

$$= 3\sqrt{5} + 5\sqrt{9}\sqrt{5}$$

$$= 3\sqrt{5} + 5(3)\sqrt{5}$$

$$= 3\sqrt{5} + 15\sqrt{5}$$

$$= 18\sqrt{5}$$

d) $5\sqrt{28} - 3\sqrt{63}$

$$= 5\sqrt{4 \cdot 7} - 3\sqrt{9 \cdot 7}$$

$$= 5\sqrt{4}\sqrt{7} - 3\sqrt{9}\sqrt{7}$$

$$= 5(2)\sqrt{7} - 3(3)\sqrt{7}$$

$$= 10\sqrt{7} - 9\sqrt{7}$$

$$= \sqrt{7}$$

e) $3\sqrt{2x^2y} + 2x\sqrt{2y}$

$$= 3\sqrt{x^2 \cdot 2y} + 2x\sqrt{2y}$$

$$= 3\sqrt{x^2}\sqrt{2y} + 2x\sqrt{2y}$$

$$= 3x\sqrt{2y} + 2x\sqrt{2y}$$

$$= 5x\sqrt{2y}$$

f) $3y\sqrt{5xy^3} - y^2\sqrt{20xy}$

$= 3y\sqrt{y^2 \cdot 5xy} - y^2\sqrt{4 \cdot 5xy}$

$= 3y\sqrt{y^2}\sqrt{5xy} - y^2\sqrt{4}\sqrt{5xy}$

$= 3y(y)\sqrt{5xy} - y^2(2)\sqrt{5xy}$

$= 3y^2\sqrt{5xy} - 2y^2\sqrt{5xy}$

$= 5y^2\sqrt{5xy}$

g) $\sqrt[3]{16} + \sqrt[3]{54}$

$= \sqrt[3]{8 \cdot 2} + \sqrt[3]{27 \cdot 2}$

$= \sqrt[3]{8}\sqrt[3]{2} + \sqrt[3]{27}\sqrt[3]{2}$

$= 2\sqrt[3]{2} + 3\sqrt[3]{2}$

$= 5\sqrt[3]{2}$

h) $\sqrt[3]{2,000x^3} - \sqrt[3]{128x^3}$

$= \sqrt[3]{1,000 \cdot 2 \cdot x^3} - \sqrt[3]{64 \cdot 2 \cdot x^3}$

$= \sqrt[3]{1,000}\sqrt[3]{x^3}\sqrt[3]{2} - \sqrt[3]{64}\sqrt[3]{x^3}\sqrt[3]{2}$

$= 10x\sqrt[3]{2} - 4x\sqrt[3]{2}$

$= 6x\sqrt[3]{2}$

23. a) $\sqrt{2}\sqrt{3} = \sqrt{2 \cdot 3} = \sqrt{6}$

b) $\left(-5\sqrt{5}\right)\left(-2\sqrt{2}\right)$

$= -5(-2)\sqrt{5}\sqrt{2}$

$= 10\sqrt{5 \cdot 2}$

$= 10\sqrt{10}$

c) $\left(3\sqrt{3x}\right)\left(4\sqrt{6x}\right)$

$= 3(4)\sqrt{3x}\sqrt{6x}$

$= 12\sqrt{3x \cdot 6x}$

$= 12\sqrt{18x^2}$

$= 12\sqrt{9 \cdot 2 \cdot x^2}$

$= 12\sqrt{9}\sqrt{x^2}\sqrt{2}$

$= 12(3)(x)\sqrt{2}$

$= 36x\sqrt{2}$

d) $\left(\sqrt[3]{4}\right)\left(2\sqrt[3]{4}\right) = 2\sqrt[3]{4}\sqrt[3]{4}$

$= 2\sqrt[3]{4 \cdot 4} = 2\sqrt[3]{16}$

$= 2\sqrt[3]{8 \cdot 2} = 2\sqrt[3]{8}\sqrt[3]{2}$

$= 2(2)\sqrt[3]{2} = 4\sqrt[3]{2}$

e) $\left(-2\sqrt[3]{32x^2}\right)\left(\sqrt[3]{2x^2}\right)$

$= -2\sqrt[3]{32x^2}\sqrt[3]{2x^2}$

$= -2\sqrt[3]{32 \cdot 2 \cdot x^2 \cdot x^2}$

$= -2\sqrt[3]{64x^4}$

$= -2\sqrt[3]{64 \cdot x^3 \cdot x}$

$= -2\sqrt[3]{64}\sqrt[3]{x^3}\sqrt[3]{x}$

$= -2(4)(x)\sqrt[3]{x}$

$= -8x\sqrt[3]{x}$

f) $\sqrt{2}\left(\sqrt{8} - \sqrt{18}\right)$

$= \sqrt{2}\sqrt{8} - \sqrt{2}\sqrt{18}$

$= \sqrt{2 \cdot 8} - \sqrt{2 \cdot 18}$

$= \sqrt{16} - \sqrt{36}$

$= 4 - 6 = -2$

g) $\sqrt{6y}\left(\sqrt{2y}+\sqrt{75}\right)$

$= \sqrt{6y}\sqrt{2y} + \sqrt{6y}\sqrt{75}$

$= \sqrt{6y \cdot 2y} + \sqrt{6y \cdot 75}$

$= \sqrt{12y^2} + \sqrt{450y}$

$= \sqrt{4 \cdot 3 \cdot y^2} + \sqrt{225 \cdot 2 \cdot y}$

$= \sqrt{4}\sqrt{y^2}\sqrt{3} + \sqrt{225}\sqrt{2y}$

$= 2y\sqrt{3} + 15\sqrt{2y}$

h) $\left(\sqrt{3}+\sqrt{5}\right)\left(\sqrt{3}-\sqrt{5}\right)$

$= \sqrt{3}\sqrt{3} - \sqrt{3}\sqrt{5}$

$+ \sqrt{3}\sqrt{5} - \sqrt{5}\sqrt{5}$

$= \sqrt{9} - \sqrt{15} + \sqrt{15} - \sqrt{25}$

$= \sqrt{9} - \sqrt{25}$

$= 3 - 5$

$= -2$

i) $\left(\sqrt{15}+3x\right)^2$

$= \left(\sqrt{15}+3x\right)\left(\sqrt{15}+3x\right)$

$= \sqrt{15}\sqrt{15} + 3x\sqrt{15}$

$+ 3x\sqrt{15} - (3x)(3x)$

$= \sqrt{225} + 6x\sqrt{15} - 9x^2$

$= 15 + 6x\sqrt{15} - 9x^2$

$= 3 - 5$

j) $\left(\sqrt[3]{3}+2\right)\left(\sqrt[3]{-1}\right)$

$= \left(\sqrt[3]{3}+2\right)(-1)$

$= -\sqrt[3]{3} - 2$

25. a) $\frac{1}{\sqrt{7}} = \frac{1}{\sqrt{7}}\frac{\sqrt{7}}{\sqrt{7}} = \frac{\sqrt{7}}{\sqrt{49}} = \frac{\sqrt{7}}{7}$

b) $\frac{\sqrt{3}}{\sqrt{7}} = \frac{\sqrt{3}}{\sqrt{7}}\frac{\sqrt{7}}{\sqrt{7}} = \frac{\sqrt{21}}{\sqrt{49}} = \frac{\sqrt{21}}{7}$

c) $\frac{3}{\sqrt{18}} = \frac{3}{\sqrt{18}}\frac{\sqrt{18}}{\sqrt{18}} = \frac{3\sqrt{18}}{\sqrt{324}} = \frac{3\sqrt{18}}{18}$

$= \frac{\sqrt{18}}{6}$

d) $\frac{8}{\sqrt[3]{16}} = \frac{8}{\sqrt[3]{16}}\frac{\sqrt[3]{4}}{\sqrt[3]{4}} = \frac{8\sqrt[3]{4}}{\sqrt[3]{64}} = \frac{8\sqrt[3]{4}}{4}$

$= 2\sqrt[3]{4}$

e) $\frac{7}{\sqrt{2}+1} = \frac{7}{\sqrt{2}+1}\frac{\sqrt{2}-1}{\sqrt{2}-1}$

$= \frac{7\left(\sqrt{2}+1\right)}{\sqrt{2}\sqrt{2}-\sqrt{2}+\sqrt{2}-1}$

$= \frac{7\left(\sqrt{2}+1\right)}{\sqrt{4}-1} = \frac{7\left(\sqrt{2}+1\right)}{2-1}$

$= 7\left(\sqrt{2}+1\right)$

f) $\frac{\sqrt{c}-4}{\sqrt{c}+4} = \frac{\sqrt{c}-4}{\sqrt{c}+4}\frac{\sqrt{c}-4}{\sqrt{c}-4}$

$= \frac{\sqrt{c}\sqrt{c}-4\sqrt{c}-4\sqrt{c}+16}{\sqrt{c}\sqrt{c}-4\sqrt{c}+4\sqrt{c}-16}$

$= \frac{\sqrt{c^2}-8\sqrt{c}+16}{\sqrt{c^2}-16} = \frac{c-8\sqrt{c}+16}{c-16}$

27.

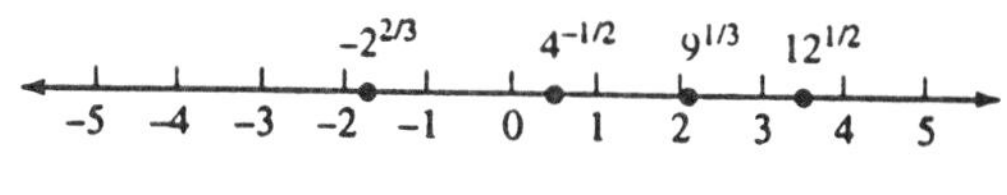

STUDY SET Section 7.1

VOCABULARY

1. prime

3. largest

5. sum

7. factoring

CONCEPTS

9. $F^2 - L^2 = (F + L) + (F - L)$

11. The distributive property.

13. GCF $= 3x$

15. 1, 4, 9, 16, 25, 36, 49, 64, 81, 100, 121, 144.

NOTATION

17. $5x^2 - 20y^2 = 5(x^2 - 4y^2)$
$= 5[x^2 - (2y)^2]$
$= 5(x - 2y)(x + 2y)$

19. The coefficient of xy is 1.

PRACTICE

21. $12 = 2{\bullet}2{\bullet}3 = 2^2{\bullet}3$

23. $15 = 3{\bullet}5$

25. $40 = 2{\bullet}2{\bullet}2{\bullet}5 = 2^3{\bullet}5$

27. $98 = 2{\bullet}7{\bullet}7 = 2{\bullet}7^2$

29. $225 = 3{\bullet}3{\bullet}5{\bullet}5 = 3^2{\bullet}5^2$

31. $288 = 2{\bullet}2{\bullet}2{\bullet}2{\bullet}2{\bullet}3{\bullet}3 = 2^5{\bullet}3^2$

33. $4a + 12 = 4(a + 3)$

35. $r^4 + r^2 = r^2(r^2 + 1)$

37. $4y^2 + 8y - 2xy = 2y(2y + 4 - x)$

39. $3x = 3{\bullet}x \qquad 6 = 2{\bullet}3$
$GCF = 3$
$3x + 6 = 3(x + 2)$

41. $xy = x{\bullet}y \qquad xz = x{\bullet}z$
$GCF = x$
$xy - xz = x(y - z)$

43. $t^3 = t{\bullet}t{\bullet}t \qquad 2t^2 = 2{\bullet}t{\bullet}t$
$GCF = t{\bullet}t = t^2$
$t^3 + 2t^2 = t^2(t + 2)$

45. $a^3 = a{\bullet}a{\bullet}a \qquad a^2 = a{\bullet}a$
$GCF = a{\bullet}a = a^2$
$a^3 - a^2 = a^2(a - 1)$

47. $24x^2y^3 = 2{\bullet}2{\bullet}2{\bullet}3{\bullet}x{\bullet}x{\bullet}y{\bullet}y$
$8xy^2 = 2{\bullet}2{\bullet}2{\bullet}x{\bullet}y{\bullet}y$
$GCF = 2{\bullet}2{\bullet}2{\bullet}x{\bullet}y{\bullet}y = 8xy^2$
$24x^2y^3 + 8xy^2 = 8xy^2(3xy + 1)$

49. $12uvw^3 = 2{\bullet}2{\bullet}3{\bullet}u{\bullet}v{\bullet}w{\bullet}w{\bullet}w$
$18uv^2w^2 = 2{\bullet}3{\bullet}3{\bullet}u{\bullet}v{\bullet}v{\bullet}w{\bullet}w$
$GCF = 2{\bullet}3{\bullet}u{\bullet}v{\bullet}w{\bullet}w = 6uvw^2$

51. $3x = 3{\bullet}x \qquad 3y = 3{\bullet}y \qquad 6z = 2{\bullet}3{\bullet}z$
$GCF = 3$
$3x + 3y - 6z = 3(x + y - 2z)$

53. $2b = 2{\bullet}b \qquad 2c = 2{\bullet}c \qquad 2d = 2{\bullet}d$
$GCF = 2$
$2b + 2c - 2d = 2(b + c - d)$

55. $12r^2 = 2{\bullet}2{\bullet}3{\bullet}r{\bullet}r \qquad 3rs = 3{\bullet}r{\bullet}s$
$9r^2s^2 = 3{\bullet}3{\bullet}r{\bullet}r{\bullet}s{\bullet}s$
$GCF = 3r$
$12r^2 - 3rs + 9r^2s^2$
$= 3r(4r - s + 3rs^2)$

57. $-a - b = -(a + b)$

59. $-2x + 5y = -(2x - 5y)$

61. $-3m - 4n + 1 = -(3m + 4n - 1)$

63. $-3ab - 5ac + 9bc$
$= -(3ab + 5ac - 9bc)$

65. $3x^2 = 3 \cdot x \cdot x \qquad 6x = 2 \cdot 3 \cdot x$
$GCF = 3x$
$-GCF = -3x$
$-3x^2 - 6x = -3x(x+2)$

67. $4a^2b^3 = 2 \cdot 2 \cdot a \cdot a \cdot b \cdot b \cdot b$
$12a^3b^2 = 2 \cdot 2 \cdot 3 \cdot a \cdot a \cdot a \cdot b \cdot b$
$GCF = 4a^2b^2$
$-GCF = -4a^2b^2$
$-4a^2b^3 + 12a^3b^2 = -4a^2b^2(b - 3a)$

69. $4a^2b^2c^2 = 2 \cdot 2 \cdot a \cdot a \cdot b \cdot b \cdot c \cdot c$
$14a^2b^2c = 2 \cdot 7 \cdot a \cdot a \cdot b \cdot b \cdot c$
$10ab^2c^2 = 2 \cdot 5 \cdot a \cdot b \cdot b \cdot c \cdot c$
$GCF = 2ab^2$
$-GCF = -2ab^2$
$-4a^2b^2c^2 + 14a^2b^2c - 10ab^2c^2$
$= -2ab^2c(2ac - 7a + 5c)$

71. $y^2 - 49 = (y+7)(y-7)$

73. $t^2 - w^2 = (t+w)(t-w)$

75. $25t^2 - 36u^2 = (5t+6u)(5t-6u)$

77. $x^2 - 16 = (x+4)(x-4)$

79. $4y^2 - 1 = (2y+1)(2y-1)$

81. $9x^2 - y^2 = (3x+y)(3x-y)$

83. $16a^2 - 25b^2 = (4a+5b)(4a-5b)$

85. $a^2 + b^2$ is prime.

87. $a^4 - 4b^2 = (a^2+2b)(a^2-2b)$

89. $8x^2 = 2 \cdot 2 \cdot 2 \cdot x \cdot x$
$32y^2 = 2 \cdot 2 \cdot 2 \cdot 2 \cdot 2 \cdot x \cdot x$
$GCF = 8$
$8x^2 - 32y^2 = 8(x^2 - 4y^2)$
$= 8(x+2y)(x-2y)$

91. $2a^2 = 2 \cdot a \cdot a \qquad 2 = 2 \cdot 1$
$GCF = 2$
$2a^2 - 2 = 2(a^2 - 1)$
$= 2(a+1)(a-1)$

93. $3r^2 = 3 \cdot r \cdot r \qquad 12s^2 = 2 \cdot 2 \cdot 3s \cdot s$
$GCF = 3$
$3r^2 - 12s^2 = 3(r^2 - 4s^2)$
$= 3(r+2s)(r-2s)$

95. $x^3 = x \cdot x \cdot x \qquad xy^2 = x \cdot y \cdot y$
$GCF = x$
$x^3 - xy^2 = x(x^2 - y^2)$
$= x(x+y)(x-y)$

97. $4a^2x = 2 \cdot 2 \cdot a \cdot a \cdot x \quad 9b^2x = 3 \cdot 3 \cdot b \cdot b \cdot x$
$GCF = x$
$4a^2x - 9b^2x = x(4a^2 - 9b^2)$
$= x(2a+3b)(2a-3b)$

99. $x^4 - 81 = (x^2+9)(x^2-9)$
$= (x^2+9)(x-3)(x+3)$

101. $a^4 - 16 = (a^2+4)(a^2-4)$
$= (a^2+4)(a+2)(a-2)$

103. $81r^4 - 256s^4$
$= (9r^2 + 16s^2)(9r^2 - 16s^2)$
$= (9r^2 + 16s^2)(3r+4s)(3r-4s)$

105. $2x^4 = 2 \cdot x \cdot x \cdot x \cdot x \quad 2y^4 = 2 \cdot y \cdot y \cdot y \cdot y$
$GCF = 2$
$2x^4 - 2y^4 = 2(x^4 - y^4)$
$= 2(x^2+y^2)(x^2-y^2)$
$= 2(x^2+y^2)(x+y)(x-y)$

APPLICATIONS

107. a) **Analyze:** We need to find the area of the entire frame. We know one side has length $2x^2$ inches and the other has length $6x$ inches.

Form: area of frame is length of frame • width of frame

$A = 2x^2 \cdot 6x$
$= 12x^3$

State: The area is $12x^3$ square inches.

b) **Analyze:** The length of the portrait is $5x$ inches and the width of the portrait is $4x$ inches.
Form: Area is length • width.

$$A = 5x \cdot 4x$$
$$= 20x^2$$

State: The area of the portrait is $20x^2$ square inches.

c) **Analyze:** The area of the mat is the area of the frame minus the area of the portrait.
Form: Area: mat is frame – portrait.

$$A = 12x^3 - 20x^2$$
$$= 4x^2(3x - 5)$$

State: The area of the mat is $4x^2(3x - 5)$ square inches.

REVIEW

111. Let $x_1 = 3,\ y_1 = 5$
$x_2 = -2,\ y_2 = -7.$
The distance formula is

$$d = \sqrt{(x_2 - x_1)^2 + (y_2 - y_1)^2}.$$

$$d = \sqrt{((-2) - (3))^2 + ((-7) - (5))^2}$$

$$d = \sqrt{(-2 - 3)^2 + (-7 - 5)^2}$$

$$d = \sqrt{(-5)^2 + (-12)^2}$$

$$d = \sqrt{25 + 144}$$

$$d = \sqrt{169}$$

$$d = 13$$

113. $4x - y = 7$
$4(3) - (5) = 7$
$12 - 5 = 7$
$7 = 7$
Yes, the point (3, 5) lies on the graph of the line $4x - y = 7$.

STUDY SET Section 7.2

VOCABULARY

1. trinomial

3. factors

5. prime

CONCEPTS

7. 4•1 and 2•2

9. $x^2 + 2xy + y^2 = (x + y)^2$

11. descending

13.

Product of factors	Sum of factors
1(8)	9
2(4)	6
-1(-8)	-9
-2(-4)	-6

15. a) The coefficient of x^2 is 1.

b) The last term is -15 and is the product of -5 and 3.

c) The coefficient of -2x is -2 and is the sum of -5 and 3.

NOTATION

17. $3x^2 + 15x + 18 = 3(x^2 + 5x + 6)$
$= 3(x + 3)(x + 2)$

PRACTICE

19. $x^2 + 3x + 2 = (x + 2)(x + 1)$

21. $t^2 - 9t + 14 = (t - 7)(t - 2)$

23. $a^2 + 6a - 16 = (a + 8)(a - 2)$

25. $z^2 + 12z + 11 = (z + 1)(z + 11)$

27. $m^2 - 5m + 6 = (m - 3)(m - 2)$

29. $a^2 - 4a - 5 = (a - 5)(a + 1)$

31. $x^2 + 5x - 24 = (x - 3)(x + 8)$

33. $a^2 - 10a - 39 = (z - 13)(a + 3)$

35. $u^2 + 10u + 15$ prime

37. $s^2 + 11s - 26 = (s + 13)(s - 2)$

39. $r^2 - 2r + 4$ prime

41. $m^2 - m - 12 = (m - 4)(m + 3)$

43. $x^2 + 6x + 9 = (x + 3)(x + 3)$
$= (x + 3)^2$

45. $y^2 - 8y + 16 = (y - 4)(y - 4)$
$= (y - 4)^2$

47. $t^2 + 20t + 100 = (t + 10)(t + 10)$
$= (t + 10)^2$

49. $u^2 - 18u + 81 = (u - 9)(u - 9)$
$= (u - 9)^2$

51. $-x^2 - 7x - 10$
$= -(x^2 + 7x + 10)$
$= -(x + 5)(x + 2)$

53. $-t^2 - 15t + 34$
$= -(t^2 + 15t - 34)$
$= -(t + 17)(t - 2)$

55. $-r^2 + 14r - 40$
$= -(r^2 - 14r + 40)$
$= -(r - 10)(r - 4)$

57. $3a^2 + 13a + 4 = (3a + 1)(a + 4)$

59. $4z^2 - 13z + 3 = (z - 3)(4z - 1)$

61. $2m^2 + 5m - 12 = (2m - 3)(m + 4)$

63. $2x^2 - 3x + 1 = (2x - 1)(x - 2)$

65. $6y^2 + 7y + 2 = (3y + 2)(2y + 1)$

67. $6x^2 - 7x + 2 = (3x - 1)(2x - 2)$

69. $2x^2 - 3x - 2 = (2x + 1)(x - 2)$

71. $10y^2 - 3y - 1 = (5y+2)(2y-1)$

73. $12y^2 - 5y - 2 = (3y-2)(4y+1)$

75. $5t^2 + 13t + 6 = (5t+3)(t+2)$

77. $16m^2 - 14m + 3 = (8m-3)(2m-1)$

79. $4 - 5x + x^2 = x^2 - 5x + 4$
$= (x-4)(x-1)$

81. $10y + 9 + y^2 = y^2 + 10y + 9$
$= (y+1)(y+9)$

83. $-r^2 + 2 + r = -r^2 + r + 2$
$= -(r^2 - r - 2)$
$= -(r-2)(r+1)$

85. $2x^2 + 10x + 12 = 2(x^2 + 5x + 6)$
$= 2(x+2)(x+3)$

87. $3y^3 + 6y^2 + 3y = 3y(y^2 + 2y + 1)$
$= 3y(y+1)(y+1)$
$= 3y(y+1)^2$

89. $-5a^2 + 25a - 30 = -5(a^2 - 5a + 6)$
$= -5(a-2)(a-3)$

91. $3z^2 - 15z + 12 = 3(z^2 - 5z + 4)$
$= 3(z-4)(z-1)$

REVIEW

97. $x - 3 > 5$
$x - 3 + 3 > 5 + 3$
$x > 8$

--------(======>
8

99. $-3x - 5 \geq 4$
$-3x - 5 + 5 \geq 4 + 5$
$-3x \geq 9$
$\frac{-3x}{-3} \leq \frac{9}{-3}$
$x \leq -3$

<=====]------
-3

STUDY SET Section 7.3

VOCABULARY

1. quadratic

CONCEPTS

3. zero; If $ab = 0$, then $a = 0$ or $b = 0$.

5. zero

7. a) $3x^2 + 4x + 2 = 0$ is quadratic.
b) $3x + 7 = 0$ is linear.
c) $2 = -16 - 4x$ is linear.
d) $-6x + 2 = x^2$ is quadratic.

9. a) If $x = 0$,
$x^2 + 6x - 16 = (0)^2 + 6(0) - 16$
$= 0 + 0 - 16$
$= -16$

b) $x^2 + 6x - 16 = (x - 2)(x + 8)$

c) $x^2 + 6x - 16 = 0$
$(x - 2)(x + 8) = 0$
$x - 2 = 0$ or $x + 8 = 0$
For $x - 2 = 0$
$x - 2 + 2 = 0 + 2$
$x = 2$
For $x + 8 = 0$
$x + 8 - 8 = 0 - 8$
$x = -8$

NOTATION

11. $7y^2 + 14y = 0$
$7y(y + 2) = 0$
$7y = 0$ or $y + 2 = 0$
For $7y = 0$
$\frac{7y}{7} = \frac{0}{7}$
$y = 0$
For $y + 2 = 0$
$y + 2 - 2 = 0 - 2$
$y = -2$

PRACTICE

13. $(x - 2)(x + 3) = 0$
$x - 2 = 0$ or $x + 3 = 0$
For $x - 2 = 0$
$x - 2 + 2 = 0 + 2$
$x = 2$
For $x + 3 = 0$
$x + 3 - 3 = 0 - 3$
$x = -3$

15. $(t - 4)(t + 1) = 0$
$t - 4 = 0$ or $t + 1 = 0$
For $t - 4 = 0$
$t - 4 + 4 = 0 + 4$
$t = 4$
For $t + 1 = 0$
$t + 1 - 1 = 0 - 1$
$t = -1$

17. $(2s - 5)(s + 6) = 0$
$2s - 5 = 0$ or $s + 6 = 0$
For $2s - 5 = 0$
$2s - 5 + 5 = 0 + 5$
$2s = 5$
$\frac{2s}{2} = \frac{5}{2}$
$s = \frac{5}{2}$
For $s + 6 = 0$
$s + 6 - 6 = 0 - 6$
$s = -6$

19. $(x - 1)(x + 2)(x - 3) = 0$
$x - 1 = 0$ or $x + 2 = 0$ or $x - 3 = 0$
For $x - 1 = 0$
$x - 1 + 1 = 0 + 1$
$x = 1$
For $x + 2 = 0$
$x + 2 - 2 = 0 - 2$
$x = -2$
For $x - 3 = 0$
$x - 3 + 3 = 0 + 3$
$x = 3$

21. $x(x - 3) = 0$
$x = 0$ or $(x - 3) = 0$
For $x = 0$
$x = 0$
For $x - 3 = 0$
$x - 3 + 3 = 0 + 3$
$x = 3$

23. $x(2x-5)=0$
$x=0 \quad \text{or} \quad 2x-5=0$
For $x=0$
$x=0$
For $2x-5=0$
$2x-5+5=0+5$
$2x=5$
$\frac{2x}{2}=\frac{5}{2}$
$x=\frac{5}{2}$

25. $w^2-7w=0$
$w(w-7)=0$
$w=0 \quad \text{or} \quad x-7=0$
For $w=0$
$w=0$
For $w-7=0$
$w-7+7=0+7$
$w=7$

27. $3x^2+8x=0$
$x(3x+8)=0$
$x=0 \quad \text{or} \quad 3x+8=0$
For $x=0$
$x=0$
For $3x+8=0$
$3x+8-8=0-8$
$3x=-8$
$\frac{3x}{3}=\frac{-8}{3}$
$x=-\frac{8}{3}$

29. $8s^2-16s=0$
$s(8s-16)=0$
$s=0 \quad \text{or} \quad 8s-16=0$
For $s=0$
$s=0$
For $8s-16=0$
$8s-16+16=0+16$
$8s=16$
$\frac{8s}{8}=\frac{16}{8}$
$s=2$

31. $x^2-25=0$
$(x+5)(x-5)=0$
$x+5=0 \quad \text{or} \quad x-5=0$
For $x+5=0$
$x+5-5=0-5$
$x=-5$
For $x-5=0$
$x-5+5=0+5$
$x=5$

33. $y^2-49=0$
$(y+7)(y-7)=0$
$y+7=0 \quad \text{or} \quad y-7=0$
For $y+7=0$
$y+7-7=0-7$
$y=-7$
For $y-7=0$
$y-7+7=0+7$
$y=7$

35. $4x^2+1=0$
$(2x+1)(2x-1)=0$
$2x+1=0 \quad \text{or} \quad 2x-1=0$
For $2x+1=0$
$2x+1-1=0-1$
$2x=-1$
$\frac{2x}{2}=\frac{-1}{2}$
$x=-\frac{1}{2}$
For $2x-1=0$
$2x-1+1=0+1$
$2x=1$
$\frac{2x}{2}=\frac{1}{2}$
$x=\frac{1}{2}$

37. $9y^2-4=0$
$(3y+2)(3y-2)=0$
$3y+2=0 \quad \text{or} \quad 3y-2=0$
For $3y+2=0$
$3y+2-2=0-2$
$3y=-2$
$\frac{3y}{3}=\frac{-2}{3}$
$y=-\frac{2}{3}$
For $3y-2=0$
$3y-2+2=0+2$
$3y=2$
$\frac{3y}{3}=\frac{2}{3}$
$y=\frac{2}{3}$

39. $x^2 - 100 = 0$

$(x+10)(x-10) = 0$

$x + 10 = 0 \quad \text{or} \quad x - 10 = 0$

For $x + 10 = 0$

$x + 10 - 10 = 0 - 10$

$x = -10$

For $x - 10 = 0$

$x - 10 + 10 = 0 + 10$

$x = 10$

41. $4x^2 = 81$

$4x^2 - 81 = 0$

$(2x+9)(2x-9) = 0$

$2x + 9 = 0 \quad \text{or} \quad 2x - 9 = 0$

For $2x + 9 = 0$

$2x + 9 - 9 = 0 - 9$

$2x = -9$

$\frac{2x}{2} = \frac{-9}{2}$

$x = -\frac{9}{2}$

For $2x - 9 = 0$

$2x - 9 + 9 = 0 + 9$

$2x = 9$

$\frac{2x}{2} = \frac{9}{2}$

$x = \frac{9}{2}$

43. $x^2 - 13x + 12 = 0$

$(x-12)(x-1) = 0$

$x - 12 = 0 \quad \text{or} \quad x - 1 = 0$

For $x - 12 = 0$

$x - 12 + 12 = 0 + 12$

$x = 12$

For $x - 1 = 0$

$x - 1 + 1 = 0 + 1$

$x = 1$

45. $a^2 - 2a + 1 = 0$

$(a-1)(a-1) = 0$

$a - 1 = 0$

$a - 1 + 1 = 0 + 1$

$a = 1$

47. $-4x - 21 + x^2 = 0$

$x^2 - 4x - 21 = 0$

$(x+3)(x-7) = 0$

$x + 3 = 0 \quad \text{or} \quad x - 7 = 0$

For $x + 3 = 0$

$x + 3 - 3 = 0 - 3$

$x = -3$

For $x - 7 = 0$

$x - 7 + 7 = 0 + 7$

$x = 7$

49. $x^2 + 8 - 9x = 0$

$x^2 - 9x + 8 = 0$

$(x-8)(x-1) = 0$

$x - 8 = 0 \quad \text{or} \quad x - 1 = 0$

For $x - 8 = 0$

$x - 8 + 8 = 0 + 8$

$x = 8$

For $x - 1 = 0$

$x - 1 + 1 = 0 + 1$

$x = 1$

51. $a^2 + 8a = -15$

$a^2 + 8a + 15 = -15 + 15$

$a^2 + 8a + 15 = 0$

$(a+5)(a+3) = 0$

$a + 5 = 0 \quad \text{or} \quad a + 3 = 0$

For $a + 5 = 0$

$a + 5 - 5 = 0 - 5$

$a = -5$

For $a + 3 = 0$

$a + 3 - 3 = 0 - 3$

$a = -3$

53. $2y - 8 = -y^2$

$y^2 + 2y - 8 = -y^2 + y^2$

$y^2 + 2y - 8 = 0$

$(y+4)(y-2) = 0$

$y + 4 = 0 \quad \text{or} \quad y - 2 = 0$

For $y + 4 = 0$

$y + 4 - 4 = 0 - 4$

$y = -4$

For $y - 2 = 0$

$y - 2 + 2 = 0 + 2$

$y = 2$

55. $x^3 + 3x^2 + 2x = 0$
$x(x^2 + 3x + 2) = 0$
$x(x + 2)(x + 1) = 0$
$x = 0$ or $x + 2 = 0$ or $x + 1 = 0$
For $x = 0$
$x = 0$
For $x + 2 = 0$
$x + 2 - 2 = 0 - 2$
$x = -2$
For $x + 1 = 0$
$x + 1 - 1 = 0 - 1$
$x = -1$

57. $k^3 - 27k - 6k^2 = 0$
$k^3 - 6k^2 - 27k = 0$
$k(k^2 - 6k - 27) = 0$
$k(k + 3)(k - 9) = 0$
$k = 0$ or $k + 3 = 0$ or $k - 9 = 0$
For $k = 0$
$k = 0$
For $k + 3 = 0$
$k + 3 - 3 = 0 - 3$
$k = -3$
For $k - 9 = 0$
$k - 9 + 9 = 0 + 9$
$k = 9$

59. $(x - 1)(x^2 + 5x + 6) = 0$
$(x - 1)(x + 2)(x + 3) = 0$
$x - 1 = 0$ or $x + 2 = 0$ or $x + 3 = 0$
For $x - 1 = 0$
$x - 1 + 1 = 0 + 1$
$x = 1$
For $x + 2 = 0$
$x + 2 - 2 = 0 - 2$
$x = -2$
For $x + 3 = 0$
$x + 3 - 3 = 0 - 3$
$x = -3$

61. $2x^2 - 5x + 2 = 0$
$(2x - 1)(x - 2) = 0$
$2x - 1 = 0$ or $x - 2 = 0$
For $2x - 1 = 0$
$2x - 1 + 1 = 0 + 1$
$2x = 1$
$\frac{2x}{2} = \frac{1}{2}$
$x = \frac{1}{2}$
For $x - 2 = 0$
$x - 2 + 2 = 0 + 2$
$x = 2$

63. $5x^2 - 6x + 1 = 0$
$(5x - 1)(x - 1) = 0$
$5x - 1 = 0$ or $x - 1 = 0$
For $5x - 1 = 0$
$5x - 1 + 1 = 0 + 1$
$5x = 1$
$\frac{5x}{5} = \frac{1}{5}$
$x = \frac{1}{5}$
For $x - 1 = 0$
$x - 1 + 1 = 0 + 1$
$x = 1$

65. $4r^2 + 4r = -1$
$4r^2 + 4r + 1 = -1 + 1$
$4r^2 + 4r + 1 = 0$
$(2r + 1)(2r + 1) = 0$
$2r + 1 = 0$
$2r + 1 - 1 = 0 - 1$
$2r = -1$
$\frac{2r}{2} = \frac{-1}{2}$
$r = -\frac{1}{2}$

67. $15x^2 - 2 = 7x$
$15x^2 + 7x - 2 = 7x - 7x$
$15x^2 + 7x - 2 = 0$
$(5x + 1)(3x - 2) = 0$
$5x + 1 = 0$ or $3x - 2 = 0$
For $5x + 1 = 0$
$5x + 1 - 1 = 0 - 1$
$5x = -1$
$\frac{5x}{5} = \frac{-1}{5}$
$x = -\frac{1}{5}$
For $3x - 2 = 0$
$3x - 2 + 2 = 0 + 2$
$3x = 2$
$\frac{3x}{3} = \frac{2}{3}$
$x = \frac{2}{3}$

69. $\frac{1}{2}x(2x-3)=10$
$2\left(\frac{1}{2}x(2x-3)\right)=2(10)$
$x(2x-3)=20$
$2x^2-3x=20$
$2x^2-3x-20=20-20$
$2x^2-3x-20=0$
$(2x+5)(x-4)=0$
$2x+5=0 \quad \text{or} \quad x-4=0$
For $2x+5=0$
$2x+5-5=0-5$
$2x=-5$
$\frac{2x}{2}=\frac{-5}{2}$
$x=-\frac{5}{2}$
For $x-4=0$
$x-4+4=0+4$
$x=4$

71. $(d+1)(8d+1)=18d$
$8d^2+d+8d+1=18d$
$8d^2+9d+1=18d$
$8d^2+9d-18d+1=18d-18d$
$8d^2-9d+1=0$
$(8d-1)(d-1)=0$
$8d-1=0 \quad \text{or} \quad d-1=0$
For $8d-1=0$
$8d-1+1=0+1$
$8d=1$
$\frac{8d}{8}=\frac{1}{8}$
$d=\frac{1}{8}$
For $d-1=0$
$d-1+1=0+1$
$d=1$

73. $2x(3x^2+10x)=-6x$
$6x^3+20x^2=-6x$
$6x^3+20x^2+6x=-6x+6x$
$6x^3+20x^2+6x=0$
$2x(3x^2+10x+3)=0$
$2x(3x+1)(x+3)=0$
$2x=0 \text{ or } 3x+1=0 \text{ or } x+3=0$
For $2x=0$
$\frac{2x}{2}=\frac{0}{2}$
$x=0$
For $3x+1=0$
$3x+1-1=0-1$
$3x=-1$
$\frac{3x}{3}=\frac{-1}{3}$
$x=-\frac{1}{3}$
For $x+3=0$
$x+3-3=0-3$
$x=-3$

75. $x^3+7x^2=x^2-9x$
$x^3+7x^2-x^2=x^2-x^2-9x$
$x^3+6x^2=-9x$
$x^3+6x^2+9x=-9x+9x$
$x^3+6x^2+9x=0$
$x(x^2+6x+9)=0$
$x(x+3)(x+3)=0$
$x=0 \quad \text{or} \quad x+3=0$
For $x=0$
$x=0$
For $x+3=0$
$x+3-3=0-3$
$x=-3$

APPLICATIONS

77. Analyze: We know $h=vt-16t^2$, where h is the height of the object above the ground after t seconds and v is the initial velocity. We have $v=$ 144 ft/sec.
Form: $h=144t-16t^2$
When the object hits the ground, its height will be 0, so we must solve
$0=144t-16t^2$
Solve:
$0=(144-16t)t$
$144-16t=0 \quad \text{or} \quad t=0$
For $144-16t=0$
$144-16t+16t=0+16t$
$144=16t$
$\frac{144}{16}=\frac{16t}{16}$
$9=t$
State: The $t=0$ solution corresponds to when the object is initially thrown. So, the object will hit the ground 9 seconds after it is thrown.

79. Analyze: We know $h = vt - 16t^2$, where h is the height of the cannonball above the ground after t seconds and v is the initial velocity. We have $v = 220$ ft/sec. We want to find t when $h = 600$.

Form: $660 = 220t - 16t^2$

Solve:

$16t^2 + 660 = 220t - 16t^2 + 16t^2$
$16t^2 + 660 = 220t$
$16t^2 - 220t + 660 = 220t - 220t$
$16t^2 - 220t + 660 = 0$
$4(4t^2 - 55t + 150) = 0$
$4(4t - 15)(t - 10) = 0$
$4t - 15 = 0 \quad \text{or} \quad t - 10 = 0$

For $4t - 15 = 0$

$4t - 15 + 15 = 0 + 15$
$4t = 15$
$\frac{4t}{4} = \frac{15}{4}$
$t = \frac{15}{4}$

For $t - 10 = 0$

$t - 10 + 10 = 0 + 10$
$t = 10$

State: The cannonball will be at 600 feet after $\frac{15}{4}$ seconds (on the way up) and after 10 seconds (on its way down).

81. Analyze: When the diver hits the water her height above the water is zero. So, we need to find the t value when $h = 0$.

Form: $h = -16t^2 + 64$

$0 = -16t^2 + 64$

Solve:

$0 = -(16t^2 - 64)$
$0 = -(4t + 8)(4t - 8)$
$4t + 8 = 0 \quad \text{or} \quad 4t - 8 = 0$

For $4t + 8 = 0$

$4t + 8 - 8 = 0 - 8$
$4t = -8$
$\frac{4t}{4} = \frac{-8}{4}$
$t = \frac{-8}{4}$
$t = -2$

For $4t - 8 = 0$

$4t - 8 + 8 = 0 + 8$
$4t = 8$
$\frac{4t}{4} = \frac{8}{4}$
$t = \frac{8}{4}$
$t = 2$

State: Since the time cannot be negative, we disregard the result $t = -2$. Thus, the dive takes 2 seconds.

83. Analyze: The area of the slab is the length times the width. We know the area is 36 square meters.

Form: area = length • width

$36 = (2w + 1) \cdot w$

Solve:

$36 = 2w^2 + w$
$36 - 36 = 2w^2 + w - 36$
$0 = 2w^2 + w - 36$
$0 = (2w + 9)(w - 4)$
$2w + 9 = 0 \quad \text{or} \quad w - 4 = 0$

For $2w + 9 = 0$

$2w + 9 - 9 = 0 - 9$
$2w = -9$
$\frac{2w}{2} = \frac{-9}{2}$
$w = -\frac{9}{2}$

For $w - 4 = 0$

$w - 4 + 4 = 0 + 4$
$w = 4$

State: Since the width cannot be negative, we disregard the solution $w = -\frac{9}{2}$. Thus, the width is 4 meters and the length is $2w + 1 = 2(4) + 1 = 8 + 1 = 9$ meters.

85. Analyze: We need to find the lengths of three sides of a right triangle. Recall the Pythagorean theorem gives the relationship between the sides of a right triangle: $a^2 + b^2 = c^2$, where c is the hypotenuse.

Form: We let a represent the "rise" of the boat ramp since the other two distances can be expressed in terms of it. The ramp is $a + 8$ meters in length and the run is $a + 7$ meters in length. We substitute these distances into the Pythagorean theorem, noting the hypotenuse is $a + 6$ meters.

$a^2 + b^2 = c^2$

$a^2 + (a + 7)^2 = (a + 8)^2$

Solve:

$a^2 + a^2 + 14a + 49 = a^2 + 16a + 64$

$2a^2 + 14a + 49 = a^2 + 16a + 64$

$2a^2 - a^2 + 14a + 49$

$= a^2 - a^2 + 16a + 64$

$a^2 + 14a + 49 = 16a + 64$

$a^2 + 14a - 16a + 49 = 16a - 16a + 64$

$a^2 - 2a + 49 = 64$

$a^2 - 2a + 49 - 64 = 64 - 64$

$a^2 - 2a - 15 = 0$

$(a + 3)(a - 5) = 0$

$a + 3 = 0$ or $a - 5 = 0$

For $a + 3 = 0$

$a + 3 - 3 = 0 - 3$

$a = -3$

For $a - 5 = 0$

$a - 5 + 5 = 0 + 5$

$a = 5$

State: We discard the result $a = -3$ since a triangle cannot have a negative number for the length of a side. The length of the rise is 5 meters. The length of the run is $a + 7 = (5) + 7 = 12$ meters. The length of the ramp is $a + 8 = (5) + 8 = 13$ meters.

Check: $5^2 + 12^2 = 13^2$

$25 + 144 = 169$

$169 = 169$

87. Analyze: We need to find the lengths of three sides of a right triangle. Recall the Pythagorean theorem gives the relationship between the sides of a right triangle: $a^2 + b^2 = c^2$, where c is the hypotenuse.

Form: We have the two sides x and $x + 1$ and the hypotenuse $x + 2$.

$a^2 + b^2 = c^2$

$(x + 1)^2 + x^2 = (x + 2)^2$

Solve:

$x^2 + 2x + 1 + x^2 = x^2 + 4x + 4$

$2x^2 + 2x + 1 = x^2 + 4x + 4$

$2a^2 - x^2 + 2x + 1 = x^2 - x^2 + 4x + 4$

$x^2 + 2x + 1 = 4x + 4$

$x^2 + 2x - 4x + 1 = 4x - 4x + 4$

$x^2 - 2x + 1 = 4$

$x^2 - 2x + 1 - 4 = 4 - 4$

$x^2 - 2x - 3 = 0$

$(x + 1)(x - 3) = 0$

$x + 1 = 0$ or $x - 3 = 0$

For $x + 1 = 0$

$x + 1 - 1 = 0 - 1$

$x = -1$

For $x - 3 = 0$

$x - 3 + 3 = 0 + 3$

$x = 3$

State: Since the length of the side of a triangle cannot be negative, we discard the result $x = -1$. Thus, the length of the back is 3 mm, the length of the cutting edge is $3 + 1 = 4$ mm, and the length of the span is $3 + 2 = 5$ mm.

Check: $3^2 + 4^2 = 5^2$

$9 + 16 = 25$

$25 = 25$

89. Analyze: We know the area of the triangle is 30 square feet and that the base of the triangle is 2 feet more than twice the height. We know the area of a triangle is $\frac{1}{2}$•base•height.

Form: Let h represent the height of the triangle. The the base is $2h + 2$ (two more than twice the height).

$$\text{Area} = \tfrac{1}{2}\text{•base•height}$$
$$\text{Area} = \tfrac{1}{2}(2h+2)h$$
$$30 = \tfrac{1}{2}(2h+2)h$$

Solve:

$$30 = \tfrac{1}{2}(2h+2)h$$
$$30 = \tfrac{1}{2}(2h^2+2h)$$
$$30 = \tfrac{1}{2}\cdot 2h^2 + \tfrac{1}{2}\cdot 2h$$
$$30 = h^2 + h$$
$$30 - 30 = h^2 + h - 30$$
$$0 = h^2 + h - 30$$
$$0 = (h+6)(h-5)$$
$$h + 6 = 0 \quad \text{or} \quad h - 5 = 0$$

For $h + 6 = 0$

$$h + 6 - 6 = 0 - 6$$
$$h = -6$$

For $h - 5 = 0$

$$h - 5 + 5 = 0 + 5$$
$$h = 5$$

State: We disregard the solution $h = -6$, since the height cannot be negative. Thus, the height of the triangle is 5 feet and the base of the triangle is $2h + 2 = 2(5) + 2$
$= 10 + 2 = 12$ feet.

Check: $30 = \frac{1}{2}(5)(12)$

$$30 = \tfrac{1}{2}(60)$$
$$30 = 30$$

91. Analyze: We know the area of the parallelogram is 200 square centimeters and that the area is base times height. We also know the base is twice the height.

Form: Let h represent the height of the parallelogram. We know the base is twice the height, so the base is $2h$ centimeters long.

$$\text{Area} = \text{base•height}$$
$$\text{Area} = 2h\text{•}h$$
$$200 = 2h^2$$

Solve:

$$200 = 2h^2$$
$$200 - 200 = 2h^2 - 200$$
$$0 = 2h^2 - 200$$
$$0 = 2(h^2 - 100)$$
$$0 = 2(h+10)(h-10)$$
$$h + 10 = 0 \quad \text{or} \quad h - 10 = 0$$

For $h + 10 = 0$

$$h + 10 - 10 = 0 - 10$$
$$h = -10$$

For $h - 10 = 0$

$$h - 10 + 10 = 0 + 10$$
$$h = 10$$

State: Since the height cannot be negative, we disregard the solution $h = -10$. Thus, the height is 10 centimeters and the base is $2(10) = 20$ centimeters.

Check: $200 = 2(10)^2$

$$200 = 2(100)$$
$$200 = 200$$

93. **Analyze:** We need to find the height of the trapezoid. We know the area of the trapezoid is $\frac{h(B+b)}{2}$ where h is the height, and b and B are the bases. We know the area is 24 square meters.
Form: We know one of the bases is 8 meters and the other is the same as the height. If h is the height, we have

$$\text{Area} = \frac{h(B+b)}{2}$$
$$24 = \frac{h(8+h)}{2}$$

Solve:

$$2(24) = 2\left(\frac{h(8+h)}{2}\right)$$
$$48 = h(8+h)$$
$$48 = 8h + h^2$$
$$48 = h^2 + 8h$$
$$48 - 48 = h^2 + 8h - 48$$
$$0 = h^2 + 8h - 48$$
$$0 = (h+12)(h-4)$$
$$h + 12 = 0 \quad \text{or} \quad h - 4 = 0$$

For $h + 12 = 0$

$$h + 12 - 12 = 0 - 12$$
$$h = -12$$

For $h - 4 = 0$

$$h - 4 + 4 = 0 + 4$$
$$h = 4$$

State: We disregard the solution $h = -12$ since the height cannot be negative. Thus, the height of the trapezoid is 4 meters.

Check: $24 = \frac{4(4+8)}{2}$

$$24 = \frac{4(12)}{2}$$
$$24 = \frac{48}{2}$$
$$24 = 24$$

REVIEW

97. Let E represent the time the patient exercises.
$15 \text{ min} \leq E \leq 30 \text{ min}$

99. **Analyze:** We know the perimeter of the rectangle is 120 cm. The equation for perimeter is $2l + 2w$, where l is the length and w is the width. We also know that the length is three times the width. We need to calculate the area. So, we must first find the length and the width.
Form:
Perimeter:

$$120 = 2l + 2w$$

Length:

$$l = 3w$$

Thus, we have the system of equations:

$$120 = 2l + 2w$$
$$l = 3w$$

Solve: Substitute the second equation into the first equation.

$$120 = 2(3w) + 2w$$
$$120 = 6w + 2w$$
$$120 = 8w$$
$$\frac{120}{8} = \frac{8w}{8}$$
$$15 = w$$

Find l:

$$l = 3w$$
$$l = 3(15)$$
$$l = 45$$

State: We have length of 45 centimeters and width of 15 centimeters. Thus, the area is $45(15) = 675 \text{ cm}^2$.

STUDY SET Section 7.4

VOCABULARY

1. completing

3. coefficients

CONCEPTS

5. $x^2 + 2bx + b^2 = (x + b)^2$

7. two

9. square; 8

11. a) Subtract 35 from both sides.
b) Add 1 to both sides.

13. Because $x^2 - 2x - 1 = 0$ does not factor.

NOTATION

15. $(y-1)^2 = 9$
$y - 1 = \sqrt{9}$ or $y - 1 = -\sqrt{9}$
$y - 1 = 3$ $\quad$ $y - 1 = -3$
$y = 4$ $\quad$ $y = -2$

17. Two solutions: $\sqrt{10}, -\sqrt{10}$

PRACTICE

19. $x^2 = 1$
$x = \sqrt{1}$ or $x = -\sqrt{1}$
$x = 1$ $\quad$ $x = -1$

21. $x^2 = 9$
$x = \sqrt{9}$ or $x = -\sqrt{9}$
$x = 3$ $\quad$ $x = -3$

23. $t^2 = 20$
$t = \sqrt{20}$ or $t = -\sqrt{20}$
$t = 2\sqrt{5}$ $\quad$ $t = -2\sqrt{5}$

25. $3m^2 = 27$
$m^2 = 9$
$m = \sqrt{9}$ or $m = -\sqrt{9}$
$m = 3$ $\quad$ $m = -3$

27. $4x^2 = 16$
$x^2 = 4$
$x = \sqrt{4}$ or $x = -\sqrt{4}$
$x = 2$ $\quad$ $x = -2$

29. $x^2 = \frac{9}{16}$
$x = \sqrt{\frac{9}{16}}$ or $x = -\sqrt{\frac{9}{16}}$
$x = \frac{3}{4}$ $\quad$ $x = -\frac{3}{4}$

31. $(x+1)^2 = 25$
$x + 1 = \sqrt{25}$ or $x + 1 = -\sqrt{25}$
$x + 1 = 5$ $\quad$ $x + 1 = -5$
$x = 4$ $\quad$ $x = -6$

33. $(x+2)^2 = 81$
$x + 2 = \sqrt{81}$ or $x + 2 = -\sqrt{81}$
$x + 2 = 9$ $\quad$ $x + 2 = -9$
$x = 7$ $\quad$ $x = -11$

35. $(x-2)^2 = 8$
$x - 2 = \sqrt{8}$ or $x - 2 = -\sqrt{8}$
$x - 2 = 2\sqrt{2}$ $\quad$ $x - 2 = -2\sqrt{2}$
$x = 2 + 2\sqrt{2}$ $\quad$ $x = 2 - 2\sqrt{2}$
$x = 2 \pm 2\sqrt{2}$

37. $x^2 = 45.82$
$x = \sqrt{45.82}$ or $x = -\sqrt{45.82}$
$x \approx 6.77$ $\quad$ $x \approx -6.77$

39. $(x+2)^2 = 90.04$
$x + 2 = \sqrt{90.04}$ $\quad$ $x + 2 = -\sqrt{90.04}$
$x + 2 \approx 9.49$ $\quad$ $x + 2 \approx -9.49$
$x \approx 7.49$ $\quad$ $x \approx -11.49$

41. $y^2 + 4y + 4 = 4$
$(y+2)^2 = 4$
$y + 2 = \sqrt{4}$ or $y + 2 = -\sqrt{4}$
$y + 2 = 2$ $\quad$ $y + 2 = -2$
$y = 0$ $\quad$ $y = -4$

43. $9x^2 - 12x + 4 = 16$

$(3x-2)^2 = 16$

$3x - 2 = \sqrt{16}$ $\quad$ $3x - 2 = -\sqrt{16}$

$3x - 2 = 4$ $\quad$ $3x - 2 = -4$

$3x = 6$ $\quad$ $3x = -2$

$x = 3$ $\quad$ $x = -\frac{2}{3}$

45. $x^2 + 2x$

$$\left(\frac{2}{2}\right)^2 = (1)^2 = 1$$

$x^2 + 2x + 1$

47. $x^2 - 4x$

$$\left(\frac{4}{2}\right)^2 = (2)^2 = 4$$

$x^2 - 4x + 4$

49. $x^2 + 7x$

$$\left(\frac{7}{2}\right)^2 = \frac{49}{4}$$

$x^2 + 7x + \frac{49}{4}$

51. $a^2 - 3a$

$$\left(\frac{3}{2}\right)^2 = \frac{9}{4}$$

$a^2 - 3a + \frac{9}{4}$

53. $b^2 + \frac{2}{3}b$

$$\left(\frac{\frac{2}{3}}{2}\right)^2 = \left(\frac{1}{3}\right)^2 = \frac{1}{9}$$

$b^2 + \frac{2}{3}b + \frac{1}{9}$

55. $x^2 + 6x + 8 = 0$

$x^2 + 6x = -8$

$$\left(\frac{6}{2}\right)^2 = (3)^2 = 9$$

$x^2 + 6x + 9 = -8 + 9$

$x^2 + 6x + 9 = 1$

$(x+3)^2 = 1$

$x + 3 = \sqrt{1}$ or $x + 3 = -\sqrt{1}$

$x + 3 = 1$ $\quad$ $x + 3 = -1$

$x = -2$ $\quad$ $x = -4$

57. $k^2 - 8k + 12 = 0$

$k^2 - 8k = -12$

$$\left(\frac{8}{2}\right)^2 = (4)^2 = 16$$

$k^2 - 8k + 16 = -12 + 16$

$k^2 - 8k + 16 = 4$

$(k-4)^2 = 4$

$k - 4 = \sqrt{4}$ or $k - 4 = -\sqrt{4}$

$k - 4 = 2$ $\quad$ $k - 4 = -2$

$k = 6$ $\quad$ $k = 2$

59. $x^2 - 2x - 15 = 0$

$x^2 - 2x = 15$

$$\left(\frac{2}{2}\right)^2 = (1)^2 = 1$$

$x^2 - 2x + 1 = 15 + 1$

$x^2 - 2x + 1 = 16$

$(x-1)^2 = 16$

$x - 1 = \sqrt{16}$ or $x - 1 = -\sqrt{16}$

$x - 1 = 4$ $\quad$ $x - 1 = -4$

$x = 5$ $\quad$ $x = -3$

61. $g^2 + 5g - 6 = 0$

$g^2 + 5g = 6$

$$\left(\frac{5}{2}\right)^2 = \frac{25}{4}$$

$g^2 + 5g + \frac{25}{4} = 6 + \frac{25}{4}$

$g^2 + 5g + \frac{25}{4} = \frac{24}{4} + \frac{25}{4}$

$g^2 + 5g + \frac{25}{4} = \frac{49}{4}$

$(g + \frac{5}{2})^2 = \frac{49}{4}$

$g + \frac{5}{2} = \sqrt{\frac{49}{4}}$ or $g + \frac{5}{2} = -\sqrt{\frac{49}{4}}$

$g + \frac{5}{2} = \frac{7}{2}$ $\quad$ $g + \frac{5}{2} = -\frac{7}{2}$

$g = \frac{2}{2}$ $\quad$ $g = -\frac{12}{2}$

$g = 1$ $\quad$ $g = -6$

63. $2x^2 = 4 - 2x$
$\frac{1}{2}(2x^2) = \frac{1}{2}(4 - 2x)$
$x^2 = 2 - x$
$x^2 + x = 2$

$$\left(\frac{1}{2}\right)^2 = \frac{1}{4}$$

$x^2 + x + \frac{1}{4} = 2 + \frac{1}{4}$
$x^2 + x + \frac{1}{4} = \frac{8}{4} + \frac{1}{4}$
$x^2 + x + \frac{1}{4} = \frac{9}{4}$
$(x + \frac{1}{2})^2 = \frac{9}{4}$
$x + \frac{1}{2} = \sqrt{\frac{9}{4}}$ or $x + \frac{1}{2} = -\sqrt{\frac{9}{4}}$
$x + \frac{1}{2} = \frac{3}{2}$ $\quad$ $x + \frac{1}{2} = -\frac{3}{2}$
$x = \frac{2}{2}$ $\quad$ $x = -\frac{4}{2}$
$x = 1$ $\quad$ $x = -2$

65. $3x^2 + 9x + 6 = 0$
$\frac{1}{3}(3x^2 + 9x + 6) = \frac{1}{3}(0)$
$x^2 + 3x + 2 = 0$
$x^2 + 3x = -2$

$$\left(\frac{3}{2}\right)^2 = \frac{9}{4}$$

$x^2 + 3x + \frac{9}{4} = -2 + \frac{9}{4}$
$x^2 + 3x + \frac{9}{4} = -\frac{8}{4} + \frac{9}{4}$
$x^2 + 3x + \frac{9}{4} = \frac{1}{4}$
$(x + \frac{3}{2})^2 = \frac{1}{4}$
$x + \frac{3}{2} = \sqrt{\frac{1}{4}}$ or $x + \frac{3}{2} = -\sqrt{\frac{1}{4}}$
$x + \frac{3}{2} = \frac{1}{2}$ $\quad$ $x + \frac{3}{2} = -\frac{1}{2}$
$x = -\frac{2}{2}$ $\quad$ $x = -\frac{4}{2}$
$x = 1$ $\quad$ $x = -2$

67. $2x^2 = 3x + 2$
$\frac{1}{2}(2x^2) = \frac{1}{2}(3x + 2)$
$x^2 = \frac{3}{2}x + 1$
$x^2 - \frac{3}{2}x = 1$

$$\left(\frac{\frac{3}{2}}{2}\right)^2 = \left(\frac{3}{4}\right)^2 = \frac{9}{16}$$

$x^2 - \frac{3}{2}x + \frac{9}{16} = 1 + \frac{9}{16}$
$x^2 - \frac{3}{2}x + \frac{9}{16} = \frac{16}{16} + \frac{9}{16}$
$x^2 - \frac{3}{2}x + \frac{9}{16} = \frac{25}{16}$
$(x - \frac{3}{4})^2 = \frac{25}{16}$
$x - \frac{3}{4} = \sqrt{\frac{25}{16}}$ or $x - \frac{3}{4} = -\sqrt{\frac{25}{16}}$
$x - \frac{3}{4} = \frac{5}{4}$ $\quad$ $x - \frac{3}{4} = -\frac{5}{4}$
$x = \frac{8}{4}$ $\quad$ $x = -\frac{2}{4}$
$x = 2$ $\quad$ $x = -\frac{1}{2}$

69. $4x^2 = 2 - 7x$
$\frac{1}{4}(4x^2) = \frac{1}{4}(2 - 7x)$
$x^2 = \frac{1}{2} - \frac{7}{4}x$
$x^2 + \frac{7}{4}x = \frac{1}{2}$

$$\left(\frac{\frac{7}{4}}{2}\right)^2 = \left(\frac{7}{8}\right)^2 = \frac{49}{64}$$

$x^2 + \frac{7}{4}x + \frac{49}{64} = \frac{1}{2} + \frac{49}{64}$
$x^2 + \frac{7}{4}x + \frac{49}{64} = \frac{32}{64} + \frac{49}{64}$
$x^2 + \frac{7}{4}x + \frac{49}{64} = \frac{81}{64}$
$(x + \frac{7}{8})^2 = \frac{81}{64}$
$x + \frac{7}{8} = \sqrt{\frac{81}{64}}$ or $x + \frac{7}{8} = -\sqrt{\frac{81}{64}}$
$x + \frac{7}{8} = \frac{9}{8}$ $\quad$ $x + \frac{7}{8} = -\frac{9}{8}$
$x = \frac{2}{8}$ $\quad$ $x = -\frac{16}{8}$
$x = \frac{1}{4}$ $\quad$ $x = -2$

71. $x^2 + 4x + 1 = 0$
$x^2 + 4x = -1$

$$\left(\frac{4}{2}\right)^2 = (2)^2 = 4$$

$x^2 + 4x + 4 = -1 + 4$
$x^2 + 4x + 4 = 3$
$(x + 2)^2 = 3$
$x + 2 = \sqrt{3}$ or $x + 2 = -\sqrt{3}$
$x = -2 + \sqrt{3}$ $\quad$ $x = -2 - \sqrt{3}$
$x \approx -0.27$ $\quad$ $x \approx -3.73$

73. $x^2 - 2x - 4 = 0$
$x^2 - 2x = 4$

$$\left(\frac{2}{2}\right)^2 = (1)^2 = 1$$

$x^2 - 2x + 1 = 4 + 1$
$x^2 - 2x + 1 = 5$
$(x - 1)^2 = 5$
$x - 1 = \sqrt{5}$ or $x - 1 = -\sqrt{5}$
$x = 1 + \sqrt{5}$ $\quad$ $x = 1 - \sqrt{5}$
$x \approx 3.24$ $\quad$ $x \approx -1.24$

75. $x^2 = 4x + 3$
$x^2 - 4x = 3$
$$\left(\frac{4}{2}\right)^2 = (2)^2 = 4$$
$x^2 - 4x + 4 = 3 + 4$
$x^2 - 4x + 4 = 7$
$(x - 2)^2 = 7$
$x - 2 = \sqrt{7}$ or $x - 2 = -\sqrt{7}$
$x = 2 + \sqrt{7}$ $\quad x = 2 - \sqrt{7}$
$x \approx 4.65$ $\quad x \approx -0.65$

77. $4x^2 + 4x + 1 = 20$
$4x^2 + 4x = 19$
$\frac{1}{4}(4x^2 + 4x) = \frac{1}{4}(19)$
$x^2 + x = \frac{19}{4}$
$$\left(\frac{1}{2}\right)^2 = \frac{1}{4}$$
$x^2 + x + \frac{1}{4} = \frac{19}{4} + \frac{1}{4}$
$x^2 + x + \frac{1}{4} = \frac{20}{4}$
$x^2 + x + \frac{1}{4} = 5$
$(x + \frac{1}{2})^2 = 5$
$x + \frac{1}{2} = \sqrt{5}$ or $x + \frac{1}{2} = -\sqrt{5}$
$x + \frac{1}{2} = \sqrt{5}$ $\quad x + \frac{1}{2} = -\sqrt{5}$
$x = -\frac{1}{2} + \sqrt{5}$ $\quad x = -\frac{1}{2} - \sqrt{5}$
$x \approx 1.74$ $\quad x \approx -2.74$

79. $2x(x + 3) = 8$
$2x^2 + 6x = 8$
$\frac{1}{2}(2x^2 + 6x) = \frac{1}{2}(8)$
$x^2 + 3x = 4$
$$\left(\frac{3}{2}\right)^2 = \frac{9}{4}$$
$x^2 + 3x + \frac{9}{4} = 4 + \frac{9}{4}$
$x^2 + 3x + \frac{9}{4} = \frac{16}{4} + \frac{9}{4}$
$x^2 + 3x + \frac{9}{4} = \frac{25}{4}$
$(x + \frac{3}{2})^2 = \frac{25}{4}$
$x + \frac{3}{2} = \sqrt{\frac{25}{4}}$ or $x + \frac{3}{2} = -\sqrt{\frac{25}{4}}$
$x + \frac{3}{2} = \frac{5}{2}$ $\quad x + \frac{3}{2} = -\frac{5}{2}$
$x = \frac{2}{2}$ $\quad x = -\frac{8}{2}$
$x = 1$ $\quad x = -4$

81. $6(x^2 - 1) = 5x$
$6x^2 - 6 = 5x$
$6x^2 - 5x - 6 = 0$
$6x^2 - 5x = 6$
$\frac{1}{6}(6x^2 - 5x) = \frac{1}{6}(6)$
$x^2 - \frac{5}{6}x = 1$
$$\left(\frac{\frac{5}{6}}{2}\right)^2 = \left(\frac{5}{12}\right) = \frac{25}{144}$$
$x^2 - \frac{5}{6}x + \frac{25}{144} = 1 + \frac{25}{144}$
$x^2 - \frac{5}{6}x + \frac{25}{144} = \frac{144}{144} + \frac{25}{144}$
$x^2 - \frac{5}{6}x + \frac{25}{144} = \frac{169}{144}$
$(x - \frac{5}{12})^2 = \frac{169}{144}$
$x - \frac{5}{12} = \sqrt{\frac{169}{144}}$ $\quad x - \frac{5}{12} = -\sqrt{\frac{169}{144}}$
$x - \frac{5}{12} = \frac{13}{12}$ $\quad x - \frac{5}{12} = -\frac{13}{12}$
$x = \frac{18}{12}$ $\quad x = -\frac{8}{12}$
$x = \frac{3}{2}$ $\quad x = -\frac{2}{3}$

83. $x(x + 3) - \frac{1}{2} = -2$
$x^2 + 3x - \frac{1}{2} = -2$
$x^2 + 3x = -2 + \frac{1}{2}$
$x^2 + 3x = -\frac{4}{2} + \frac{1}{2}$
$x^2 + 3x = -\frac{3}{2}$
$$\left(\frac{3}{2}\right)^2 = \frac{9}{4}$$
$x^2 + 3x + \frac{9}{4} = -\frac{3}{2} + \frac{9}{4}$
$x^2 + 3x + \frac{9}{4} = -\frac{6}{4} + \frac{9}{4}$
$x^2 + 3x + \frac{9}{4} = \frac{3}{4}$
$(x + \frac{3}{2})^2 = \frac{3}{4}$
$x + \frac{3}{2} = \sqrt{\frac{3}{4}}$ $\quad x + \frac{3}{2} = -\sqrt{\frac{3}{4}}$
$x + \frac{3}{2} = \frac{\sqrt{3}}{2}$ $\quad x + \frac{3}{2} = -\frac{\sqrt{3}}{2}$
$x = -\frac{3}{2} + \frac{\sqrt{3}}{2}$ $\quad x = -\frac{3}{2} - \frac{\sqrt{3}}{2}$
$x = \frac{-3+\sqrt{3}}{2}$ $\quad x = \frac{-3-\sqrt{3}}{2}$
$$x = \frac{-3\pm\sqrt{3}}{2}$$

85. **Analyze:** We know the area of the course is 800 square feet, and area is length times width.

Form: Area = (length) • (width)

$800 = (x + 20)(x)$

Solve:

$800 = x^2 + 20x$

$x^2 + 20x = 800$

$\left(\frac{20}{2}\right)^2 = (10)^2 = 100$

$x^2 + 20x + 100 = 800 + 100$

$x^2 + 20x + 100 = 900$

$(x + 10)^2 = 900$

$x + 10 = \sqrt{900}$ $\quad x + 10 = -\sqrt{900}$

$x + 10 = 30$ $\quad x + 10 = -30$

$x = 20$ $\quad x = -40$

State: Since the width of the rectangle cannot be negative, we disregard the solution $x = -40$. Thus, the width of the rectangle is 20 feet and the length is $20 + 20 = 40$ feet.

Check: $800 = (20)(40)$

$800 = 800$

91. $(y - 1)^2 = (y - 1)(y - 1)$

$= y^2 - y - y + 1$

$= y^2 - 2y + 1$

93. $(x + y)^2 = (x + y)(x + y)$

$= x^2 + xy + xy + y^2$

$= x^2 + 2xy + y^2$

95. $(2z)^2 = (2z)(2z)$

$= 2(2)z \cdot z$

$= 4z^2$

STUDY SET Section 7.5

VOCABULARY

1. quadratic

3. solve

CONCEPTS

5. $a \neq 0$

7. $3x^2 - 5 = 3x^2 + 0x - 5$
So, $a = 3$, $b = 0$, and $c = -5$.

9. $A = lw$

11. Evaluate 5^2.

13. a) 2 solutions, because 15 is positive in the square root and the $\pm$.

b) $\frac{-3+\sqrt{15}}{2}$ and $\frac{-3-\sqrt{15}}{2}$

c) $\frac{-3+\sqrt{15}}{2} \approx 0.44$
$\frac{-3-\sqrt{15}}{2} \approx -3.44$

15. $\frac{-1+\sqrt{15}}{2} \approx 0.62$
$\frac{-1-\sqrt{15}}{2} \approx -1.62$

$\frac{-1-\sqrt{5}}{2}$ $\frac{-1+\sqrt{5}}{2}$

-3 -2 -1 0 1 2 3

NOTATION

17.
$$x = \frac{-b\pm\sqrt{b^2-4ac}}{2a}$$
$$= \frac{-(-5)\pm\sqrt{(-5)^2-4(1)(-6)}}{2(1)}$$
$$= \frac{5\pm\sqrt{25+24}}{2}$$
$$= \frac{5\pm\sqrt{49}}{2}$$
$$= \frac{5\pm 7}{2}$$
$\frac{5+7}{2} = 6$ or $\frac{5-7}{2} = -1$

19. The student did not extend the fraction bar so that it underlines the complete numerator. The -4 is also in the numerator.

PRACTICE

21. $x^2 + 4x + 3 = 0$
$a = 1,\ b = 4,\ c = 3$

23. $3x^2 - 2x + 7 = 0$
$a = 3,\ b = -2,\ c = 7$

25. $4y^2 = 2y - 1$
$4y^2 - 2y + 1 = 0$
$a = 4,\ b = -2,\ c = 1$

27. $x(3x - 5) = 2$
$3x^2 - 5x = 2$
$3x^2 - 5x - 2 = 0$
$a = 3,\ b = -5,\ c = -2$

29. $7(x^2 + 3) = -14x$
$7x^2 + 21 = -14x$
$7x^2 + 14x + 21 = 0$
$a = 7,\ b = 14,\ c = 21$

31. $x^2 - 5x + 6 = 0$
$a = 1,\ b = -5,\ c = 6$
$$x = \frac{-b\pm\sqrt{b^2-4ac}}{2a}$$
$$= \frac{-(-5)\pm\sqrt{(-5)^2-4(1)(6)}}{2(1)}$$
$$= \frac{5\pm\sqrt{25-24}}{2}$$
$$= \frac{5\pm\sqrt{1}}{2}$$
$$= \frac{5\pm 1}{2}$$
$x = \frac{5+1}{2} = \frac{6}{2} = 3$
or $x = \frac{5-1}{2} = \frac{4}{2} = 2$

33. $x^2 + 7x + 12 = 0$

$a = 1,\ b = 7,\ c = 12$

$$x = \frac{-b \pm \sqrt{b^2 - 4ac}}{2a} = \frac{-(7) \pm \sqrt{(7)^2 - 4(1)(12)}}{2(1)} = \frac{-7 \pm \sqrt{49 - 48}}{2} = \frac{-7 \pm \sqrt{1}}{2} = \frac{-7 \pm 1}{2}$$

$$x = \frac{-7+1}{2} = \frac{-6}{2} = -3$$

or $$x = \frac{-7-1}{2} = \frac{-8}{2} = -4$$

35. $2x^2 - x - 1 = 0$

$a = 2,\ b = -1,\ c = -1$

$$x = \frac{-b \pm \sqrt{b^2 - 4ac}}{2a} = \frac{-(-1) \pm \sqrt{(-1)^2 - 4(2)(-1)}}{2(2)} = \frac{1 \pm \sqrt{1+8}}{4} = \frac{1 \pm \sqrt{9}}{4} = \frac{1 \pm 3}{4}$$

$$x = \frac{1+3}{4} = \frac{4}{4} = 1$$

or $$x = \frac{1-3}{4} = \frac{-2}{4} = -\frac{1}{2}$$

37. $3x^2 + 5x + 2 = 0$

$a = 3,\ b = 5,\ c = 2$

$$x = \frac{-b \pm \sqrt{b^2 - 4ac}}{2a} = \frac{-(5) \pm \sqrt{(5)^2 - 4(3)(2)}}{2(3)} = \frac{-5 \pm \sqrt{25 - 24}}{6} = \frac{-5 \pm \sqrt{1}}{6} = \frac{-5 \pm 1}{6}$$

$$x = \frac{-5+1}{6} = \frac{-4}{6} = -\frac{2}{3}$$

or $$x = \frac{-5-1}{6} = \frac{-6}{6} = -1$$

39. $4x^2 + 4x - 3 = 0$

$a = 4,\ b = 4,\ c = -3$

$$x = \frac{-b \pm \sqrt{b^2 - 4ac}}{2a} = \frac{-(4) \pm \sqrt{(4)^2 - 4(4)(-3)}}{2(4)} = \frac{-4 \pm \sqrt{16+48}}{8} = \frac{-4 \pm \sqrt{64}}{8} = \frac{-4 \pm 8}{8}$$

$$x = \frac{-4+8}{8} = \frac{4}{8} = \frac{1}{2}$$

or $$x = \frac{-4-8}{8} = \frac{-12}{8} = -\frac{3}{2}$$

41. $x^2 + 3x + 1 = 0$

$a = 1,\ b = 3,\ c = 1$

$$x = \frac{-b \pm \sqrt{b^2 - 4ac}}{2a} = \frac{-(3) \pm \sqrt{(3)^2 - 4(1)(1)}}{2(1)} = \frac{-3 \pm \sqrt{9-4}}{2} = \frac{-3 \pm \sqrt{5}}{2}$$

$$x = \frac{-3+\sqrt{5}}{2}$$

or $$x = \frac{-3-\sqrt{5}}{2}$$

43. $3x^2 - x = 3$

$3x^2 - x - 3 = 0$

$a = 3,\ b = -1,\ c = -3$

$$x = \frac{-b \pm \sqrt{b^2 - 4ac}}{2a} = \frac{-(-1) \pm \sqrt{(-1)^2 - 4(3)(-3)}}{2(3)} = \frac{1 \pm \sqrt{1+36}}{6} = \frac{1 \pm \sqrt{37}}{6}$$

$$x = \frac{1+\sqrt{37}}{6}$$

or $$x = \frac{1-\sqrt{37}}{6}$$

45. $x^2 + 5 = 2x$

$x^2 - 2x + 5 = 0$

$a = 1,\ b = -2,\ c = 5$

$$x = \frac{-b \pm \sqrt{b^2 - 4ac}}{2a} = \frac{-(-2) \pm \sqrt{(-2)^2 - 4(1)(5)}}{2(1)} = \frac{2 \pm \sqrt{4-20}}{2} = \frac{2 \pm \sqrt{-16}}{2}$$

No real solutions since $\sqrt{-16}$ is not a real number.

47. $x^2 = 1 - 2x$
$x^2 + 2x - 1 = 0$
$a = 1,\ b = 2,\ c = -1$
$$x = \frac{-b\pm\sqrt{b^2-4ac}}{2a}$$
$$= \frac{-(2)\pm\sqrt{(2)^2-4(1)(-1)}}{2(1)}$$
$$= \frac{-2\pm\sqrt{4+4}}{2}$$
$$= \frac{-2\pm\sqrt{8}}{2}$$
$$= \frac{-2\pm2\sqrt{2}}{2}$$
$$= -1\pm\sqrt{2}$$
$$x = -1+\sqrt{2}$$
or $$x = -1-\sqrt{2}$$

49. $3x^2 = 6x + 2$
$3x^2 - 6x - 2 = 0$
$a = 3,\ b = -6,\ c = -2$
$$x = \frac{-b\pm\sqrt{b^2-4ac}}{2a}$$
$$= \frac{-(-6)\pm\sqrt{(-6)^2-4(3)(-2)}}{2(3)}$$
$$= \frac{6\pm\sqrt{36+24}}{6}$$
$$= \frac{6\pm\sqrt{60}}{6}$$
$$= \frac{6\pm2\sqrt{15}}{6}$$
$$= \frac{3\pm\sqrt{15}}{3}$$
$$x = \frac{3+\sqrt{15}}{3}$$
or $$x = \frac{3-\sqrt{15}}{3}$$

51. Use square root method.
$(2y-1)^2 = 25$
$2y - 1 = \sqrt{25}$ or $2y - 1 = -\sqrt{25}$
$2y - 1 = 5$ $\quad$ $2y - 1 = -5$
$2y = 6$ $\quad$ $2y = -4$
$y = 3$ $\quad$ $y = -2$

53. Use quadratic formula (doesn't look like it can be easily factored).
$2x^2 + x = 5$
$2x^2 + x - 5 = 0$
$a = 2,\ b = 1,\ c = -5$
$$x = \frac{-b\pm\sqrt{b^2-4ac}}{2a}$$
$$= \frac{-(1)\pm\sqrt{(1)^2-4(2)(-5)}}{2(2)}$$
$$= \frac{-1\pm\sqrt{1+40}}{4}$$
$$= \frac{-1\pm\sqrt{41}}{4}$$
$$= \frac{-1\pm\sqrt{41}}{4}$$
$$x = \frac{-1+\sqrt{41}}{4} \approx 1.35$$
or $$x = \frac{-1-\sqrt{41}}{4} \approx -1.85$$

55. Use quadratic formula.
$x^2 - 2x - 1 = 0$
$a = 1,\ b = -2,\ c = -1$
$$x = \frac{-b\pm\sqrt{b^2-4ac}}{2a}$$
$$= \frac{-(-2)\pm\sqrt{(-2)^2-4(1)(-1)}}{2(1)}$$
$$= \frac{2\pm\sqrt{4+4}}{2}$$
$$= \frac{2\pm\sqrt{8}}{2}$$
$$= \frac{2\pm2\sqrt{2}}{2}$$
$$= 1\pm\sqrt{2}$$
$$x = 1+\sqrt{2} \approx 2.41$$
or $$x = 1-\sqrt{2} \approx -.41$$

57. Use factor method.
$x^2 - 2x - 35 = 0$
$(x-7)(x+5) = 0$
$x - 7 = 0$ or $x + 5 = 0$
$x = 7$ $\quad$ $x = -5$

59. Use quadratic formula.
$x^2 + 2x + 7 = 0$
$a = 1,\ b = 2,\ c = 7$
$$x = \frac{-b\pm\sqrt{b^2-4ac}}{2a}$$
$$= \frac{-(2)\pm\sqrt{(2)^2-4(1)(7)}}{2(1)}$$
$$= \frac{-2\pm\sqrt{4-28}}{2}$$
$$= \frac{-2\pm\sqrt{-24}}{2}$$
No real solutions since $\sqrt{-24}$ is not a real number.

61. Use factor method.

$4c^2 + 16c = 0$
$4c(c + 4) = 0$
$4c = 0$ or $c + 4 = 0$
$c = 0$ $\quad c = -4$

63. Isolate y.
$18 = 3y^2$
$6 = y^2$
$y^2 = 6$
$y = \sqrt{6}$ or $y = -\sqrt{6}$
$y \approx 2.45$ $\quad y \approx -2.45$

65. Use quadratic formula.
$2.4x^2 - 9.5x + 6.2 = 0$
$a = 2.4,\ b = -9.5,\ c = 6.2$

$$x = \frac{-b \pm \sqrt{b^2 - 4ac}}{2a}$$
$$= \frac{-(-9.5) \pm \sqrt{(-9.5)^2 - 4(2.4)(6.2)}}{2(2.4)}$$
$$= \frac{9.5 \pm \sqrt{90.25 - 59.52}}{4.8}$$
$$= \frac{9.5 \pm \sqrt{30.73}}{4.8}$$
$$x = \frac{9.5 + \sqrt{30.73}}{4.8} \approx 3.1$$
or
$$x = \frac{9.5 - \sqrt{30.73}}{4.8} \approx 0.8$$

APPLICATIONS

67. **Analyze:** We know the area is 32 square feet and the length is 4 feet longer than the width.
Form: Let w represent the width of the rectangle.
Area = (length) • (width)
$32 = (w + 4)(w)$
Solve:
$32 = w^2 + 4w$
$w^2 + 4w = 32$
$w^2 + 4w - 32 = 0$
$(w - 4)(w + 8) = 0$
$w - 4 = 0$ or $w + 8 = 0$
$w = 4$ $\quad w = -8$
State: We disregard the solution $w = -8$ since the width cannot be negative. Thus, the width of the mural is 4 feet and the length is $4 + 4 = 8$ feet.

Check: $32 = (8)(4)$
$32 = 32$

69. **Analyze:** Recall that the area of a triangle is one-half the base times the height. We are given that the height of the triangle is x and the base is $x + 4$. The area is 30 square inches.
Form:
Area $= \frac{1}{2}$(base) • (height)
$30 = \frac{1}{2}(x + 4)(x)$
Solve:
$30 = \frac{1}{2}(x^2 + 4x)$
$2(30) = 2\left(\frac{1}{2}(x^2 + 4x)\right)$
$60 = x^2 + 4x$
$x^2 + 4x = 60$
$x^2 + 4x - 60 = 0$
$(x + 10)(x - 6) = 0$
$x + 10 = 0$ or $x - 6 = 0$
$x = -10$ $\quad x = 6$
State: The height of the triangle is 6 inches.

71. **Analyze:** We need the dimensions of the rectangular garden. We are given that the area is 180 sq. feet. We know the dimensions of the rectangle than includes the garden and the walkway of uniform width.
Form: Let x represent the width of the walkway. The width of the garden is $16 - 2x\,a$ and the length is $24 - 2x$. So, we have
Area $= l$•w
$180 = (16 - 2x)(24 - 2x)$
Solve:
$180 = 384 - 32x - 48x + 4x^2$
$180 = 384 - 80x + 4x^2$
$180 = 4x^2 - 80x + 384$
$0 = 4x^2 - 80x + 204$
$\frac{1}{4}(0) = \frac{1}{4}(4x^2 - 80x + 204)$
$0 = x^2 - 20x - 51$
$x^2 - 20x - 51 = 0$
$(x - 17)(x - 3) = 0$
$x - 17 = 0$ or $x - 3 = 0$
$x = 17$ $\quad x = 3$

State: The solution $x = 17$ does not make sense since the width of the garden is $16 - 2x$. So, if $x = 3$ then the width of the garden is $16 - 2(3) = 16 - 6$ $= 10$ feet. The length of the garden is $24 - 2(3) = 24 - 6 = 18$ feet. The length of the garden is 18 feet and the width is 10 feet.

Check: $180 = (10)(18)$

$180 = 180$

73. **Analyze:** We know that an object will fall s feet in t seconds, where $s = 16t^2$. The workman is 1454 feet above the ground. We need to find t when $s = 1454$.

Form:

$s = 16t^2$

$1454 = 16t^2$

Solve:

$1454 = 16t^2$

$\frac{1454}{16} = t^2$

$t^2 = \frac{1454}{16}$

$t = \pm\sqrt{\frac{1454}{16}}$

$t = \sqrt{\frac{1454}{16}} \approx 9.5$

or $t = -\sqrt{\frac{1454}{16}} \approx -9.5$

State: We disregard the negative solution because time is not negative. Thus, it takes 9.5 seconds for the hammer to fall.

75. **Analyze:** Since we know the diagonal of the picture, we can use the right triangle and Pythagorean's theorem to find the dimensions.

Form: We know the hypotenuse of the triangle is 10 inches and the sides are h and $h + 2$ inches.

$a^2 + b^2 = c^2$

$h^2 + (h + 2)^2 = 10^2$

Solve:

$h^2 + h^2 + 4h + 4 = 100$

$2h^2 + 4h + 4 = 100$

$2h^2 + 4h = 96$

$\frac{1}{2}(2h^2 + 4h) = \frac{1}{2}(96)$

$h^2 + 2h = 48$

$\left(\frac{2}{2}\right)^2 = (1)^2 = 1$

$h^2 + 2h + 1 = 48 + 1$

$h^2 + 2h + 1 = 49$

$(h + 1)^2 = 49$

$h + 1 = \pm\sqrt{49}$

$h + 1 = \pm 7$

$h + 1 = 7$ or $h + 1 = -7$

$h = 6$ $\quad h = -8$

State: The solution $h = -8$ does not make sense since the height cannot be negative. Thus, the height is 6 inches and the length is $(6) + 2 = 8$ inches.

Check: $6^2 + 8^2 = 100$

$36 + 64 = 100$

$100 = 100$

77. **Analyze:** Since one boat is traveling north and one boat is traveling east, the lines they are traveling create a right angle. We know that one boat has sailed 10 nautical miles farther than the other and that the distance between the boats is 50 nautical miles. We can use the Pythagorean theorem to calculate how far each boat has sailed.

Form: Let x represent the distance one boat has traveled. Then the other has traveled $x + 10$. The distance between the boats is the hypotenuse of the right triangle.

$a^2 + b^2 = c^2$

$x^2 + (x + 10)^2 = 50^2$

Solve:

$x^2 + x^2 + 20x + 100 = 2500$

$2x^2 + 20x + 100 = 2500$

$2x^2 + 20x = 2400$

$\frac{1}{2}(2x^2 + 20x) = \frac{1}{2}(2400)$

$x^2 + 10x = 1200$

$x^2 + 10x - 1200 = 0$

$(x - 30)(x + 40) = 0$

$x - 30 = 0$ or $x + 40 = 0$

$x = 30$ $\quad x = -40$

State: We disregard the solution $x = -40$ since distance traveled cannot be negative. Thus, one boat has traveled 30 nautical miles and the other has traveled $(30) + 10 = 40$ nautical miles.
Check: $30^2 + 40^2 = 2500$

$900 + 1600 = 2500$

$2500 = 2500$

79. **Analyze:** We are given that the amount invested is \$5000 and the account balance is \$5624.50 after 2 years. We need to find the interest rate, r.
Form: We have $P = 5000$ and $A = 5624.50$.

$A = P(1+r)^2$

$5624.50 = 5000(1+r)^2$

Solve:

$1.1249 = (1+r)^2$

$\pm\sqrt{1.1249} = 1 + r$

$-1 \pm \sqrt{1.1249} = r$

$r = -1 + \sqrt{1.1249} \approx 0.061$

$r = -1 - \sqrt{1.1249} \approx -2.06$

State: We disregard the solution $r \approx -2.06$ since interest rates cannot be negative. Thus, the interest rate is 6.1%.

81. **Analyze:** We are given that the revenue is $R = -\frac{1}{6}x^2 + 450x$, where x is the number of television sets manufactured and sold. We want to find R when $x = 600$.
Form:

$R = -\frac{1}{6}x^2 + 450x$

$R = -\frac{1}{6}(600)^2 + 450(600)$

Solve:

$R = -\frac{1}{6}(360{,}000) + 270{,}000$

$R = -60,000 + 270,000$

$R = 210,000$

State: \$210,000 in revenue will be earned.

83. **Analyze:** The tin is 12 inches on each side before the corner squares are cut out. The length of the side after the corner squares are cut out will be $12 - 2x$, where x is the length of the square that is removed. So, the length of the side of the box $12 - 2x$. We are given that the area of the bottom of the box is 64 square inches.
Form: Area of a square is the length of a side squared.

$A = (\text{length of a side})^2$

$64 = (12 - 2x)^2$

Solve:

$64 = (12 - 2x)^2$

$\pm\sqrt{64} = 12 - 2x$

$\pm 8 = 12 - 2x$

$8 = 12 - 2x$ or $-8 = 12 - 2x$

$-4 = -2x \qquad -20 = -2x$

$2 = x \qquad 10 = x$

State: We disregard the solution $x = 10$ since we cannot cut two 10-inch squares from each corner of a 12-inch square piece of tin. Thus, the length of the square that is removed is 2 inches. This means that the depth of the box is 2 inches.

REVIEW

89. $A = p + prt$

$A - p = prt$

$\frac{A-p}{pt} = r$

91. slope $= \frac{3}{5}$ and $(0,\ 12)$ is the y-intercept. Use the point-intercept form:

$y = mx + b$

$y = \frac{3}{5}x + 12$

$5(y) = 5\left(\frac{3}{5}x + 12\right)$

$5y = 3x + 60$

$-60 = 3x - 5y$

93. $\sqrt{80} = \sqrt{16 \cdot 5} = \sqrt{16}\sqrt{5} = 4\sqrt{5}$

95. $\frac{x}{\sqrt{7x}} = \frac{x}{\sqrt{7x}}\frac{\sqrt{7x}}{\sqrt{7x}} = \frac{x\sqrt{7x}}{7x} = \frac{\sqrt{7x}}{7}$

STUDY SET Section 7.6

VOCABULARY

1. quadratic

3. y-intercept

5. symmetry

CONCEPTS

7. $a > 0$; $a < 0$

9. $(0,\, c)$

11. a) a parabola
    b) $(1, 0)$ and $(3, 0)$
    c) (0, -3)
    d) (2, 1)

13. zero x-intercepts

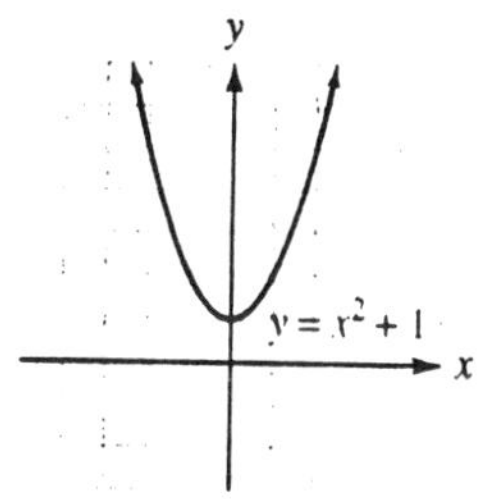

one x-intercept

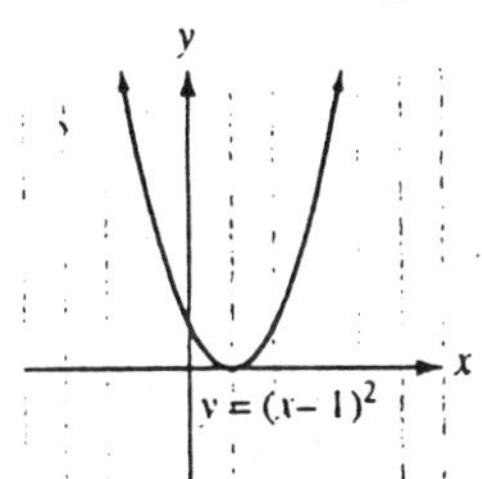

two x-intercepts

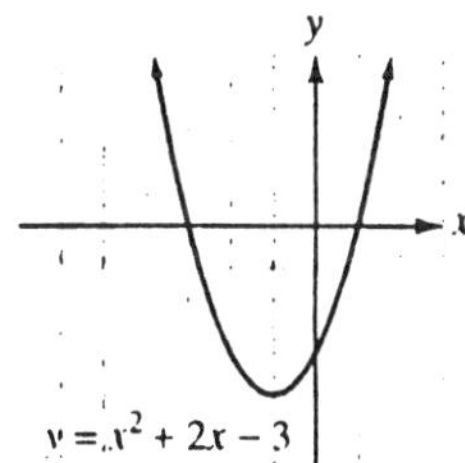

15. The vertex tells us that the most cases of flu (25) were reported in the fifth week.

NOTATION

17. $y = x^2 - 4x + 6$
    $= x^2 - 4x + 4 - 4 + 6$
    $= (x - 2)^2 + 2$
    The vertex is at (2, 2).

PRACTICE

19. moved up 1

21. opens the opposite direction

23.

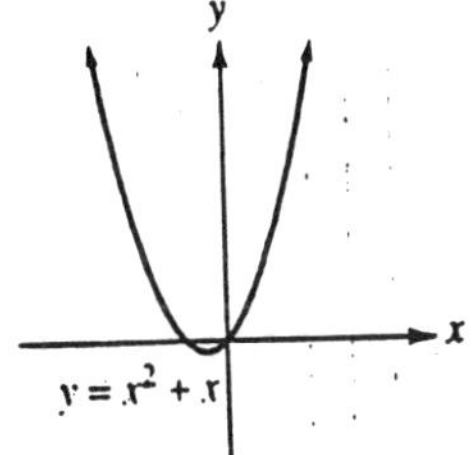

25.

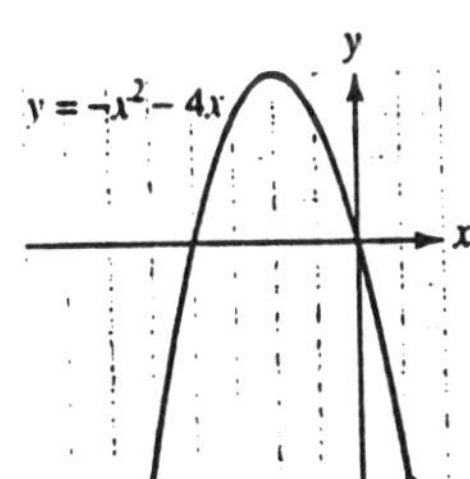

27.

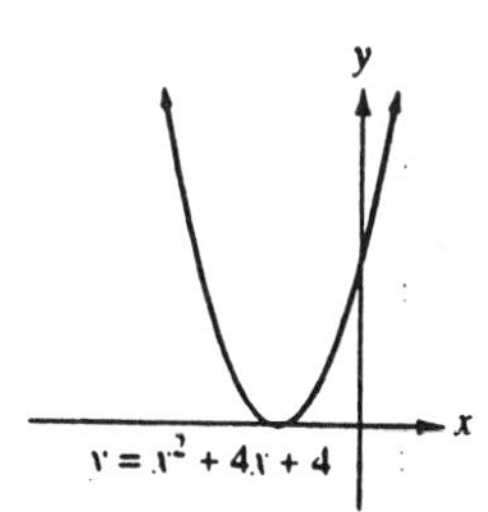

29.

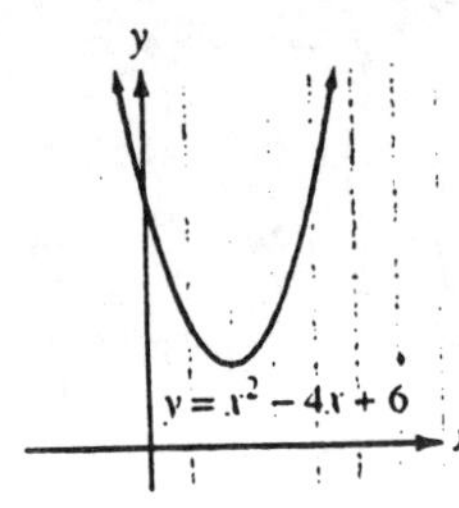

31. $y = -3(x-2)^2 + 4$
The vertex is at (2, 4).

33. $y = 4(x-3)^2 + 2$
The vertex is at (3, 2).

35. $y = (x-1)^2$
$= (x-1)^2 + 0$
The vertex is at (1, 0).

37. $y = -7x^2 + 4$
$= -7(x-0)^2 + 4$
The vertex is (0, 4).

39. $y = x^2 + 2x + 5$
The x-coordinate of the vertex is at
$x = -\frac{b}{2a}$. $a = 1$ and $b = 2$.
$x = -\frac{(2)}{2(1)} = -\frac{2}{2} = -1.$
The y-coordinate is
$y = (-1)^2 + 2(-1) + 5$
$= 1 - 2 + 5 = 4.$
The vertex is $(-1, 4)$.

41. $y = x^2 - 6x - 12$
The x-coordinate of the vertex is at
$x = -\frac{b}{2a}$. $a = 1$ and $b = -6$.
$x = -\frac{(-6)}{2(1)} = \frac{6}{2} = 3.$
The y-coordinate is
$y = (3)^2 - 6(3) - 12$
$= 9 - 18 - 12 = -21.$
The vertex is $(3, -21)$.

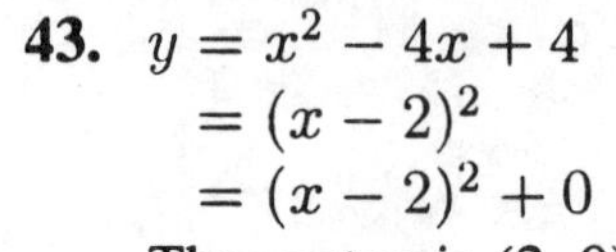

43. $y = x^2 - 4x + 4$
$= (x-2)^2$
$= (x-2)^2 + 0$
The vertex is (2, 0).

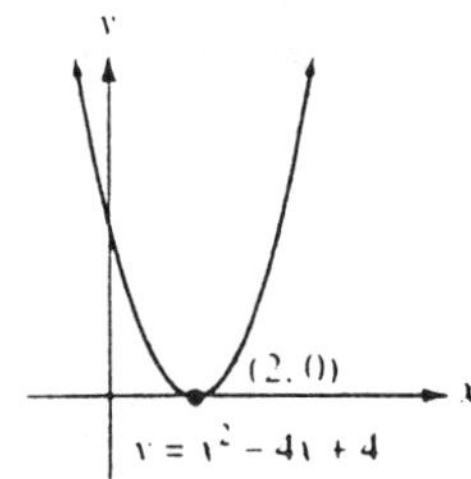

45. $y = -x^2 - 2x - 1$
$= -(x^2 + 2x + 1)$
$= -(x+1)^2$
$= -(x+1)^2 + 0$
The vertex is $(-1, 0)$.

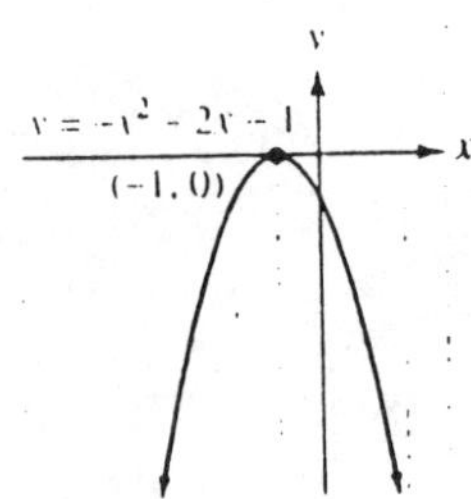

47. $y = x^2 + 2x - 3$
$= x^2 + 2x + 1 - 1 - 3$
$= (x+1)^2 - 4$
The vertex is $(-1, -4)$.

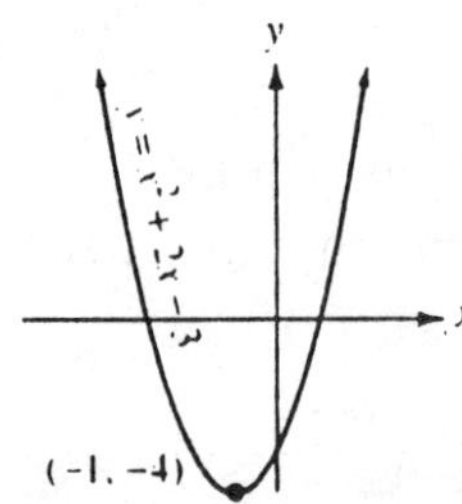

49. $y = -x^2 - 6x - 7$
$= -(x^2 + 6x + 7)$
$= -(x^2 + 6x + 9 - 9 + 7)$
$= -((x + 3)^2 - 2)$
$= -(x + 3)^2 + 2$
The vertex is $(-3, 2)$.

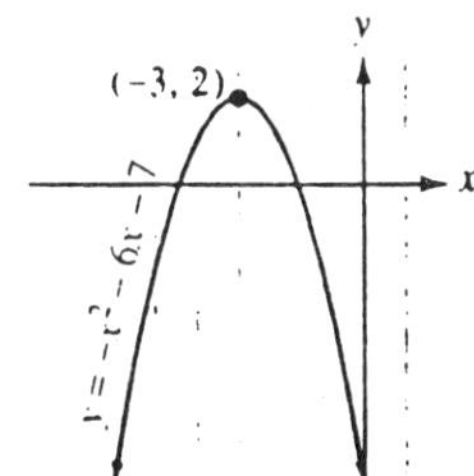

51. $y = 2x^2 + 8x + 6$
$= 2(x^2 + 4x + 3)$
$= 2(x^2 + 4x + 4 - 4 + 3)$
$= 2((x + 2)^2 - 1)$
$= 2(x + 2)^2 - 2$
The vertex is $(-2, -2)$.

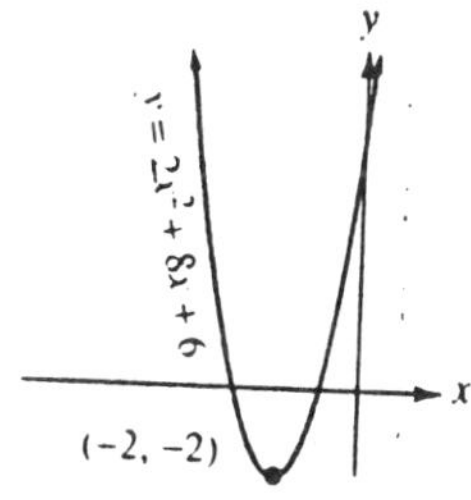

53. $y = -3x^2 + 6x - 2$
$= -3(x^2 - 2x) - 2$
$= -3(x^2 - 2x + 1 - 1) - 2$
$= -3((x - 1)^2 - 1) - 2$
$= -3(x - 1)^2 + 3 - 2$
$= -3(x - 1)^2 + 1$
The vertex is $(1, 1)$.

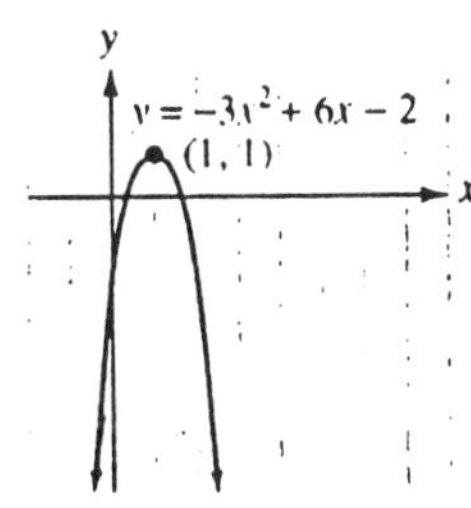

55. $y = x^2 - x - 2$
$a = 1,\ b = -1,\ c = -2$
The x-coordinate of the vertex is at
$x = -\frac{b}{2a} = -\frac{(-1)}{2(1)} = \frac{1}{2}$.
The y-coordinate of the vertex is at
$y = (\frac{1}{2})^2 - (\frac{1}{2}) - 2$
$= \frac{1}{4} - \frac{1}{2} - 2$
$= \frac{1}{4} - \frac{2}{4} - \frac{8}{4} = -\frac{9}{4}$
The vertex is $\left(\frac{1}{2}, -\frac{9}{4}\right)$.
To find the x-intercepts, we set $y = 0$ and solve for x :
$0 = x^2 - x - 2$
$0 = (x - 2)(x + 1)$
$x - 2 = 0$ or $x + 1 = 0$
$x = 2$ $\quad x = -1$
So, the x-intercepts are $(2, 0)$ and $(-1, 0)$.
The y-intercept is $(0, c) = (0, -2)$.

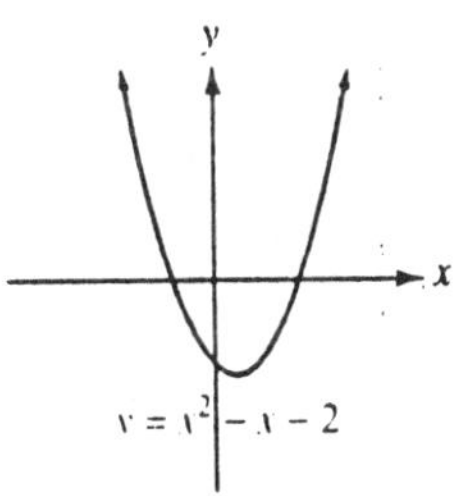

57. $y = -x^2 + 2x + 3$
$a = -1,\ b = 2,\ c = 3$
The x-coordinate of the vertex is at
$x = -\frac{b}{2a} = -\frac{(2)}{2(-1)} = \frac{2}{2} = 1.$
The y-coordinate of the vertex is at
$y = -(1)^2 + 2(1) + 3$
$= -1 + 2 + 3 = 4$
The vertex is $(1, 4)$.
To find the x-intercepts, we set $y = 0$ and solve for x :
$0 = -x^2 + 2x + 3$
$-1(0) = -1(-x^2 + 2x + 3)$
$0 = x^2 - 2x - 3$
$0 = (x - 3)(x + 1)$
$x - 3 = 0$ or $x + 1 = 0$
$x = 3$ $\quad x = -1$
So, the x-intercepts are $(3, 0)$ and $(-1, 0)$.
The y-intercept is $(0, c) = (0, 3)$.

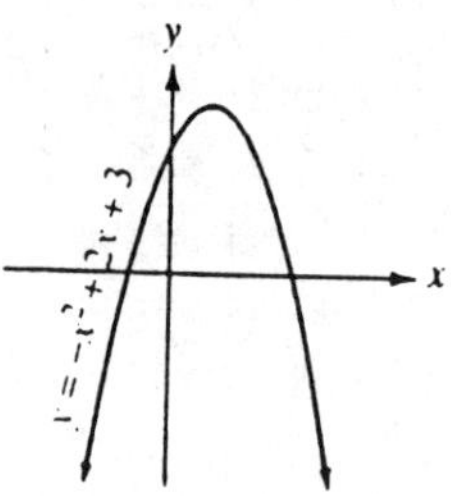

59. $y = 2x^2 + 3x - 2$
$a = 2,\ b = 3,\ c = -2$
The x-coordinate of the vertex is at
$x = -\frac{b}{2a} = -\frac{(3)}{2(2)} = -\frac{3}{4}.$
The y-coordinate of the vertex is at
$y = 2(-\frac{3}{4})^2 + 3(-\frac{3}{4}) - 2$
$= 2(\frac{9}{16}) - \frac{9}{4} - 2$
$= \frac{9}{8} - \frac{9}{4} - 2$
$= \frac{9}{8} - \frac{18}{8} - \frac{16}{8} = -\frac{25}{8}$
The vertex is $\left(-\frac{3}{4}, -\frac{25}{8}\right)$.
To find the x-intercepts, we set $y = 0$ and solve for x :
$0 = 2x^2 + 3x - 2$
$0 = (2x - 1)(x + 2)$
$2x - 1 = 0$ or $x + = 0$
$2x = 1$ $\quad x = -2$
$x = \frac{1}{2}$
So, the x-intercepts are $(\frac{1}{2}, 0)$ and $(-2, 0)$.
The y-intercept is $(0, c) = (0, -2)$.

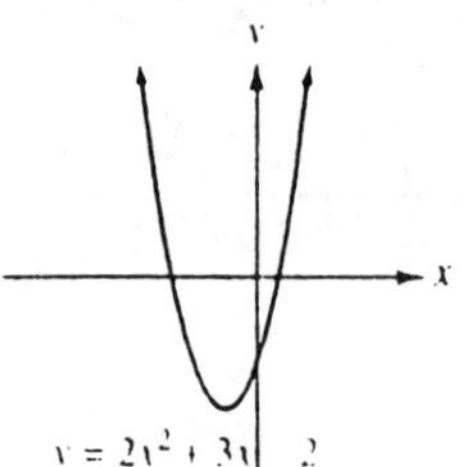

APPLICATIONS

61.

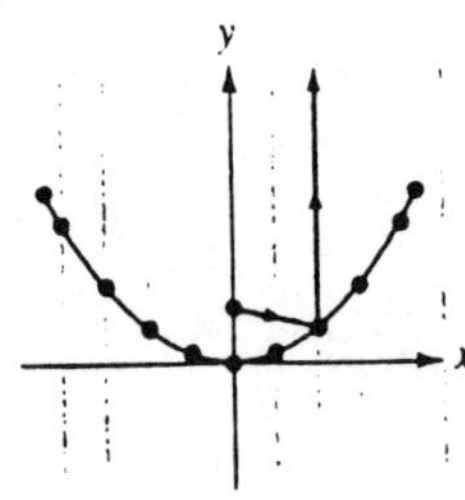

63. For $x = -80$,
$y = 0.005x^2 = 0.005(-80)^2$
$= 0.005(6400) = 32.$
For $x = -60$,
$y = 0.005x^2 = 0.005(-60)^2$
$= 0.005(3600) = 18.$
For $x = -40$,
$y = 0.005x^2 = 0.005(-40)^2$
$= 0.005(1600) = 8.$
For $x = -20$,
$y = 0.005x^2 = 0.005(-20)^2$
$= 0.005(400) = 2.$
For $x = 0$,
$y = 0.005x^2 = 0.005(0)^2$
$= 0.005(0) = 0.$
For $x = 20$,
$y = 0.005x^2 = 0.005(20)^2$
$= 0.005(400) = 2.$
For $x = 40$,
$y = 0.005x^2 = 0.005(40)^2$
$= 0.005(1600) = 8.$
For $x = 60$,
$y = 0.005x^2 = 0.005(60)^2$
$= 0.005(3600) = 18.$
For $x = 80$,
$y = 0.005x^2 = 0.005(80)^2$
$= 0.005(6400) = 32.$

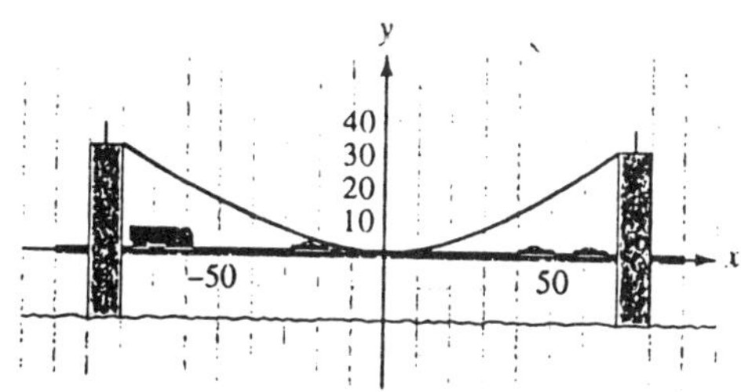

65. a) The revenue is the product of the number of TV's sold, (x), and the price of each TV, ($450 - \frac{1}{6}x$). Thus, revenue is given by

$$R = x(450 - \tfrac{1}{6}x)$$
$$R = 450x - \tfrac{1}{6}x^2$$

We need the x-coordinate of the vertex in order to know the number of TV's that will maximize the revenue the x-coordinate of the vertex is

$$x = -\frac{b}{2a} = -\frac{(450)}{2(-\frac{1}{6})} = 1350.$$

The company must sell 1350 TV's to maximize its revenue.

$$R = 450(1350) - \tfrac{1}{6}(1350)^2$$
$$= 607,500 - 303,750$$
$$= 303,750$$

The maximum revenue is \$303,750.

b) The maximum revenue is given by the y-coordinate of the vertex.

REVIEW

71. $\sqrt{12} + \sqrt{27} = \sqrt{4 \cdot 3} + \sqrt{9 \cdot 3}$
$= \sqrt{4}\sqrt{3} + \sqrt{9}\sqrt{3}$
$= 2\sqrt{3} + 3\sqrt{3}$

73. $\left(\sqrt{3}+1\right)\left(\sqrt{3}-1\right)$
$= \sqrt{3}\sqrt{3} - \sqrt{3} + \sqrt{3} - 1$
$- \sqrt{9} - 1$
$= 3 - 1$
$= 2$

CHAPTER 7 REVIEW

1. a) $35 = 5 \cdot 7$
b) $45 = 3^2 \cdot 5$
c) $96 = 2^5 \cdot 3$
d) $99 = 3^2 \cdot 11$
e) $2050 = 2 \cdot 5^2 \cdot 41$
f) $4096 = 2^{12}$

3. $-a - 7 = -(a + 7)$

5. a) $6x^2y - 24y^2 = 6y(x^2 - 4y^2)$
$= 6y(x - 2y)(x + 2y)$

b) $2x^4 - 162 = 2(x^4 - 81)$
$= 2(x^2 - 9)(x^2 + 9)$
$= 29x - 3)(x + 3)(x^2 + 9)$

c) $-m^2 + 100 = -(m^2 - 100)$
$= -(m - 10)(m + 10)$

7. We can check by multiplying.
$(3x - 4)(2x + 5)$
$= 6x^2 + 15x - 8x - 20$
$= 6x^2 + 7x - 20$

9. a) $x^2 + 2x = 0$
$x(x + 2) = 0$
$x = 0$ or $x + 2 = 0$
$x = -2$
Solutions are $x = 0$ and $x = -2$.

b) $2x^2 - 6x = 0$
$2x(x - 3) = 0$
$2x = 0$ or $x - 3 = 0$
$x = 0$ $x = 3$
Solutions are $x = 0$ and $x = 3$.

c) $x^2 - 9 = 0$
$(x + 3)(x - 3) = 0$
$x + 3 = 0$ or $x - 3 = 0$
$x = -3$ $x = 3$
Solutions are $x = 3$ and $x = -3$.

d) $36p^2 = 25$
$p^2 = \frac{25}{36}$
$p^2 - \frac{25}{36} = 0$
$(p + \frac{5}{6})(p - \frac{5}{6}) = 0$
$p + \frac{5}{6} = 0$ or $p - \frac{5}{6} = 0$
$p = -\frac{5}{6}$ $p = \frac{5}{6}$
Solutions are $p = -\frac{5}{6}$ and $p = \frac{5}{6}$.

e) $a^2 - 7a + 12 = 0$
$(a - 3)(a - 4) = 0$
$a - 3 = 0$ or $a - 4 = 0$
$a = 3$ $a = 4$
Solutions are $a = 3$ and $a = 4$.

f) $t^2 + 4t + 4 = 0$
$(t + 2)^2 = 0$
$t + 2 = 0$
$t = -2$
Solution is $t = -2$.

g) $2x - x^2 + 24 = 0$
$-x^2 + 2x + 24 = 0$
$x^2 - 2x - 24 = 0$
$(x + 4)(x - 6) = 0$
$x + 4 = 0$ or $x - 6 = 0$
$x = -4$ $x = 6$
Solutions are $x = -4$ and $x = 6$.

h) $(t + 1)(8t + 1) = 18t$
$8t^2 + t + 8t + 1 = 18t$
$8t^2 + 9t + 1 = 18t$
$8t^2 - 9t + 1 = 0$
$(8t - 1)(t - 1) = 0$
$8t - 1 = 0$ or $t - 1 = 0$
$8t = 1$ $t = 1$
$t = \frac{1}{8}$
Solutions are $t = \frac{1}{8}$ and $t = 1$.

i) $x^2 - 11x = 12$
$x^2 - 11x - 12 = 0$
$(x + 1)(x - 12) = 0$
$x + 1 = 0$ or $x - 12 = 0$
$x = -1$ $x = 12$
Solutions are $x = -1$ and $x = 12$.

j) $2p^3 = 2p(p+2)$
$2p^3 = 2p^2 + 4p$
$2p^3 - 2p^2 - 4p = 0$
$2p(p^2 - p - 2) = 0$
$p(p+1)(p-2) = 0$
$p = 0 \quad p+1 = 0 \quad p-2 = 0$
$p = -1 \quad p = 2$
Solutions are $p = 0$, $p = -1$, and $p = 2$.

11. **Analyze:** The bomb is starting from a height of 3000 feet and when it hits the target the height of the bomb will be zero. We need to find the time (t) when the height is zero.
Form: $h = 3000 + 40t - 16t^2$
$0 = 3000 + 40t - 16t^2$
Solve:
$0 = -8(-375 - 5t + 2t^2)$
$0 = -375 - 5t + 2t^2$
$0 = 2t^2 - 5t - 375$
$0 = (2t+25)(t-15)$
$2t + 25 = 0 \quad t - 15 = 0$
$2t = -25 \quad t = 15$
$t = -\frac{25}{2}$
State: We disregard the solution $t = -\frac{25}{2}$ since time cannot be negative. Thus, the bomb takes 15 seconds to reach its target.

13. a) $x^2 = 25$
$x = \pm\sqrt{25}$
$x = \pm 5$

b) $x^2 = 400$
$x = \pm\sqrt{400}$
$x = \pm 20$

c) $2x^2 = 18$
$x^2 = 9$
$x = \pm\sqrt{9}$
$x = \pm 3$

d) $4y^2 = 9$
$y^2 = \frac{9}{4}$
$y = \pm\sqrt{\frac{9}{4}}$
$y = \pm\frac{3}{2}$

e) $t^2 = 8$
$t = \pm\sqrt{8}$
$t = \pm\sqrt{4}\sqrt{2}$
$t = \pm 2\sqrt{2}$

f) $2x^2 - 1 = 149$
$2x^2 = 150$
$x^2 = 75$
$x = \pm\sqrt{75}$
$x = \pm\sqrt{25}\sqrt{3}$
$x = \pm 5\sqrt{3}$

15. a) $x^2 = 12$
$x = \pm\sqrt{12}$
$x \approx \pm 3.46$

b) $(x-1)^2 = 55$
$x - 1 = \pm\sqrt{55}$
$x = 1 \pm \sqrt{55}$
$x = 1 + \sqrt{55} \quad x = 1 - \sqrt{55}$
$x \approx 8.42 \quad x \approx -6.42$

17. $x^2 + 4x + 1$ does not factor.

19. $x^2 + 4x + 1 = 0$
$x^2 + 4x = -1$
$\left(\frac{4}{2}\right)^2 = (2)^2 = 4$
$x^2 + 4x + 4 = -1 + 4$
$(x+2)^2 = 3$
$x + 2 = \pm\sqrt{3}$
$x = -2 \pm \sqrt{3}$
$x = -2 + \sqrt{3} \quad x = -2 - \sqrt{3}$
$x \approx -0.27 \quad x \approx -3.73$

21. $3x^2 + 2x - 2 = 0$
$a = 3,\ b = 2,\ c = -2$

$$x = \frac{-b \pm \sqrt{b^2 - 4ac}}{2a}$$

$$x = \frac{-(2) \pm \sqrt{(2)^2 - 4(3)(-2)}}{2(3)}$$

$$x = \frac{-2 \pm \sqrt{4+24}}{6}$$

$$x = \frac{-2 \pm \sqrt{28}}{6}$$

$$x = \frac{-2 \pm \sqrt{4}\sqrt{7}}{6}$$

$$x = \frac{-2 \pm 2\sqrt{7}}{6}$$

$$x = \frac{-1 \pm \sqrt{7}}{3}$$

$x = \frac{-1 + \sqrt{7}}{3}$ $\qquad$ $x = \frac{-1 - \sqrt{7}}{3}$

$x \approx 0.55$ $\qquad$ $x \approx -1.22$

23. **Analyze:** We need to find two sides of a right triangle. We are given that one side is 14 feet longer than the other and that the hypotenuse is 26 feet.
Form: Use Pythagorean theorem.
$a^2 + b^2 = c^2$
$w^2 + (w + 14)^2 = 26^2$
Solve:
$w^2 + w^2 + 28w + 196 = 676$
$2w^2 + 28w + 196 = 676$
$2w^2 + 28w - 480 = 0$
$w^2 + 28w - 240$
$(w - 24)(w + 10) = 0$
$w - 24 = 0$ $\qquad$ $w + 10 = 0$
$w = 24$ $\qquad$ $w = -10$
State: We disregard the solution $w = -10$ since the side length cannot be negative. Thus, the width of the gate frame is 24 feet and the length is $w + 14 = (24) + 14 = 38$ feet.

25. a) The x-intercepts are $(-3, 0)$ and $(1, 0)$.

b) The y-intercept is $(0, -3)$.

c) The vertex is $(-1, -4)$.

d)

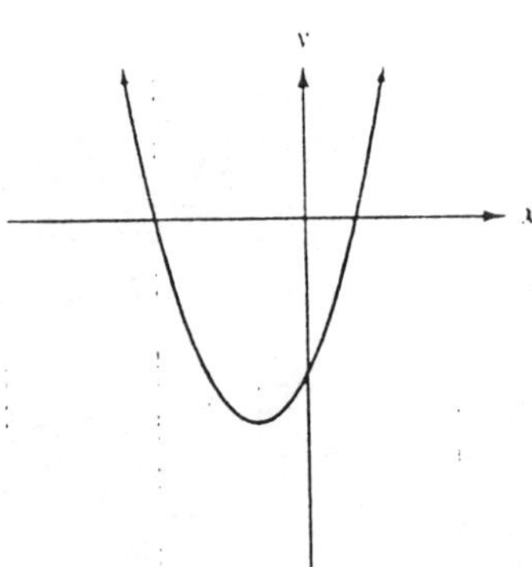

27. a) $y = x^2 + 8x + 10$
$= x^2 + 8x + 16 - 16 + 10$
$= x^2 + 8x + 16 - 6$
$= (x + 4)^2 - 6$
Parabola opens up with a vertex of $(-4, -6)$.

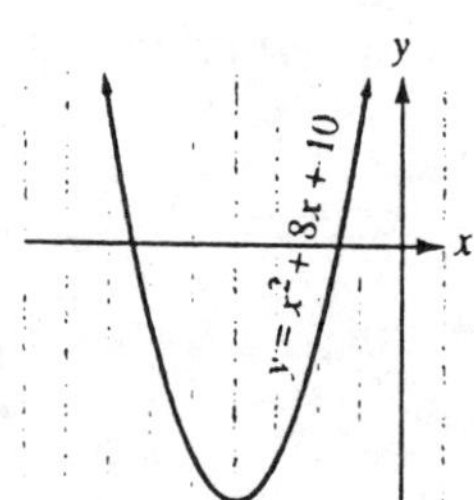

b) $y = -2x^2 - 4x - 6$
$= -2(x^2 + 2x) - 6$
$= -2(x^2 + 2x + 1 - 1) - 6$
$= -2(x^2 + 2x + 1) + 2 - 6$
$= -2(x^2 + 2x + 1) - 4$
$= -2(x + 1)^2 - 4$
Parabola opens down with a vertex of $(-1, -4)$.

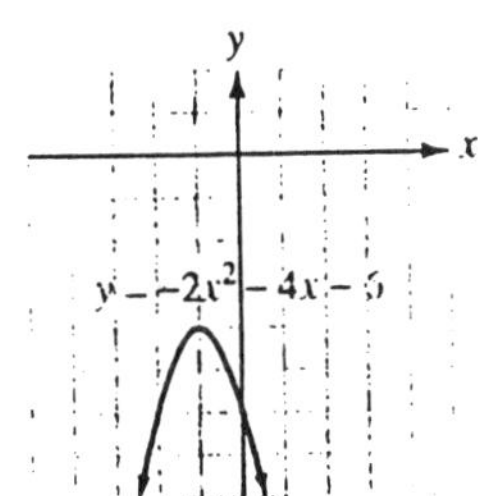

29. $y = x^2 + 6x + 5$
$a = 1, \ b = 6, \ c = 5$
To find the x-intercepts, we set $y = 0$ and solve for x :
$0 = x^2 + 6x + 5$
$0 = (x + 1)(x + 5)$
$x + 1 = 0 \qquad x + 5 = 0$
$x = -1 \qquad x = -5$
The x-intercepts are $(-1, 0)$ and $(-5, 0)$.
The y-intercept is $(0, c) = (0, 5)$.

STUDY SET Section 8.1

VOCABULARY

1. comparison; quotient

3. equal

CONCEPTS

5. speed of a car: $\frac{\text{miles}}{\text{hour}}$;
team's winning percentage: $\frac{\text{wins}}{\text{games}}$;
GPA: $\frac{\text{credits earned}}{\text{credits attempted}}$.

NOTATION

7. a) $\frac{3}{11}$
b) $\frac{27}{32}$

PRACTICE

9. $\frac{5}{7}$

11. $\frac{17}{34} = \frac{17}{2 \cdot 17} = \frac{1}{2}$

13. $\frac{22}{33} = \frac{2 \cdot 11}{3 \cdot 11} = \frac{2}{3}$

15. $\frac{7}{24.5} = \frac{7 \cdot 10}{24.5 \cdot 10} = \frac{70}{245} = \frac{2 \cdot 35}{7 \cdot 35} = \frac{2}{7}$

17. $\frac{4 \text{ ounces}}{12 \text{ ounces}} = \frac{4 \cdot 1}{4 \cdot 3} = \frac{1}{3}$

19. $\frac{12 \text{ minutes}}{1 \text{ hour}} = \frac{12 \text{ minutes}}{60 \text{ minutes}} = \frac{12 \cdot 1}{12 \cdot 5} = \frac{1}{5}$

21. $\frac{3 \text{ days}}{1 \text{ week}} = \frac{3 \text{ days}}{7 \text{ days}} = \frac{3}{7}$

23. $\frac{18 \text{ months}}{2 \text{ years}} = \frac{18 \text{ months}}{24 \text{ months}} = \frac{6 \cdot 3}{6 \cdot 4} = \frac{3}{4}$

25. The total amount of the budget is found by adding Rent, Food, Gas and Electric, Phone, and Entertainment.
Total Budget = \$750 + \$652 + \$188 + \$125 + \$110 = \$1825

27. $\frac{\text{Entertainment}}{\text{Total Budget}} = \frac{\$110}{\$1825} = \frac{5 \cdot 22}{5 \cdot 365} = \frac{22}{365}$

29. The total amount of deductions is found by adding Medical, Real Estate Taxes, Contributions, Mortgage Deduction, and Union Dues.
Total Budget = \$995 + \$1245 + \$1680 + \$4580 + \$225 = \$8725

31. $\frac{\text{Contributions}}{\text{Total Deductions}} = \frac{\$1680}{\$8725} = \frac{5 \cdot 336}{5 \cdot 1745} = \frac{336}{1745}$

APPLICATIONS

33. The ratio of faculty to students is 125 to 2,000.

$$\frac{\text{faculty}}{\text{students}} = \frac{125}{2000} = \frac{125 \cdot 1}{125 \cdot 16} = \frac{1}{16}$$

The ratio of faculty to students is $\frac{1}{16}$.

35. Infant blood pressure:

$$\frac{\text{infant Systolic}}{\text{infant Diastolic}} = \frac{80}{45}.$$

Age 30 blood pressure:

$$\frac{\text{age 30 Systolic}}{\text{age 30 Diastolic}} = \frac{120}{80}.$$

Age 40 blood pressure:

$$\frac{\text{age 40 Systolic}}{\text{age 40 Diastolic}} = \frac{140}{85}.$$

37. The ratio of dollars to gallons is

$$\frac{\$21.59}{17 \text{ gallons}} = \frac{\$21.59 \cdot 100}{17 \cdot 100 \text{ gallons}} = \frac{\$2159}{1700 \text{ gallons}}$$

$= \$1.27$ per gallon.

39. The ratio of cost to rides is

$\frac{\$36}{63 \text{ rides}} = \frac{3600¢}{63 \text{ rides}} = 57¢$ per ride.

41. The unit cost of a 6-ounce can is

$\frac{89¢}{6 \text{ ounces}} = 14.83¢$ per ounce.

The unit cost of an 8-ounce can is

$\frac{\$1.19}{8 \text{ ounces}} = \frac{119¢}{8 \text{ ounces}} = 14.88¢$ per ounce.

The 6-ounce can is a better buy.

43. The first student's reading speed is

$\frac{54 \text{ pages}}{40 \text{ minutes}} = 1.35$ pages per minute.

The second student's reading speed is

$\frac{80 \text{ pages}}{62 \text{ minutes}} = 1.29$ pages per minute.

The first student reads faster.

45. The ratio of gallons to minutes is

$\frac{11{,}880 \text{ gallons}}{27 \text{ minutes}} = 440$ gallons per minute.

47. We can write the sales tax rate as the ratio

$$\text{sales tax} = \frac{\text{amount of sales tax}}{\text{cost of sweater, without tax}}$$

The amount of sales tax is

$\$36.75 - \$35.00 = \$1.75$,

so, the sales tax rate $= \frac{\$1.75}{\$35} = .05$

The tax rate is .05, or 5%.

49. The rate of speed is

$\frac{325 \text{ miles}}{5 \text{ hours}} = \frac{5 \cdot 65 \text{ miles}}{5 \cdot 1 \text{ hours}} = \frac{65 \text{ miles}}{1 \text{ hour}}$.

The car's rate of speed is 65 mph.

51. The first car's mpg rating is

$\frac{1{,}235 \text{ miles}}{51.3 \text{ gallons}} = 24.1$ mpg.

The second car's mpg rating is

$\frac{1{,}456 \text{ miles}}{55.78 \text{ gallons}} = 26.1$ mpg.

The second car has the better mpg rating.

53. a) The 1920 marriage/divorce ratio is 7.5:1. This is written as a fraction as

$\frac{7.5 \text{ marriages}}{1 \text{ divorce}}$.

This means that for every 7.5 marriages, there is 1 divorce.

b) 2, 1

c) In 1920, for every 7.5 marriages there was 1 divorce. In 1996, for every 2 marriages there was 1 divorce. This means that the number of divorces for a given number of marriages is increasing.

55. The scale 1:1,000,000 means that one inch on the map represents 1,000,000 inches on the earth.

REVIEW

59. $0.2x + 4 = 3.8$

$0.2x + 4 - 4 = 3.8 - 4$

$0.2x = -0.2$

$\frac{0.2x}{0.2} = \frac{-0.2}{0.2}$

$x = -1$

61. $3(x + 2) = 24$

$3x + 6 = 24$

$3x + 6 - 6 = 24 - 6$

$3x = 18$

$\frac{3x}{3} = \frac{18}{3}$

$x = 6$

63. $\sqrt{16} = 4$

65. $\sqrt{81x^4y^8} = \sqrt{81}\sqrt{x^4}\sqrt{y^8} = 9x^2y^4$

STUDY SET Section 8.2
VOCABULARY

1. proportion

3. means

5. shape

CONCEPTS

7. ad; bc

9. a) For United States:

$\frac{\text{defense spending}}{\text{gross domestic product}} = \frac{\$278}{\$7{,}316}$

$= \frac{2 \cdot 139}{2 \cdot 3658} = \frac{139}{3{,}658}$

For China:

$\frac{\text{defense spending}}{\text{gross domestic product}} = \frac{\$32}{\$561}$

$= \frac{32}{561}$

b) $\frac{139}{3{,}658} = \frac{32}{561}$

The product of the extremes is

$139 \cdot 561 = 32 \cdot 3{,}658$

$77{,}979 \neq 117{,}056$

Since the products are not equal, the ratios of the defense spending to the gross domestic product are not a proportion.

The ratio for the U.S. is $\frac{139}{3{,}658} = 0.038$.

The ratio for China is $\frac{32}{561} = 0.057$.

China's defense spending is larger than the United States'.

NOTATION

11. $\frac{12}{18} = \frac{x}{24}$

$12 \cdot 24 = 18 \cdot x$

$288 = 18x$

$\frac{288}{18} = \frac{18x}{18}$

$16 = x$

13. triangle

PRACTICE

15. $\frac{9}{7} = \frac{81}{70}$

$9 \cdot 70 = 7 \cdot 81$

$630 \neq 560$

The statement is not a proportion.

17. $\frac{7}{3} = \frac{14}{6}$

$7 \cdot 6 = 3 \cdot 14$

$42 = 42$

The statement is a proportion.

19. $\frac{9}{19} = \frac{38}{80}$

$9 \cdot 80 = 19 \cdot 38$

$720 \neq 722$

The statement is not a proportion.

21. $\frac{10.4}{3.6} = \frac{41.6}{14.4}$

$10.4 \cdot 14.4 = 3.6 \cdot 41.6$

$149.76 = 149.76$

The statement is a proportion.

23. $\frac{2}{3} = \frac{x}{6}$

$2 \cdot 6 = 3 \cdot x$

$12 = 3x$

$\frac{12}{3} = \frac{3x}{3}$

$4 = x$

25. $\frac{5}{10} = \frac{3}{c}$

$5 \cdot c = 10 \cdot 3$

$5c = 30$

$\frac{5c}{5} = \frac{30}{5}$

$c = 6$

27. $\frac{6}{x} = \frac{8}{4}$

$6 \cdot 4 = x \cdot 8$

$24 = 8x$

$\frac{24}{8} = \frac{8x}{8}$

$3 = x$

29. $\frac{x}{3} = \frac{9}{3}$

$x \cdot 3 = 3 \cdot 9$

$3x = 27$

$\frac{3x}{3} = \frac{27}{3}$

$x = 9$

31. $\frac{x+1}{5} = \frac{3}{15}$

$(x+1) \cdot 15 = 5 \cdot 3$

$15x + 15 = 15$

$15x + 15 - 15 = 15 - 15$

$15x = 0$

$\frac{15x}{15} = \frac{0}{15}$

$x = 0$

33. $\frac{x+3}{12} = \frac{-7}{6}$

$(x+3) \cdot 6 = 12 \cdot (-7)$

$6x + 18 = -84$

$6x + 18 - 18 = -84 - 18$

$6x = -102$

$\frac{6x}{6} = \frac{-102}{6}$

$x = -17$

35. $\frac{4-x}{13} = \frac{11}{26}$

$(4-x) \cdot 26 = 13 \cdot 11$

$104 - 26x = 143$

$104 - 104 - 26x = 143 - 104$

$-26x = 39$

$\frac{-26x}{-26} = \frac{39}{-26}$

$x = -\frac{3 \cdot 13}{2 \cdot 13} = -\frac{3}{2}$

37. $\frac{2x+1}{18} = \frac{14}{3}$

$(2x+1) \cdot 3 = 18 \cdot 14$

$6x + 3 = 252$

$6x + 3 - 3 = 252 - 3$

$6x = 249$

$\frac{6x}{6} = \frac{249}{6}$

$x = \frac{3 \cdot 83}{3 \cdot 2} = \frac{83}{2}$.

APPLICATIONS

39. **Analyze:** We know the cost of 3 pints of yogurt, we are to find the cost of 51 pints of yogurt.

Form: Let c represent the cost of 51 pints of yogurt. The ratios of the numbers of pints to their costs are equal.

$\frac{3}{1} = \frac{51}{c}$

Solve: $3 \cdot c = 1 \cdot 51$

$3c = 51$

$\frac{3c}{3} = \frac{51}{3}$

$c = 17$

State: 51 pints of yogurt will cost \$17.

Check: $3(17) = 51$

$51 = 51$

41. **Analyze:** We know the cost of a 30-second TV ad, we are to find the cost of a 45-second ad.

Form: Let c represent the cost of a 45-second ad. The ratios of the length and the cost of the ads are equal.

$\frac{30}{1.2} = \frac{45}{c}$

Solve: $30 \cdot c = 1.2 \cdot 45$

$30c = 54$

$\frac{30c}{30} = \frac{54}{30}$

$c = 1.8$

State: The cost of a 45-second ad is \$1.8 million.

Check: $30(1.8) = 54$

$54 = 54$

43. **Analyze:** We want to maintain the same concentration of pure essence.

Form: Let d represent the amount of pure essence to be added to 56 drops of alcohol.

The ratios of the drops of pure essence and drops of alcohol are equal.

$$\frac{3}{7} = \frac{d}{56}$$

Solve: $3 \bullet 56 = 7 \bullet d$

$$168 = 7d$$
$$\frac{168}{7} = \frac{7d}{7}$$
$$24 = d$$

State: The mixture should contain 24 drops of pure essence.

Check: $7(24) = 168$

$$168 = 168$$

45. **Analyze:** The cook needs to maintain the same ratios between each ingredient and the number of servings in the original recipe.

Form: Let

$c =$ cups of chicken broth,

$r =$ cups of rice,

$x =$ cups of onions,

s = cups of shredded carrots,

$l =$ cups of light cream,

$f =$ tablespoons of flour, and

$p =$ teaspoons of pepper.

The ratios of the amount of each ingredient to the number of servings are equal.

Chicken: $\frac{3}{6} = \frac{c}{15}$ Rice: $\frac{2/3}{6} = \frac{r}{15}$

Onions: $\frac{1/4}{6} = \frac{x}{15}$ Carrots: $\frac{1/2}{6} = \frac{s}{15}$

Light Cream: $\frac{1}{6} = \frac{l}{15}$ Flour: $\frac{2}{6} = \frac{f}{15}$

Pepper: $\frac{1/8}{6} = \frac{p}{15}$

Solve:

Chicken:

$$\frac{3}{6} = \frac{c}{15}$$
$$3 \bullet 15 = 6 \bullet c$$
$$45 = 6c$$
$$\frac{45}{6} = \frac{6c}{6}$$
$$\frac{3 \bullet 15}{3 \bullet 2} = c$$
$$\frac{15}{2} = c$$

Rice:

$$\frac{2/3}{6} = \frac{r}{15}$$
$$\frac{2}{3} \bullet 15 = 6 \bullet r$$
$$10 = 6r$$
$$\frac{10}{6} = \frac{6r}{6}$$
$$\frac{2 \bullet 5}{2 \bullet 3} = r$$
$$\frac{5}{3} = r$$

Onions:

$$\frac{1/4}{6} = \frac{x}{15}$$
$$\frac{1}{4} \bullet 15 = 6 \bullet x$$
$$\frac{15}{4} = 6x$$
$$\frac{15/4}{6} = \frac{6x}{6}$$
$$\frac{15}{4} \bullet \frac{1}{6} = x$$
$$\frac{15}{24} = x$$
$$\frac{3 \bullet 5}{3 \bullet 8} = x$$
$$\frac{5}{8} = x$$

Carrots:

$$\frac{1/2}{6} = \frac{s}{15}$$
$$\frac{1}{2} \bullet 15 = 6 \bullet s$$
$$\frac{15}{2} = 6s$$
$$\frac{15/2}{6} = \frac{6s}{6}$$
$$\frac{15}{2} \bullet \frac{1}{6} = s$$
$$\frac{15}{12} = s$$
$$\frac{3 \bullet 5}{3 \bullet 4} = s$$
$$\frac{5}{4} = s$$

Light Cream:

$$\frac{1}{6} = \frac{l}{15}$$
$$1 \bullet 15 = 6 \bullet l$$
$$15 = 6l$$
$$\frac{15}{6} = \frac{6l}{6}$$
$$\frac{3 \bullet 5}{3 \bullet 2} = l$$
$$\frac{5}{2} = l$$

Flour:

$$\frac{2}{6} = \frac{f}{15}$$
$$2 \bullet 15 = 6 \bullet f$$
$$30 = 6f$$
$$\frac{30}{6} = \frac{6f}{6}$$
$$5 = f$$

Pepper:

$$\frac{1/8}{6} = \frac{p}{15}$$
$$\frac{1}{8} \bullet 15 = 6 \bullet p$$
$$\frac{15}{8} = 6p$$
$$\frac{15/8}{6} = \frac{6p}{6}$$
$$\frac{15}{8} \bullet \frac{1}{6} = p$$
$$\frac{15}{48} = p$$
$$\frac{3 \bullet 5}{3 \bullet 16} = p$$
$$\frac{5}{16} = p$$

State: The cook should use $7\frac{1}{2}$ cups of chicken broth, $1\frac{2}{3}$ cups of rice, $\frac{5}{8}$ cups of onions, $1\frac{1}{4}$ cups of shredded carrots, $2\frac{1}{2}$ cups of light cream, 5 tablespoons of flour, and $\frac{5}{16}$ teaspoons of pepper.

47. **Analyze:** 95% is 95 out of 100. This means 95 out of 100 parts are not defective. Therefore, 5 out of 100 parts will be defective. We are to find how many defective parts there will be if there are 940 total parts.
Form: Let d represent the number of defective parts out of 940.
The ratios of defective parts and total parts are equal.

$$\frac{5}{100} = \frac{d}{940}$$

Solve: $5 \cdot 940 = 100 \cdot d$

$$4700 = 100d$$
$$\frac{4700}{100} = \frac{100d}{100}$$
$$47 = d$$

State: There will be 47 defective parts in a run of 940 pieces.
Check: $47(100) = 4700$
$4700 = 4700$

49. **Analyze:** We know the number of miles the car can travel on 1 gallon of gas. We are to find how much gas is needed to go 315 miles.
Form: Let g represent the number of gallons of gas needed to go 315 miles.
The ratios of the number of gallons of gas and miles traveled are equal.

$$\frac{42 \text{ miles}}{1 \text{ gallon}} = \frac{315 \text{ miles}}{g \text{ gallons}}$$

Solve: $42 \cdot g = 1 \cdot 315$

$$42g = 315$$
$$\frac{42g}{42} = \frac{315}{42}$$
$$g = \frac{15 \cdot 21}{2 \cdot 21} = \frac{15}{2} = 7\frac{1}{2}$$

State: $7\frac{1}{2}$ gallons of gas is needed to travel 315 miles.
Check: $42(\frac{15}{2}) = 315$
$\frac{630}{2} = 315$
$315 = 315$

51. **Analyze:** If Bill missed 10 hours, he worked 30 hours. We know the amount Bill earns for a 40-hour week. We are to find the amount bill earned for a 30-hour week.
Form: Let a represent the amount Bill earned for a 30-hour week.
The ratios of the number of hours worked and the amount earned are equal.

$$\frac{40 \text{ hours}}{\$412} = \frac{30 \text{ hours}}{\$a}$$

Solve: $40 \cdot a = 412 \cdot 30$

$$40a = 12,360$$
$$\frac{40a}{40} = \frac{12,360}{40}$$
$$a = 309$$

State: Bill earned \$309 last week.
Check: $40(309) = 12,360$
$12,360 = 12,360$

53. **Analyze:** We know the length of an N-scale caboose. We are to find the length of the real caboose.
Form: Let l represent the length of the real caboose.
The ratios of the real length and the scale length are equal.

$$\frac{169 \text{ feet}}{1 \text{ feet}} = \frac{l \text{ feet}}{3.5 \text{ inches}}$$

Convert inches to feet:

$$\frac{169 \text{ feet}}{1 \text{ feet}} = \frac{l}{\frac{3.5}{12} \text{ feet}}$$

Solve: $\frac{169}{1} = \frac{l}{\frac{350}{120}}$

$$169 \cdot \frac{350}{120} = 1 \cdot l$$
$$\frac{5915}{120} = l$$
$$49.29 = l$$

To covert 0.29 feet into inches,

$$\frac{0.29 \text{ ft}}{1} \cdot \frac{12 \text{ in.}}{1 \text{ ft}} \approx 3.48$$

State: The real caboose is about 49 ft $3\frac{1}{2}$ in.
Check: $49.29 = 169(\frac{35}{120})$
$49.29 = 49.29$

55. Analyze: We know the calories fat and protein in a 10-oz. shake. We are to find the calories, fat, and protein in a 16-oz. shake.

Form: Let

c = amount of calories in a 16-oz. shake,

f = amount of fat in a 16-oz shake, and

p = amount of protein in a 16-oz. shake.

The ratios of the amount of calories, fat, and protein in a 10-oz. shake and a 16-oz. shake are equal.

calories: $\frac{355}{10} = \frac{c}{16}$

fat: $\frac{8}{10} = \frac{f}{16}$

protein: $\frac{9}{10} = \frac{p}{16}$

Solve:

calories:

$$\frac{355}{10} = \frac{c}{16}$$
$$355 \cdot 16 = 10 \cdot c$$
$$5680 = 10c$$
$$\frac{5680}{10} = \frac{10c}{10}$$
$$568 = c$$

fat:

$$\frac{8}{10} = \frac{f}{16}$$
$$8 \cdot 16 = 10 \cdot f$$
$$128 = 10f$$
$$\frac{128}{10} = \frac{10f}{10}$$
$$12.8 = f$$
$$13 \approx f$$

protein:

$$\frac{9}{10} = \frac{p}{16}$$
$$9 \cdot 16 = 10 \cdot p$$
$$144 = 10p$$
$$\frac{144}{10} = \frac{10p}{10}$$
$$14.4 = p$$
$$14 \approx p$$

State: There are 568 calories, 13 gm of fat, and 14 gm of protein in a 16-oz. shake.

57. If the directions are correct, then we would want 6 gallons to be 50 times 16 oz. First convert 6 gallons to ounces.

$$6 \text{ gal.} \cdot \frac{128 \text{ oz}}{1 \text{ gal}} = 768 \text{ oz.}$$

Now, multiply 16 by 50.

$$16(50) = 800 \text{ oz.}$$

768 oz. is not exactly 800 oz., but it's close.

59. Analyze: Since the triangles have the same shape, they are similar. So, the lengths of their corresponding sides are in proportion.

Form: Let h represent the height of the tree.

$$\frac{h \text{ feet}}{6 \text{ feet}} = \frac{26 \text{ feet}}{4 \text{ feet}}$$

Solve: $\frac{h}{6} = \frac{26}{4}$

$$h \cdot 4 = 6 \cdot 26$$
$$4h = 156$$
$$\frac{4h}{4} = \frac{156}{4}$$
$$h = 39$$

State: The height of the tree is 39 feet.

Check: $\frac{39}{6} = 6.5$

$$\frac{156}{4} = 6.5$$

So, the ratios are the same.

61. Analyze: Since the triangles are similar, the lengths of their corresponding sides are in proportion.

Form: We are given that w is the width of the river. By similar triangles:

$$\frac{w}{20} = \frac{75}{32}$$

Solve: $w \cdot 32 = 20 \cdot 75$

$$32w = 1500$$
$$\frac{32w}{32} = \frac{1500}{32} = \frac{375 \cdot 4}{8 \cdot 4} = \frac{375}{8}$$
$$w = 46\tfrac{7}{8}$$

State: The river is $46\frac{7}{8}$ feet wide.

Check: $\frac{46\frac{7}{8}}{20} = 2.34$

$$\frac{75}{32} = 2.34$$

So, the ratios are the same.

63. **Analyze:** The two situations can be drawn with right triangles with descent as the height and horizontal distance as the base. Since the triangles are similar, the lengths of their corresponding sides are in proportion.
Form: Let x represent the amount of altitude lost (descent) when the plane's horizontal distance is 5 miles.
By similar triangles:

$$\frac{1350}{x} = \frac{1}{5}$$

Solve: $1350 \cdot 5 = x \cdot 1$

$6,750 = x$

State: The plane will descend 6,750 ft.
Check: $\frac{1350}{6{,}750} = 0.2$

$\frac{1}{5} = 0.2$

So, the ratios are the same.

REVIEW

67. $\frac{9}{10} = \frac{9 \cdot 10}{10 \cdot 10} = \frac{90}{100} = 90\%$

69. $33\frac{1}{3}\% = 33.\overline{3}\% = .33\overline{3} = .\overline{3} = \frac{1}{3}$

71. 30% of 1,600 is x.
$30\% \cdot 1600 = x$
$0.30 \cdot 1600 = x$
$480 = x$

73. **Analyze:** We know the original price and the percent of discount. We are to find the sale price.
Form: Let $x =$ the sale price. Then,

sale price	=	original price	–	discount

$x = \$98 - 0.25(\$98)$

Solve: $x = 98 - 0.25(98)$

$x = 98 - 24.5$

$x = 73.5$

State: The dress was \$73.50.

STUDY SET Section 8.3

VOCABULARY

1. numerator

3. 0

5. negatives

CONCEPTS

7. $\frac{ac}{bc} = \frac{a}{b}$

9. factor; common

NOTATION

11. $\frac{x^2+5x-6}{x^2-1} = \frac{(x+6)(x-1)}{(x+1)(x-1)}$

$= \frac{x+6}{x+1}$

PRACTICE

13. $\frac{8}{10} = \frac{4 \cdot 2}{5 \cdot 2} = \frac{4}{5}$

15. $\frac{28}{35} = \frac{4 \cdot 7}{5 \cdot 7} = \frac{4}{5}$

17. $\frac{8}{52} = \frac{2 \cdot 4}{13 \cdot 4} = \frac{2}{13}$

19. $\frac{10}{45} = \frac{2 \cdot 5}{9 \cdot 5} = \frac{2}{9}$

21. $-\frac{18}{54} = -\frac{1 \cdot 18}{3 \cdot 18} = -\frac{1}{3}$

23. $\frac{4x}{2} = \frac{2 \cdot 2x}{2 \cdot 1} = \frac{2x}{1} = 2x$

25. $-\frac{6x}{18} = -\frac{6 \cdot x}{6 \cdot 3} = -\frac{x}{3}$

27. $\frac{45}{9a} = \frac{9 \cdot 5}{9 \cdot a} = \frac{5}{a}$

29. $\frac{5+5}{5z} = \frac{10}{5z} = \frac{5 \cdot 2}{5 \cdot z} = \frac{2}{z}$

31. $\frac{(3+4)a}{24-3} = \frac{7a}{21} = \frac{7 \cdot a}{7 \cdot 3} = \frac{a}{3}$

33. $\frac{2x}{3x} = \frac{2 \cdot x}{3 \cdot x} = \frac{2}{3}$

35. $\frac{6x^2}{4x^2} = \frac{2 \cdot 3 \cdot x \cdot x}{2 \cdot 2 \cdot x \cdot x} = \frac{3}{2}$

37. $\frac{2x^2}{3y}$ is in lowest terms.

39. $\frac{15x^2y}{5xy^2} = \frac{3 \cdot 5 \cdot x \cdot x \cdot y}{1 \cdot 5 \cdot x \cdot y \cdot y} = \frac{3x}{y}$

41. $\frac{28x}{32y} = \frac{4 \cdot 7 \cdot x}{4 \cdot 8 \cdot y} = \frac{7x}{8y}$

43. $\frac{x+3}{3(x+3)} = \frac{1 \cdot (x+3)}{3 \cdot (x+3)} = \frac{1}{3}$

45. $\frac{5x+35}{x+7} = \frac{5 \cdot (x+7)}{1 \cdot (x+7)} = \frac{5}{1} = 5$

47. $\frac{x^2+3x}{2x+6} = \frac{x \cdot (x+3)}{2 \cdot (x+3)} = \frac{x}{2}$

49. $\frac{15-3x^2}{25y-5xy} = \frac{3 \cdot x \cdot (5-x)}{5 \cdot y \cdot (5-x)} = \frac{3x}{5y}$

51. $\frac{6a-6b+6c}{9a-9b+9c} = \frac{6 \cdot (a-b+c)}{9 \cdot (a-b+c)} = \frac{6}{9}$

$= \frac{2 \cdot 3}{3 \cdot 3} = \frac{2}{3}$

53. $\frac{x-7}{7-x} = \frac{x-7}{-x+7} = \frac{1 \cdot (x-7)}{-1 \cdot (x-7)}$

$= \frac{1}{-1} = -1$

55. $\frac{6x-3y}{3y-6x} = \frac{3 \cdot (2x-y)}{3 \cdot (y-2x)} = \frac{2x-y}{y-2x} = \frac{2x-y}{-2x+y}$

$= \frac{1 \cdot (2x-y)}{-1 \cdot (2x-y)} = \frac{1}{-1} = -1$

57. $\frac{a+b-c}{c-a-b} = \frac{a+b-c}{-a-b+c} = \frac{1 \cdot (a+b-c)}{-1 \cdot (a+b-c)}$

$= \frac{1}{-1} = -1$

59. $\frac{x^2+3x+2}{x^2+x-2} = \frac{(x+2) \cdot (x+1)}{(x+2) \cdot (x-1)} = \frac{x+1}{x-1}$

61. $\frac{x^2-8x+15}{x^2-x-6} = \frac{(x-5) \cdot (x-3)}{(x+2) \cdot (x-3)} = \frac{x-5}{x+2}$

63. $\frac{2x^2-8x}{x^2-6x+8} = \frac{2x \cdot (x-4)}{(x-2) \cdot (x-4)} = \frac{2x}{x-2}$

65. $\frac{xy+2x^2}{2xy+y^2} = \frac{x\bullet(y+2x)}{y\bullet(2x+y)} = \frac{x\bullet(y+2x)}{y\bullet(y+2x)}$

$= \frac{x}{y}$

67. $\frac{x^2+3x+2}{x^3+x^2} = \frac{(x+2)\bullet(x+1)}{x^2\bullet(x+1)} = \frac{x+2}{x^2}$

69. $\frac{x^2-8x+16}{x^2-16} = \frac{(x-4)\bullet(x-4)}{(x+4)\bullet(x-4)} = \frac{x-4}{x+4}$

71. $\frac{2x^2-8}{x^2-3x+2} = \frac{2\bullet(x^2-4)}{(x-1)\bullet(x-2)}$

$= \frac{2\bullet(x-2)\bullet(x-2)}{(x-1)\bullet(x-2)} = \frac{2(x-2)}{x-1}$

73. $\frac{x^2-2x-15}{x^2+2x-15} = \frac{(x-5)\bullet(x+3)}{(x+5)\bullet(x-3)}$

Since no factors cancel, it is already in lowest terms.

75. $\frac{x^2-3(2x-3)}{9-x^2} = \frac{x^2-6x+9}{-x^2+9}$

$= \frac{(x-3)\bullet(x-3)}{-1\bullet(x^2-9)}$

$= \frac{(x-3)\bullet(x-3)}{-1\bullet(x-3)\bullet(x+3)}$

$= -\frac{x-3}{x+3}$

77. $\frac{4(x+3)+4}{3(x+2)+6} = \frac{4x+12+4}{3x+6+6} = \frac{4x+16}{3x+12}$

$= \frac{4\bullet(x+4)}{3\bullet(x+4)} = \frac{4}{3}$

79. $\frac{x^2-9}{(2x+3)-(x+6)} = \frac{x^2-9}{2x+3-x-6}$

$= \frac{x^2-9}{x-3} = \frac{(x-3)\bullet(x+3)}{x-3}$

$= \frac{x+3}{1} = x+3$

APPLICATIONS

81. pitch $= \frac{\text{rise}}{\text{run}} = \frac{x^2+4x+4}{x^2-4}$

$= \frac{(x+2)\bullet(x+2)}{(x+2)\bullet(x-2)} = \frac{x+2}{x-2}$

REVIEW

85. $(a+b)+c = a+(b+c)$

87. One of them is zero.

$(0)\bullet b = 0$

$a\bullet(0) = 0$

89. $\frac{5}{3}$

91. The number. $x+0=x$

STUDY SET Section 8.4

VOCABULARY

1. numerator

CONCEPTS

3. numerators; denominators

5. 1

7. divisor; multiply

NOTATION

9. $\frac{x^2+x}{3x-6} \cdot \frac{x-2}{x+1} = \frac{(x^2+x)(x-2)}{(3x-6)(x+1)}$

$= \frac{x(x+1)(x-2)}{3(x-2)(x+1)}$

$= \frac{x}{3}$

PRACTICE

11. $\frac{5}{7} \cdot \frac{9}{13} = \frac{5 \cdot 9}{7 \cdot 13} = \frac{45}{91}$

13. $\frac{25}{35} \cdot \frac{-21}{55} = \frac{25 \cdot (-21)}{35 \cdot 55}$

$= -\frac{5 \cdot 5 \cdot 3 \cdot 7}{5 \cdot 7 \cdot 5 \cdot 11}$

$= -\frac{\cancel{5} \cdot \cancel{5} \cdot 3 \cdot \cancel{7}}{\cancel{5} \cdot \cancel{7} \cdot \cancel{5} \cdot 11}$

$= -\frac{3}{11}$

15. $\frac{2}{3} \cdot \frac{15}{2} \cdot \frac{1}{7} = \frac{2 \cdot 15 \cdot 1}{3 \cdot 2 \cdot 7} = \frac{2 \cdot 3 \cdot 5}{2 \cdot 3 \cdot 7}$

$= \frac{\cancel{2} \cdot \cancel{3} \cdot 5}{\cancel{2} \cdot \cancel{3} \cdot 7} = \frac{5}{7}$

17. $\frac{3x}{y} \cdot \frac{y}{2} = \frac{3 \cdot x \cdot y}{y \cdot 2} = \frac{3 \cdot x \cdot \cancel{y}}{\cancel{y} \cdot 2} = \frac{3x}{2}$

19. $\frac{5y}{7} \cdot \frac{7x}{5z} = \frac{5 \cdot y \cdot 7 \cdot x}{7 \cdot 5 \cdot z}$

$= \frac{\cancel{5} \cdot \cancel{7} \cdot y \cdot x}{\cancel{7} \cdot \cancel{5} \cdot z} = \frac{yx}{z}$

21. $\frac{7z}{9z} \cdot \frac{4z}{2z} = \frac{7z \cdot 4z}{9z \cdot 2z} = \frac{7 \cdot z \cdot 2 \cdot 2 \cdot z}{3 \cdot 3 \cdot z \cdot 2 \cdot z}$

$= \frac{7 \cdot \cancel{z} \cdot \cancel{2} \cdot 2 \cdot \cancel{z}}{3 \cdot 3 \cdot \cancel{z} \cdot \cancel{2} \cdot \cancel{z}}$

$= \frac{7 \cdot 2}{3 \cdot 3} = \frac{14}{9}$

23. $\frac{2x^2y}{3xy} \cdot \frac{3xy^2}{2} = \frac{2x^2y \cdot 3xy^2}{3xy \cdot 2}$

$= \frac{2 \cdot x \cdot x \cdot y \cdot 3 \cdot x \cdot y \cdot y}{3 \cdot x \cdot y \cdot 2}$

$= \frac{\cancel{2} \cdot \cancel{x} \cdot x \cdot \cancel{y} \cdot \cancel{3} \cdot x \cdot y \cdot y}{\cancel{3} \cdot \cancel{x} \cdot \cancel{y} \cdot \cancel{2}}$

$= \frac{x \cdot x \cdot y \cdot y}{1} = x^2y^2$

25. $\frac{8x^2y^2}{4x^2} \cdot \frac{2xy}{2y} = \frac{8x^2y^2 \cdot 2xy}{4x^2 \cdot 2y}$

$= \frac{2 \cdot 2 \cdot 2 \cdot x \cdot x \cdot y \cdot y \cdot 2 \cdot x \cdot y}{2 \cdot 2 \cdot x \cdot x \cdot 2 \cdot y}$

$= \frac{\cancel{2} \cdot \cancel{2} \cdot \cancel{2} \cdot \cancel{x} \cdot \cancel{x} \cdot \cancel{y} \cdot y \cdot 2 \cdot x \cdot y}{\cancel{2} \cdot \cancel{2} \cdot \cancel{x} \cdot \cancel{x} \cdot \cancel{2} \cdot \cancel{y}}$

$= \frac{2 \cdot x \cdot y \cdot y}{1} = 2xy^2$

27. $-\frac{2xy}{x^2} \cdot \frac{3xy}{2} = -\frac{2xy \cdot 3xy}{x^2 \cdot 2}$

$= -\frac{2 \cdot x \cdot y \cdot 3 \cdot x \cdot y}{x \cdot x \cdot 2}$

$= -\frac{\cancel{2} \cdot \cancel{x} \cdot y \cdot 3 \cdot \cancel{x} \cdot y}{\cancel{x} \cdot \cancel{x} \cdot 2}$

$= -\frac{3 \cdot y \cdot y}{1} = -3xy^2$

29. $\frac{ab^2}{a^2b} \bullet \frac{b^2c^2}{abc} \bullet \frac{abc^2}{a^3c^2} = \frac{ab^2 \cdot b^2c^2 \cdot abc^2}{a^2b \cdot abc \cdot a^3c^2}$

$= \frac{a \cdot b \cdot b \cdot b \cdot b \cdot c \cdot c \cdot a \cdot b \cdot c \cdot c \cdot c}{a \cdot a \cdot b \cdot a \cdot b \cdot c \cdot a \cdot a \cdot a \cdot c \cdot c}$

$= \frac{\cancel{a} \cdot \cancel{b} \cdot \cancel{b} \cdot b \cdot b \cdot \cancel{c} \cdot \cancel{c} \cdot \cancel{a} \cdot b \cdot \cancel{c} \cdot c \cdot c}{\cancel{a} \cdot \cancel{a} \cdot \cancel{b} \cdot a \cdot \cancel{b} \cdot \cancel{c} \cdot a \cdot a \cdot a \cdot \cancel{c} \cdot \cancel{c}}$

$= \frac{b \cdot b \cdot b \cdot c \cdot c}{a \cdot a \cdot a \cdot a} = \frac{b^3c^2}{a^4}$

31. $\frac{10r^2st^3}{6rs^2} \bullet \frac{3r^3t}{2rst} \bullet \frac{2s^3t^4}{5s^2t^3}$

$= \frac{10r^2st^3 \cdot 3r^3t \cdot 2s^3t^4}{6rs^2 \cdot 2rst \cdot 5s^2t^3}$

$= \frac{2 \cdot 5 \cdot r \cdot r \cdot s \cdot t \cdot t \cdot t \cdot 3 \cdot r \cdot r \cdot r \cdot t \cdot 2 \cdot s \cdot s \cdot s \cdot t \cdot t \cdot t \cdot t}{2 \cdot 3 \cdot r \cdot s \cdot s \cdot 2 \cdot r \cdot s \cdot t \cdot 5 \cdot s \cdot s \cdot t \cdot t \cdot t}$

$= \frac{\cancel{2} \cdot \cancel{5} \cdot \cancel{r} \cdot \cancel{r} \cdot \cancel{s} \cdot \cancel{t} \cdot \cancel{t} \cdot \cancel{t} \cdot \cancel{3} \cdot r \cdot r \cdot r \cdot \cancel{t} \cdot \cancel{2} \cdot \cancel{s} \cdot \cancel{s} \cdot \cancel{s} \cdot t \cdot t \cdot t \cdot t}{\cancel{2} \cdot \cancel{3} \cdot \cancel{r} \cdot \cancel{s} \cdot \cancel{s} \cdot \cancel{2} \cdot \cancel{r} \cdot \cancel{s} \cdot \cancel{t} \cdot \cancel{5} \cdot \cancel{s} \cdot s \cdot \cancel{t} \cdot \cancel{t} \cdot \cancel{t}}$

$= \frac{r \cdot r \cdot r \cdot t \cdot t \cdot t \cdot t}{s} = \frac{r^3t^4}{s}$

33. $\frac{z+7}{7} \bullet \frac{z+2}{z} = \frac{(z+7) \cdot (z+2)}{7 \cdot z}$

$= \frac{(z+7)(z+2)}{7z}$

35. $\frac{x-2}{2} \bullet \frac{2x}{x-2} = \frac{(x-2) \cdot 2x}{2 \cdot (x-2)}$

$= \frac{(x-2) \cdot \cancel{2} \cdot x}{\cancel{2} \cdot (x-2)} = \frac{x}{1} = x$

37. $\frac{x+5}{5} \bullet \frac{x}{x+5} = \frac{(x+5) \cdot x}{5 \cdot (x+5)} = \frac{x}{5}$

39. $\frac{(x+1)^2}{x+1} \bullet \frac{x+2}{x+1} = \frac{(x+1)^2 \cdot (x+2)}{(x+1) \cdot (x+1)}$

$= \frac{(x+1) \cdot (x+1) \cdot (x+2)}{(x+1) \cdot (x+1)}$

$= \frac{x+2}{1} = x + 2$

41. $\frac{2x+6}{x+3} \bullet \frac{3}{4x} = \frac{(2x+6) \cdot 3}{(x+3) \cdot 4x}$

$= \frac{2 \cdot (x+3) \cdot 3}{(x+3) \cdot 2 \cdot 2 \cdot x} = \frac{\cancel{2} \cdot (x+3) \cdot 3}{(x+3) \cdot \cancel{2} \cdot 2 \cdot x}$

$= \frac{3}{2x}$

43. $\frac{x^2-x}{x} \bullet \frac{3x-6}{3x-3} = \frac{(x^2-x) \cdot (3x-6)}{x \cdot (3x-3)}$

$= \frac{x \cdot (x-1) \cdot 3 \cdot (x-2)}{x \cdot 3 \cdot (x-1)}$

$= \frac{\cancel{x} \cdot (x-1) \cdot \cancel{3} \cdot (x-2)}{\cancel{x} \cdot \cancel{3} \cdot (x-1)} = \frac{x-2}{1}$

$= x - 2$

45. $\frac{7y-14}{y-2} \bullet \frac{x^2}{7x} = \frac{(7y-14) \cdot x^2}{(y-2) \cdot 7x}$

$= \frac{7 \cdot (y-2) \cdot x \cdot x}{(y-2) \cdot 7 \cdot x}$

$= \frac{\cancel{7} \cdot (y-2) \cdot \cancel{x} \cdot x}{(y-2) \cdot \cancel{7} \cdot \cancel{x}} = \frac{x}{1} = x$

47. $\frac{x^2+x-6}{5x} \bullet \frac{5x-10}{x+3} = \frac{(x^2+x-6) \cdot (5x-10)}{5x \cdot (x+3)}$

$= \frac{(x+3) \cdot (x-2) \cdot 5 \cdot (x-2)}{5 \cdot x \cdot (x+3)}$

$= \frac{(x+3) \cdot (x-2) \cdot \cancel{5} \cdot (x-2)}{\cancel{5} \cdot x \cdot (x+3)}$

$= \frac{(x-2)(x-2)}{x} = \frac{(x-2)^2}{x}$

49. $\frac{m^2-2m-3}{2m+4} \bullet \frac{m^2-4}{m^2+3m+2}$

$= \frac{(m^2-2m-3) \cdot (m^2-4)}{(2m+4) \cdot (m^2+3m+2)}$

$= \frac{(m-3) \cdot (m+1) \cdot (m-2) \cdot (m+2)}{2 \cdot (m+2) \cdot (m+2) \cdot (m+1)}$

$= \frac{(m-3)(m-2)}{2(m+2)}$

51. $\frac{abc^2}{a+1}\bullet\frac{c}{a^2b^2}\bullet\frac{a^2+a}{ac} = \frac{abc^2\bullet c\bullet(a^2+a)}{(a+1)\bullet a^2b^2\bullet ac}$

$= \frac{a\bullet b\bullet c\bullet c\bullet c\bullet a\bullet(a+1)}{(a+1)\bullet a\bullet a\bullet b\bullet b\bullet a\bullet c}$

$= \frac{\cancel{a}\bullet\cancel{b}\bullet\cancel{c}\bullet c\bullet c\bullet\cancel{a}\bullet(a+1)}{(a+1)\bullet\cancel{a}\bullet\cancel{a}\bullet\cancel{b}\bullet b\bullet a\bullet\cancel{c}}$

$= \frac{c\bullet c}{a\bullet b} = \frac{c^2}{ab}$

53. $\frac{3x^2+5x+2}{x^2-9}\bullet\frac{x-3}{x^2-4}\bullet\frac{x^2+5x-6}{6x+4}$

$= \frac{(3x^2+5x+2)\bullet(x-3)\bullet(x^2+5x-6)}{(x^2-9)\bullet(x^2-4)\bullet(6x+4)}$

$= \frac{(3x+2)\bullet(x+1)\bullet(x-3)\bullet(x+3)\bullet(x+2)}{(x-3)\bullet(x+3)\bullet(x-2)\bullet(x+2)\bullet 2\bullet(3x+2)}$

$= \frac{(x+1)}{2(x-2)}$

55. $\frac{1}{3}\div\frac{1}{2} = \frac{1}{3}\bullet\frac{2}{1} = \frac{1\bullet 2}{3\bullet 1} = \frac{2}{3}$

57. $\frac{21}{14}\div\frac{5}{2} = \frac{21}{14}\bullet\frac{2}{5} = \frac{21\bullet 2}{14\bullet 5}$

$= \frac{3\bullet 7\bullet 2}{2\bullet 7\bullet 5} = \frac{3\bullet\cancel{7}\bullet\cancel{2}}{\cancel{2}\bullet\cancel{7}\bullet 5}$

$= \frac{3}{5}$

59. $\frac{2}{y}\div\frac{4}{3} = \frac{2}{y}\bullet\frac{3}{4} = \frac{2\bullet 3}{y\bullet 4}$

$= \frac{2\bullet 3}{y\bullet 2\bullet 2} = \frac{\cancel{2}\bullet 3}{y\bullet\cancel{2}\bullet 2} = \frac{3}{2y}$

61. $\frac{3x}{2}\div\frac{x}{2} = \frac{3x}{2}\bullet\frac{2}{x} = \frac{3x\bullet 2}{2\bullet x}$

$= \frac{3\bullet x\bullet 2}{2\bullet x} = \frac{3\bullet\cancel{x}\bullet\cancel{2}}{\cancel{2}\bullet\cancel{x}} = \frac{3}{1} = 3$

63. $\frac{3x}{y}\div\frac{2x}{4} = \frac{3x}{y}\bullet\frac{4}{2x} = \frac{3x\bullet 4}{y\bullet 2x}$

$= \frac{3\bullet x\bullet 2\bullet 2}{y\bullet 2\bullet x} = \frac{3\bullet\cancel{x}\bullet\cancel{2}\bullet 2}{y\bullet\cancel{2}\bullet\cancel{x}}$

$= \frac{3\bullet 2}{y} = \frac{6}{y}$

65. $\frac{4x}{3x}\div\frac{2y}{9y} = \frac{4x}{3x}\bullet\frac{9y}{2y} = \frac{4x\bullet 9y}{3x\bullet 2y}$

$= \frac{4\bullet x\bullet 3\bullet 3\bullet y}{3\bullet x\bullet 2\bullet y} = \frac{\cancel{2}\bullet 2\bullet\cancel{x}\bullet\cancel{3}\bullet 3\bullet\cancel{y}}{\cancel{3}\bullet\cancel{x}\bullet\cancel{2}\bullet\cancel{y}}$

$= \frac{2\bullet 3}{1} = \frac{6}{1} = 6$

67. $\frac{x^2}{3}\div\frac{2x}{4} = \frac{x^2}{3}\bullet\frac{4}{2x} = \frac{x^2\bullet 4}{3\bullet 2x}$

$= \frac{x\bullet x\bullet 2\bullet 2}{3\bullet 2\bullet x} = \frac{\cancel{x}\bullet x\bullet\cancel{2}\bullet 2}{3\bullet\cancel{2}\bullet\cancel{x}}$

$= \frac{2\bullet x}{3} = \frac{2x}{3}$

69. $\frac{x^2y}{3xy}\div\frac{xy^2}{6y} = \frac{x^2y}{3xy}\bullet\frac{6y}{xy^2} = \frac{x^2y\bullet 6y}{3xy\bullet xy^2}$

$= \frac{x\bullet x\bullet y\bullet 6\bullet y}{3\bullet x\bullet y\bullet x\bullet y\bullet y} = \frac{\cancel{x}\bullet\cancel{x}\bullet\cancel{y}\bullet 2\bullet\cancel{3}\bullet\cancel{y}}{\cancel{3}\bullet\cancel{x}\bullet\cancel{y}\bullet\cancel{x}\bullet\cancel{y}\bullet y}$

$= \frac{2}{y}$

71. $\frac{x+2}{3x}\div\frac{x+2}{2} = \frac{x+2}{3x}\bullet\frac{2}{x+2} = \frac{(x+2)\bullet 2}{3x\bullet(x+2)}$

$= \frac{(x+2)\bullet 2}{3\bullet x\bullet(x+2)} = \frac{2}{3x}$

73. $\frac{(z-2)^2}{3z^2}\div\frac{z-2}{6z} = \frac{(z-2)^2}{3z^2}\bullet\frac{6z}{z-2}$

$= \frac{(z-2)^2\bullet 6z}{3z^2\bullet(z-2)}$

$= \frac{(z-2)\bullet(z-2)\bullet 2\bullet 3\bullet z}{3\bullet z\bullet z\bullet(z-2)}$

$= \frac{(z-2)\bullet(z-2)\bullet 2\bullet\cancel{3}\bullet\cancel{z}}{\cancel{3}\bullet\cancel{z}\bullet z\bullet(z-2)}$

$= \frac{2(z-2)}{z}$

75. $\frac{(z-7)^2}{z+2} \div \frac{z(z-7)}{5z^2} = \frac{(z-7)^2}{z+2} \bullet \frac{5z^2}{z(z-7)}$

$= \frac{(z-7)^2 \bullet 5z^2}{(z+2) \bullet z(z-7)}$

$= \frac{(z-7) \bullet (z-7) \bullet 5 \bullet z \bullet z}{(z+2) \bullet z \bullet (z-7)}$

$= \frac{(z-7) \bullet (z-7) \bullet 5 \bullet \cancel{z} \bullet z}{(z+2) \bullet \cancel{z} \bullet (z-7)}$

$= \frac{5z(z-7)}{z+2}$

77. $\frac{x^2-4}{3x+6} \div \frac{x-2}{x+2} = \frac{x^2-4}{3x+6} \bullet \frac{x+2}{x-2}$

$= \frac{(x^2-4) \bullet (x+2)}{(3x+6) \bullet (x-2)}$

$= \frac{(x-2) \bullet (x+2) \bullet (x+2)}{3 \bullet (x+2) \bullet (x-2)} \quad = \frac{x+2}{3}$

79. $\frac{x^2-1}{3x-3} \div \frac{x+1}{3} = \frac{x^2-1}{3x-3} \bullet \frac{x+1}{3}$

$= \frac{(x^2-1) \bullet (x+1)}{(3x-3) \bullet 3}$

$= \frac{(x-1) \bullet (x+1) \bullet (x+1)}{3 \bullet (x-1) \bullet 3}$

$= \frac{(x+1) \bullet (x+1)}{3 \bullet 3} = \frac{(x+1)^2}{9}$

81. $\frac{x^2-2x-35}{3x^2+27x} \div \frac{x^2+7x+10}{6x^2+12x}$

$= \frac{x^2-2x-35}{3x^2+27x} \bullet \frac{6x^2+12x}{x^2+7x+10}$

$= \frac{(x^2-2x-35) \bullet (6x^2+12x)}{(3x^2+27x) \bullet (x^2+7x+10)}$

$= \frac{(x-7) \bullet (x+5) \bullet 6x \bullet (x+2)}{3x \bullet (x+9) \bullet (x+2) \bullet (x+5)}$

$= \frac{(x-7) \bullet (x+5) \bullet 2 \bullet 3 \bullet x \bullet (x+2)}{3 \bullet x \bullet (x+9) \bullet (x+2) \bullet (x+5)}$

$= \frac{(x-7) \bullet (x+5) \bullet 2 \bullet \cancel{3} \bullet \cancel{x} \bullet (x+2)}{\cancel{3} \bullet \cancel{x} \bullet (x+9) \bullet (x+2) \bullet (x+5)}$

$= \frac{2(x-7)}{x+9}$

83. $\frac{2x^2+8x-42}{x-3} \div \frac{2x^2+14x}{x^2+5x}$

$= \frac{2x^2+8x-42}{x-3} \bullet \frac{x^2+5x}{2x^2+14x}$

$= \frac{(2x^2+8x-42) \bullet (x^2+5x)}{(x-3) \bullet (2x^2+14x)}$

$= \frac{2 \bullet (x^2+4x-21) \bullet x \bullet (x+5)}{(x-3) \bullet 2x \bullet (x+7)}$

$= \frac{2 \bullet (x+7) \bullet (x-3) \bullet x \bullet (x+5)}{2 \bullet x \bullet (x-3) \bullet (x+7)}$

$= \frac{\cancel{2} \bullet (x+7) \bullet (x-3) \bullet \cancel{x} \bullet (x+5)}{\cancel{2} \bullet \cancel{x} \bullet (x-3) \bullet (x+7)}$

$= \frac{x+5}{1} = x + 5$

85. $\frac{x}{3} \bullet \frac{9}{4} \div \frac{x^2}{6} = \frac{x}{3} \bullet \frac{9}{4} \bullet \frac{6}{x^2} = \frac{x \bullet 9 \bullet 6}{3 \bullet 4 \bullet x^2}$

$= \frac{x \bullet 3 \bullet 3 \bullet 2 \bullet 3}{3 \bullet 2 \bullet 2 \bullet x \bullet x} = \frac{\cancel{x} \bullet \cancel{3} \bullet 3 \bullet \cancel{2} \bullet 3}{\cancel{3} \bullet \cancel{2} \bullet 2 \bullet \cancel{x} \bullet x}$

$= \frac{3 \bullet 3}{2 \bullet x} = \frac{9}{2x}$

87. $\frac{x^2}{18} \div \frac{x^3}{6} \div \frac{12}{x^2} = \frac{x^2}{18} \bullet \frac{6}{x^3} \div \frac{12}{x^2}$

$= \frac{x^2}{18} \bullet \frac{6}{x^3} \bullet \frac{x^2}{12} = \frac{x^2 \bullet 6 \bullet x^2}{18 \bullet x^3 \bullet 12}$

$= \frac{x \bullet x \bullet 3 \bullet 2 \bullet x \bullet x}{2 \bullet 3 \bullet 3 \bullet x \bullet x \bullet x \bullet 2 \bullet 2 \bullet 3}$

$= \frac{\cancel{x} \bullet \cancel{x} \bullet \cancel{3} \bullet \cancel{2} \bullet \cancel{x} \bullet x}{\cancel{2} \bullet \cancel{3} \bullet 3 \bullet \cancel{x} \bullet \cancel{x} \bullet \cancel{x} \bullet 2 \bullet 2 \bullet 3}$

$= \frac{x}{3 \bullet 2 \bullet 2 \bullet 3} = \frac{x}{36}$

89. $\frac{x^2-4}{2x+6} \div \frac{x+2}{4} \bullet \frac{x+3}{x-2}$

$= \frac{x^2-4}{2x+6} \bullet \frac{4}{x+2} \bullet \frac{x+3}{x-2}$

$= \frac{(x^2-4) \bullet 4 \bullet (x+3)}{(2x+6) \bullet (x+2) \bullet (x-2)}$

$= \frac{(x-2) \bullet (x+2) \bullet 2 \bullet 2 \bullet (x+3)}{2 \bullet (x+3) \bullet (x+2) \bullet (x-2)}$

$= \frac{(x-2) \bullet (x+2) \bullet \cancel{2} \bullet 2 \bullet (x+3)}{\cancel{2} \bullet (x+3) \bullet (x+2) \bullet (x-2)}$

$= \frac{2}{1} = 2$

91. $\frac{x-x^2}{x^2-4} \left(\frac{2x+4}{x+2} \div \frac{5}{x+2} \right)$

$= \frac{x-x^2}{x^2-4} \left(\frac{2x+4}{x+2} \bullet \frac{x+2}{5} \right)$

$= \frac{(x-x^2)}{x^2-4} \bullet \frac{2x+4}{x+2} \bullet \frac{x+2}{5}$

$= \frac{(x-x^2) \bullet (2x+4) \bullet (x+2)}{(x^2-4) \bullet (x+2) \bullet 5}$

$= \frac{x \bullet (1-x) \bullet 2 \bullet (x+2) \bullet (x+2)}{(x-2) \bullet (x+2) \bullet (x+2) \bullet 5}$

$= \frac{x \bullet (1-x) \bullet 2}{(x-2) \bullet 5} = \frac{2x(1-x)}{5(x-2)}$

93. $\frac{y^2}{x+1} \bullet \frac{x^2+2x+1}{x^2-1} \div \frac{3y}{xy-y}$

$= \frac{y^2}{x+1} \bullet \frac{x^2+2x+1}{x^2-1} \bullet \frac{xy-y}{3y}$

$= \frac{y^2 \bullet (x^2+2x+1) \bullet (xy-y)}{(x+1) \bullet (x^2-1) \bullet 3y}$

$= \frac{y \bullet y \bullet (x+1) \bullet (x+1) \bullet y \bullet (x-1)}{(x+1) \bullet (x-1) \bullet (x+1) \bullet 3 \bullet y}$

$= \frac{\cancel{y} \bullet y \bullet (x+1) \bullet (x+1) \bullet y \bullet (x-1)}{(x+1) \bullet (x-1) \bullet (x+1) \bullet 3 \bullet \cancel{y}}$

$= \frac{y \bullet y}{3} = \frac{y^2}{3}$

95. $\frac{x^2+x-6}{x^2-4} \bullet \frac{x^2+2x}{x-2} \div \frac{x^2+3x}{x+2}$

$= \frac{x^2+x-6}{x^2-4} \bullet \frac{x^2+2x}{x-2} \bullet \frac{x+2}{x^2+3x}$

$= \frac{(x^2+x-6) \bullet (x^2+2x) \bullet (x+2)}{(x^2-4) \bullet (x-2) \bullet (x^2+3x)}$

$= \frac{(x+3) \bullet (x-2) \bullet x \bullet (x+2) \bullet (x+2)}{(x-2) \bullet (x+2) \bullet (x-2) \bullet x \bullet (x+3)}$

$= \frac{(x+3) \bullet (x-2) \bullet \cancel{x} \bullet (x+2) \bullet (x+2)}{(x-2) \bullet (x+2) \bullet (x-2) \bullet \cancel{x} \bullet (x+3)}$

$= \frac{x+2}{x-2}$

APPLICATIONS

97. width $= 2\left(\frac{2x+1}{2} \right) = \frac{2}{1} \bullet \frac{2x+1}{2}$

$= \frac{2 \bullet (2x+1)}{2} = \frac{\cancel{2} \bullet (2x+1)}{\cancel{2}}$

$= \frac{2x+1}{1} = 2x + 1$

length $= 3\left(\frac{(2x+1)}{2} \right) = \frac{3}{1} \bullet \frac{2x+1}{2}$

$= \frac{3 \bullet (2x+1)}{2} = \frac{3(2x+1)}{2}$

Area = length•width

$= \frac{3(2x+1)}{2} \bullet (2x+1)$

$= \frac{3(2x+1)}{2} \bullet \frac{(2x+1)}{1}$

$= \frac{3(2x+1)(2x+1)}{2}$

$= \frac{3(4x^2+4x+1)}{2}$

$= \frac{12x^2+12x+3}{2}$

The area is $\frac{12x^2+12x+3}{2}$ in^2.

REVIEW

103. $2x^3y^2(-3x^2y^4) = -3 \bullet 2 \bullet x^3 \bullet x^2 \bullet y^2 \bullet y^4$

$= -6x^{3+2}y^{2+4}$

$= -6x^5y^6$

105. $(3y)^{-4} = \frac{1}{(3y)^4} = \frac{1}{3^4y^4} = \frac{1}{81y^4}$

107. $-4(y^3 - 4y^2 + 3y - 2)$
$-4(-2y^3 - y)$
$= -4y^3 + 16y^2 - 12y + 8$
$+ 8y^3 + 4y$
$= -4y^3 + 8y^3 + 16y^2$
$- 12y + 4y + 8$
$= 4y^3 + 16y^2 - 8y + 8$

STUDY SET Section 8.5

VOCABULARY

1. LCD

CONCEPTS

3. numerators; common denominators

NOTATION

5. $\frac{6a-1}{4a+1} + \frac{2a+3}{4a+1} = \frac{6a-1+2a+3}{4a+1}$

$= \frac{8a+2}{4a+1}$

$= \frac{2(4a+1)}{4a+1}$

$= 2$

PRACTICE

7. $\frac{1}{3} + \frac{1}{3} = \frac{1+1}{3} = \frac{2}{3}$

9. $\frac{2}{9} + \frac{1}{9} = \frac{2+1}{9} = \frac{3}{9} = \frac{3 \cdot 1}{3 \cdot 3} = \frac{1}{3}$

11. $\frac{2x}{y} + \frac{2x}{y} = \frac{2x+2x}{y} = \frac{4x}{y}$

13. $\frac{4}{7y} + \frac{10}{7y} = \frac{4+10}{7y} = \frac{14}{7y} = \frac{2 \cdot 7}{7y} = \frac{2}{y}$

15. $\frac{y+2}{5z} + \frac{y+4}{5z} = \frac{y+2+y+4}{5z}$

$= \frac{2y+6}{5z} = \frac{2(y+3)}{5z}$

17. $\frac{3x-5}{x-2} + \frac{6x-13}{x-2} = \frac{3x-5+6x-13}{x-2}$

$= \frac{9x-18}{x-2} = \frac{9(x-2)}{x-2} = \frac{9}{1} = 9$

19. $\frac{5}{7} - \frac{4}{7} = \frac{5-4}{7} = \frac{1}{7}$

21. $\frac{35}{72} - \frac{44}{72} = \frac{35-44}{72} = \frac{9}{72} = \frac{9 \cdot 1}{9 \cdot 8} = \frac{1}{8}$

23. $\frac{2x}{y} - \frac{x}{y} = \frac{2x-x}{y} = \frac{x}{y}$

25. $\frac{9y}{3x} - \frac{6y}{3x} = \frac{9y-6y}{3x} = \frac{3y}{3x} = \frac{3 \cdot x}{3 \cdot y} = \frac{x}{y}$

27. $\frac{6x-5}{3xy} - \frac{3x-5}{3xy} = \frac{(6x-5)-(3x-5)}{3xy}$

$= \frac{6x-5-3x+5}{3xy} = \frac{3x}{3xy} = \frac{1}{y}$

29. $\frac{3y-2}{y+3} - \frac{2y-5}{y+3} = \frac{(3y-2)-(2y-5)}{y+3}$

$= \frac{3y-2-2y+5}{y+3} = \frac{y+3}{y+3} = 1$

31. $\frac{13x}{15} + \frac{12x}{15} - \frac{5x}{15} = \frac{13x+12x-5x}{15}$

$= \frac{20x}{15} = \frac{4 \cdot 5 \cdot x}{3 \cdot 5} = \frac{4x}{3}$

33. $\frac{x}{3y} + \frac{2x}{3y} - \frac{x}{3y} = \frac{x+2x-x}{3y} = \frac{2x}{3y}$

35. $\frac{3x}{y+2} - \frac{3y}{y+2} + \frac{x+y}{y+2} = \frac{3x-3y+x+y}{y+2}$

$= \frac{4x-2y}{y+2}$

37. $\frac{x+1}{x-2} - \frac{2(x-3)}{x-2} + \frac{3(x-3)}{x-2}$

$= \frac{(x+1)-2(x-3)+3(x-3)}{x-2}$

$= \frac{x+1-2x+6+3x-9}{x-2}$

$= \frac{2x+10}{x-2}$

39. $\frac{25}{4} = \frac{25 \cdot 5}{4 \cdot 5} = \frac{125}{20}$

41. $\frac{8}{x} = \frac{8 \cdot xy}{x \cdot xy} = \frac{8xy}{x^2y}$

43. $\frac{3x}{x+1} = \frac{3x \cdot (x+1)}{(x+1) \cdot (x+1)} = \frac{3x(x+1)}{(x+1)^2}$

45. $\frac{2y}{x} = \frac{2y \cdot (x+1)}{x \cdot (x+1)} = \frac{2y(x+1)}{x^2+x}$

47. $\frac{z}{z-1} = \frac{z \cdot (z+1)}{(z-1) \cdot (z+1)} = \frac{z(z+1)}{z^2-1}$

49. $\frac{2}{x+1} = \frac{2 \cdot (x+2)}{(x+1) \cdot (x+2)} = \frac{2(x+2)}{x^2+3x+2}$

51. $2x = 2 \cdot x$
$6x = 2 \cdot 3 \cdot x$
$\text{LCD} = 2 \cdot x \cdot 3 = 6x$

53. $3x = 3 \cdot x$
$6y = 2 \cdot 3 \cdot y$
$9xy = 3 \cdot 3 \cdot x \cdot y$
$\text{LCD} = 3 \cdot x \cdot 2 \cdot y \cdot 3 = 18xy$

55. $x^2 - 1 = (x-1)(x+1)$
$x + 1 = x + 1$
$\text{LCD} = (x-1)(x+1) = x^2 - 1$

57. $x^2 + 6x = x(x+6)$
$x + 6 = x + 6$
$\text{LCD} = x(x+6) = x^2 + 6x$

59. $x^2 - 4x - 5 = (x-5)(x+1)$
$x^2 - 25 = (x-5)(x+5)$
$\text{LCD} = (x-5)(x+1)(x+5)$

61. $\frac{1}{2} + \frac{2}{3} = \frac{1 \cdot 3}{2 \cdot 3} + \frac{2 \cdot 2}{3 \cdot 2} = \frac{3}{6} + \frac{4}{6} = \frac{7}{6}$

63. $\frac{2y}{9} + \frac{y}{3} = \frac{2y}{9} + \frac{y \cdot 3}{3 \cdot 3} = \frac{2y}{9} + \frac{3y}{9}$

$= \frac{2y+3y}{9} = \frac{5y}{9}$

65. $\frac{21x}{14} - \frac{5x}{21} = \frac{21x}{2 \cdot 7} - \frac{5x}{3 \cdot 7}$

$= \frac{21x \cdot 3}{2 \cdot 7 \cdot 3} - \frac{5x \cdot 2}{3 \cdot 7 \cdot 2}$

$= \frac{63x}{42} - \frac{10x}{42} = \frac{63x - 10x}{42}$

$= \frac{53x}{42}$

67. $\frac{4x}{3} + \frac{2x}{y} = \frac{4x \cdot y}{3 \cdot y} + \frac{2x \cdot 3}{y \cdot 3}$

$= \frac{4xy}{3y} + \frac{6x}{3y} = \frac{4xy+6x}{3y}$

69. $\frac{2}{x} - 3x = \frac{2}{x} - \frac{3x}{1}$

$= \frac{2}{x} - \frac{3x \cdot x}{1 \cdot x} = \frac{2}{x} - \frac{3x^2}{x}$

$= \frac{2-3x^2}{x}$

71. $\frac{y+2}{5y} + \frac{y+4}{15y} = \frac{(y+2) \cdot 3}{5y \cdot 3} + \frac{y+4}{15y}$

$= \frac{3y+6}{15y} + \frac{y+4}{15y}$

$= \frac{3y+6+y+4}{15y} = \frac{4y+10}{15y}$

73. $\frac{x+5}{xy} - \frac{x-1}{x^2y} = \frac{(x+5) \cdot x}{xy \cdot x} - \frac{x-1}{x^2y}$

$= \frac{x^2+5x}{x^2y} - \frac{x-1}{x^2y}$

$= \frac{x^2+5x-(x-1)}{x^2y}$

$= \frac{x^2+5x-x+1}{x^2y}$

$= \frac{x^2+4x+1}{x^2y}$

75. $\frac{x}{x+1} + \frac{x-1}{x} = \frac{x \cdot x}{(x+1) \cdot x} + \frac{(x-1) \cdot x}{x \cdot (x+1)}$

$= \frac{x^2}{x(x+1)} + \frac{x^2-x}{x(x+1)}$

$= \frac{x^2+x^2-x}{x(x+1)} = \frac{2x^2-x}{x(x+1)}$

77. $\frac{x-1}{x} + \frac{y+1}{y} = \frac{(x-1) \cdot y}{x \cdot y} + \frac{(y+1) \cdot x}{y \cdot x}$

$= \frac{xy-y}{xy} + \frac{xy+x}{xy}$

$= \frac{xy-y+xy+x}{xy} = \frac{2xy-y+x}{xy}$

79. $\frac{x}{x-2}+\frac{4+2x}{x^2-4}=\frac{x}{x-2}+\frac{4+2x}{(x-2)(x+2)}$

$=\frac{x\cdot(x+2)}{(x-2)\cdot(x+2)}+\frac{4+2x}{(x-2)\cdot(x+2)}$

$=\frac{x^2+2x}{(x-2)(x+2)}+\frac{4+2x}{(x-2)(x+1)}$

$=\frac{x^2+2x+4+2x}{(x-2)(x+2)}=\frac{x^2+4x+4}{(x-2)(x+2)}$

$=\frac{(x+2)(x+2)}{(x-2)(x+2)}=\frac{x+2}{x-2}$

81. $\frac{x+1}{x-1}+\frac{x-1}{x+1}$

$=\frac{(x+1)\cdot(x+1)}{(x-1)\cdot(x+1)}+\frac{(x-1)\cdot(x-1)}{(x+1)\cdot(x-1)}$

$=\frac{x^2+2x+1}{(x-1)(x+1)}+\frac{x^2-2x+1}{(x+1)(x-1)}$

$=\frac{x^2+2x+1+x^2-2x+1}{(x-1)(x+1)}$

$=\frac{2x^2+2}{(x-1)(x+1)}$

83. $\frac{2x+2}{x-2}-\frac{2x}{2-x}=\frac{2x+2}{x-2}-\frac{2x}{(-1)(x-2)}$

$=\frac{2x+2}{x-2}+\frac{2x}{x-2}=\frac{2x+2+2x}{x-2}$

$=\frac{4x+2}{x-2}$

85. $\frac{2x}{x^2-3x+2}+\frac{2x}{x-1}-\frac{x}{x-2}$

$=\frac{2x}{(x-2)(x-1)}+\frac{2x}{x-1}-\frac{x}{x-2}$

$=\frac{2x}{(x-2)(x-1)}+\frac{2x\cdot(x-2)}{(x-1)\cdot(x-2)}$

$-\frac{x\cdot(x-1)}{(x-2)\cdot(x-1)}$

$=\frac{2x}{(x-2)(x-1)}+\frac{2x^2-4x}{(x-1)\cdot(x-2)}$

$-\frac{x^2-x}{(x-2)\cdot(x-1)}$

$=\frac{2x+2x^2-4x-(x^2-x)}{(x-2)(x-1)}$

$=\frac{2x+2x^2-4x-x^2+x}{(x-2)(x-1)}$

$=\frac{x^2-x}{(x-2)(x-1)}=\frac{x(x-1)}{(x-2)(x-1)}$

$=\frac{x}{(x-2)}$

87. $\frac{2x}{x-1}+\frac{3x}{x+1}-\frac{x+3}{x^2-1}$

$=\frac{2x}{x-1}+\frac{3x}{x+1}-\frac{x+3}{(x-1)(x+1)}$

$=\frac{2x\cdot(x+1)}{(x-1)\cdot(x+1)}+\frac{3x\cdot(x-1)}{(x+1)\cdot(x-1)}$

$-\frac{x+3}{(x-1)(x+1)}$

$=\frac{2x^2+2x}{(x-1)(x+1)}+\frac{3x^2-3x}{(x+1)(x-1)}$

$-\frac{x+3}{(x-1)\cdot(x+1)}$

$=\frac{2x^2+2x+3x^2-3x-(x+3)}{(x-1)(x+1)}$

$=\frac{2x^2+2x+3x^2-3x-x-3}{(x-1)(x+1)}$

$=\frac{5x^2-2x-3}{(x-1)(x+1)}=\frac{(5x+3)(x-1)}{(x-1)(x+1)}$

$=\frac{5x+3}{x+1}$

89. $\frac{x+1}{2x+4} - \frac{x^2}{2x^2-8}$

$$= \frac{x+1}{2(x+2)} - \frac{x^2}{2(x^2-4)}$$

$$= \frac{x+1}{2(x+2)} - \frac{x^2}{2(x-2)(x+2)}$$

$$= \frac{(x+1)\cdot(x-2)}{2(x+2)\cdot(x-2)} - \frac{x^2}{2(x-2)(x+2)}$$

$$= \frac{x^2-x-2}{2(x+2)(x-2)} - \frac{x^2}{2(x-2)(x+2)}$$

$$= \frac{x^2-x-2-x^2}{2(x-2)(x+2)}$$

$$= \frac{-x-2}{2(x-2)(x+2)}$$

$$= \frac{-1(x+2)}{2(x-2)(x+2)}$$

$$= -\frac{1}{2(x-2)}$$

APPLICATIONS

91. height of the funnel $= \frac{3}{3x^2} + \frac{10}{3x}$

$$= \frac{3}{3x^2} + \frac{10\cdot x}{3x\cdot x}$$

$$= \frac{3}{3x^2} + \frac{10x}{3x^2}$$

$$= \frac{3+10x}{3x^2}$$

The height of the funnel is $\frac{3+10x}{3x^2}$ cm.

REVIEW

99. $49 = 7\cdot 7 = 7^2$

101. $136 = 2\cdot 2\cdot 2\cdot 17 = 2^3\cdot 17$

103. $102 = 2\cdot 3\cdot 17$

105. $144 = 2\cdot 2\cdot 2\cdot 2\cdot 3\cdot 3 = 2^4\cdot 3^2$

STUDY SET Section 8.6

VOCABULARY

1. complex fraction

CONCEPTS

3. single; divide

NOTATION

5.
$$\frac{\frac{2}{a}-\frac{1}{b}}{\frac{1}{a}+\frac{2}{b}} = \frac{\frac{2b-a}{ab}}{\frac{b+2a}{ab}}$$

$$= \frac{2b-a}{ab} \div \frac{b+2a}{ab}$$

$$= \frac{2b-a}{ab} \bullet \frac{ab}{b+2a}$$

$$= \frac{(2b-a)ab}{ab(b+2a)}$$

$$= \frac{2b-a}{b+2a}$$

PRACTICE

7.
$$\frac{\frac{2}{3}}{\frac{3}{4}} = \frac{2}{3} \div \frac{3}{4} = \frac{2}{3} \bullet \frac{4}{3} = \frac{2 \cdot 4}{3 \cdot 3} = \frac{8}{9}$$

9.
$$\frac{\frac{4}{5}}{\frac{32}{15}} = \frac{4}{5} \div \frac{32}{15} = \frac{4}{5} \bullet \frac{15}{32} = \frac{4 \cdot 15}{5 \cdot 32}$$

$$= \frac{4 \cdot 3 \cdot 5}{5 \cdot 4 \cdot 8} = \frac{\cancel{4} \cdot 3 \cdot \cancel{5}}{\cancel{5} \cdot \cancel{4} \cdot 8} = \frac{3}{8}$$

11.
$$\frac{\frac{2}{3}+1}{\frac{1}{3}+1} = \frac{\frac{2}{3}+\frac{3}{3}}{\frac{1}{3}+\frac{3}{3}} = \frac{\frac{2+3}{3}}{\frac{1+3}{3}} = \frac{\frac{5}{3}}{\frac{4}{3}}$$

$$= \frac{5}{3} \div \frac{4}{3} = \frac{5}{3} \bullet \frac{3}{4} = \frac{5 \cdot 3}{3 \cdot 4}$$

$$= \frac{5 \cdot \cancel{3}}{\cancel{3} \cdot 4} = \frac{5}{4}$$

13.
$$\frac{\frac{1}{2}+\frac{3}{4}}{\frac{3}{2}+\frac{1}{4}} = \frac{\frac{1 \cdot 2}{2 \cdot 2}+\frac{3}{4}}{\frac{3 \cdot 2}{2 \cdot 2}+\frac{1}{4}} = \frac{\frac{2}{4}+\frac{3}{4}}{\frac{6}{4}+\frac{1}{4}}$$

$$= \frac{\frac{2+3}{4}}{\frac{6+1}{4}} = \frac{\frac{5}{4}}{\frac{7}{4}} = \frac{5}{4} \div \frac{7}{4}$$

$$= \frac{5}{4} \bullet \frac{4}{7} = \frac{5 \cdot 4}{4 \cdot 7} = \frac{5 \cdot \cancel{4}}{\cancel{4} \cdot 7} = \frac{5}{7}$$

15.
$$\frac{\frac{x}{y}}{\frac{1}{x}} = \frac{x}{y} \div \frac{1}{x} = \frac{x}{y} \bullet \frac{x}{1} = \frac{x \cdot x}{y \cdot 1} = \frac{x^2}{y}$$

17.
$$\frac{\frac{5t^2}{9x^2}}{\frac{3t}{x^2t}} = \frac{5t^2}{9x^2} \div \frac{3t}{x^2t} = \frac{5t^2}{9x^2} \bullet \frac{x^2t}{3t}$$

$$= \frac{5t^2 \cdot x^2t}{9x^2 \cdot 3t} = \frac{5 \cdot t \cdot t \cdot x \cdot x \cdot t}{9 \cdot x \cdot x \cdot 3 \cdot t}$$

$$= \frac{5 \cdot t \cdot t \cdot \cancel{x} \cdot \cancel{x} \cdot \cancel{t}}{9 \cdot \cancel{x} \cdot \cancel{x} \cdot 3 \cdot \cancel{t}} = \frac{5 \cdot t \cdot t}{9 \cdot 3} = \frac{5t^2}{27}$$

19.
$$\frac{\frac{1}{x}-3}{\frac{5}{x}+2} = \frac{\frac{1}{x}-\frac{3 \cdot x}{1 \cdot x}}{\frac{5}{x}+\frac{2 \cdot x}{1 \cdot x}} = \frac{\frac{1-3x}{x}}{\frac{5+2x}{x}}$$

$$= \frac{1-3x}{x} \div \frac{5+2x}{x} = \frac{1-3x}{x} \bullet \frac{x}{5+2x}$$

$$= \frac{(1-3x) \cdot x}{x \cdot (5+2x)} = \frac{(1-3x) \cdot \cancel{x}}{\cancel{x} \cdot (5+2x)} = \frac{1-3x}{5+2x}$$

21.
$$\frac{\frac{2}{x}+2}{\frac{4}{x}+2} = \frac{\frac{2}{x}+\frac{2 \cdot x}{1 \cdot x}}{\frac{4}{x}+\frac{2 \cdot x}{1 \cdot x}} = \frac{\frac{2+2x}{x}}{\frac{4+2x}{x}}$$

$$= \frac{2+2x}{x} \div \frac{4+2x}{x} = \frac{2+2x}{x} \bullet \frac{x}{4+2x}$$

$$= \frac{(2+2x) \cdot x}{x \cdot (4+2x)} = \frac{2 \cdot (1+x) \cdot x}{x \cdot 2 \cdot (2+x)}$$

$$= \frac{\cancel{2} \cdot (1+x) \cdot \cancel{x}}{\cancel{x} \cdot \cancel{2} \cdot (2+x)} = \frac{1+x}{2+x}$$

23. $\dfrac{\frac{3y}{x}-y}{y-\frac{y}{x}} = \dfrac{\frac{3y}{x}-\frac{y\cdot x}{1\cdot x}}{\frac{y\cdot x}{1\cdot x}-\frac{y}{x}} = \dfrac{\frac{3y-xy}{x}}{\frac{xy-y}{x}}$

$= \dfrac{3y-xy}{x} \div \dfrac{xy-y}{x} = \dfrac{3y-xy}{x} \bullet \dfrac{x}{xy-y}$

$= \dfrac{(3y-xy)\cdot x}{x\cdot(xy-y)} = \dfrac{y\cdot(3-x)\cdot x}{x\cdot y\cdot(x-1)}$

$= \dfrac{\not{y}\cdot(3-x)\cdot\not{x}}{\not{x}\cdot\not{y}\cdot(x-1)} = \dfrac{3-x}{x-1}$

25. $\dfrac{\frac{1}{x+1}}{1+\frac{1}{x+1}} = \dfrac{\frac{1}{x+1}}{\frac{1\cdot(x+1)}{1\cdot(x+1)}+\frac{1}{x+1}} = \dfrac{\frac{1}{x+1}}{\frac{x+1+1}{x+1}}$

$= \dfrac{1}{x+1} \div \dfrac{x+2}{x+1} = \dfrac{1}{x+1} \bullet \dfrac{x+1}{x+2}$

$= \dfrac{1\cdot(x+1)}{(x+1)\cdot(x+2)} = \dfrac{1}{x+2}$

27. $\dfrac{\frac{x}{x+2}}{\frac{x}{x+2}+x} = \dfrac{\frac{x}{x+2}}{\frac{x}{x+2}+\frac{x\cdot(x+2)}{1\cdot(x+2)}}$

$= \dfrac{\frac{x}{x+2}}{\frac{x+x^2+2x}{x+2}} = \dfrac{x}{x+2} \div \dfrac{x^2+3x}{x+2}$

$= \dfrac{x}{x+2} \bullet \dfrac{x+2}{x^2+3x} = \dfrac{x\cdot(x+2)}{(x+2)\cdot(x^2+3x)}$

$= \dfrac{x\cdot(x+2)}{(x+2)\cdot x\cdot(x+3)}$

$= \dfrac{\not{x}\cdot(x+2)}{(x+2)\cdot\not{x}\cdot(x+3)} = \dfrac{1}{x+3}$

29. $\dfrac{1}{\frac{1}{x}+\frac{1}{y}} = \dfrac{1}{\frac{1\cdot y}{x\cdot y}+\frac{1\cdot x}{y\cdot x}} = \dfrac{1}{\frac{y+x}{xy}}$

$= 1 \div \dfrac{y+x}{xy} = 1\bullet\dfrac{xy}{y+x}$

$= \dfrac{xy}{y+x}$

31. $\dfrac{\frac{2}{x}}{\frac{2}{y}-\frac{4}{x}} = \dfrac{\frac{2}{x}}{\frac{2\cdot x}{y\cdot x}-\frac{4\cdot y}{x\cdot y}} = \dfrac{\frac{2}{x}}{\frac{2x-4y}{xy}}$

$= \dfrac{2}{x} \div \dfrac{2x-4y}{xy} = \dfrac{2}{x}\bullet\dfrac{xy}{2x-4y}$

$= \dfrac{2\cdot x\cdot y}{x\cdot(2x-4y)} = \dfrac{2\cdot x\cdot y}{x\cdot 2\cdot(x-2y)}$

$= \dfrac{\not{2}\cdot\not{x}\cdot y}{\not{x}\cdot\not{2}\cdot(x-2y)} = \dfrac{y}{x-2y}$

33. $\dfrac{3+\frac{3}{x-1}}{3-\frac{3}{x}} = \dfrac{\frac{3\cdot(x-1)}{1\cdot(x-1)}+\frac{3}{x-1}}{\frac{3\cdot x}{1\cdot x}-\frac{3}{x}} = \dfrac{\frac{3x-3+3}{x-1}}{\frac{3x-3}{x}}$

$= \dfrac{3x}{x-1} \div \dfrac{3x-3}{x} = \dfrac{3x}{x-1}\bullet\dfrac{x}{3x-3}$

$= \dfrac{3x\cdot x}{(x-1)\cdot(3x-3)} = \dfrac{3\cdot x\cdot x}{(x-1)\cdot 3\cdot(x-1)}$

$= \dfrac{\not{3}\cdot x\cdot x}{(x-1)\cdot\not{3}\cdot(x-1)} = \dfrac{x^2}{(x-1)^2}$

35. $\dfrac{\frac{3}{x}+\frac{4}{x+1}}{\frac{2}{x+1}-\frac{3}{x}} = \dfrac{\frac{3\cdot(x+1)}{x\cdot(x+1)}+\frac{4\cdot x}{(x+1)\cdot x}}{\frac{2\cdot x}{(x+1)\cdot x}-\frac{3\cdot(x+1)}{x\cdot(x+1)}}$

$= \dfrac{\frac{3x+3}{x(x+1)}+\frac{4x}{x(x+1)}}{\frac{2x}{x(x+1)}-\frac{3x+3}{x(x+1)}} = \dfrac{\frac{7x+3}{x(x+1)}}{\frac{-x-3}{x(x+1)}}$

$= \dfrac{7x+3}{x(x+1)} \div \dfrac{-x-3}{x(x+1)}$

$= \dfrac{7x+3}{x(x+1)}\bullet\dfrac{x(x+1)}{-x-3}$

$= \dfrac{(7x+3)\cdot x\cdot(x+1)}{x\cdot(x+1)\cdot(-x-3)} = \dfrac{7x+3}{-x-3}$

37. $$\frac{\frac{2}{x}-\frac{3}{x+1}}{\frac{2}{x+1}-\frac{3}{x}} = \frac{\frac{2\cdot(x+1)}{x\cdot(x+1)}-\frac{3\cdot x}{(x+1)\cdot x}}{\frac{2\cdot x}{(x+1)\cdot x}-\frac{3\cdot(x+1)}{x\cdot(x+1)}}$$

$$= \frac{\frac{2x+2}{x(x+1)}-\frac{3x}{x(x+1)}}{\frac{2x}{x(x+1)}-\frac{3x+3}{x(x+1)}} = \frac{\frac{-x+2}{x(x+1)}}{\frac{-x-3}{x(x+1)}}$$

$$= \frac{-x+2}{x(x+1)} \div \frac{-x-3}{x(x+1)}$$

$$= \frac{-x+2}{x(x+1)} \bullet \frac{x(x+1)}{-x-3}$$

$$= \frac{(-x+2)\cdot x\cdot(x+1)}{x\cdot(x+1)\cdot(-x-3)} = \frac{-x+2}{-x-3}$$

$$= \frac{(-1)(x-2)}{(-1)(x+3)} = \frac{x-2}{x+3}$$

39. $$\frac{\frac{1}{y^2+y}-\frac{1}{xy+x}}{\frac{1}{xy+x}-\frac{1}{y^2+y}} = \frac{\frac{1}{y(y+1)}-\frac{1}{x(y+1)}}{\frac{1}{x(y+1)}-\frac{1}{y(y+1)}}$$

$$= \frac{xy(y+1)\left(\frac{1}{y(y+1)}-\frac{1}{x(y+1)}\right)}{xy(y+1)\left(\frac{1}{x(y+1)}-\frac{1}{y(y+1)}\right)}$$

$$= \frac{\frac{xy(y+1)}{y(y+1)}-\frac{xy(y+1)}{x(y+1)}}{\frac{xy(y+1)}{x(y+1)}-\frac{xy(y+1)}{y(y+1)}}$$

$$= \frac{\frac{x\not{y}(y+1)}{\not{y}(y+1)}-\frac{\not{x}y(y+1)}{\not{x}(y+1)}}{\frac{\not{x}y(y+1)}{\not{x}(y+1)}-\frac{x\not{y}(y+1)}{\not{y}(y+1)}} = \frac{x-y}{y-x}$$

$$= \frac{(x-y)}{(-1)(x-y)} = -1$$

41. $$\frac{x^{-2}}{y^{-1}} = \frac{\frac{1}{x^2}}{\frac{1}{y}} = \frac{1}{x^2} \div \frac{1}{y} = \frac{1}{x^2} \div \frac{y}{1}$$

$$= \frac{y}{x^2}$$

43. $$\frac{1+x^{-1}}{x^{-1}-1} = \frac{1+\frac{1}{x}}{\frac{1}{x}-1}$$

$$= \frac{x\left(1+\frac{1}{x}\right)}{x\left(\frac{1}{x}-1\right)} = \frac{x\bullet 1+\frac{x}{x}}{\frac{x}{x}-x\bullet 1}$$

$$= \frac{x+1}{1-x}$$

45. $$\frac{a^{-2}+a}{a} = \frac{\frac{1}{a^2}+a}{a} = \frac{\frac{1}{a^2}+\frac{a\cdot a^2}{1\cdot a^2}}{a} = \frac{\frac{1+a^3}{a^2}}{a}$$

$$= \frac{1+a^3}{a^2} \div a = \frac{1+a^3}{a^2} \bullet \frac{1}{a}$$

$$= \frac{1+a^3}{a^2\cdot a} = \frac{1+a^3}{a^3}$$

47. $$\frac{2x^{-1}+4x^{-2}}{2x^{-2}+x^{-1}} = \frac{\frac{2}{x}+\frac{4}{x^2}}{\frac{2}{x^2}+\frac{1}{x}}$$

$$= \frac{x^2\left(\frac{2}{x}+\frac{4}{x^2}\right)}{x^2\left(\frac{2}{x^2}+\frac{1}{x}\right)} = \frac{\frac{2x^2}{x}+\frac{4x^2}{x^2}}{\frac{2x^2}{x^2}+\frac{x^2}{x}}$$

$$= \frac{2x+4}{2+x} = \frac{2(x+2)}{x+2} = 2$$

49. $\dfrac{1-25y^{-2}}{1+10y^{-1}+25y^{-2}} = \dfrac{1-\frac{25}{y^2}}{1+\frac{10}{y}+\frac{25}{y^2}}$

$= \dfrac{y^2\left(1-\frac{25}{y^2}\right)}{y^2\left(1+\frac{10}{y}+\frac{25}{y^2}\right)}$

$= \dfrac{y^2-\frac{25y^2}{y^2}}{y^2+\frac{10y^2}{y}+\frac{25y^2}{y^2}}$

$= \dfrac{y^2-25}{y^2+10y+25} = \dfrac{(y+5)(y-5)}{(y+5)(y+5)}$

$= \dfrac{y-5}{y+5}$

APPLICATIONS

51. $\dfrac{\text{opening of blades}}{\text{opening of handles}} = \dfrac{\frac{x}{2}}{\frac{7x}{3}} = \dfrac{x}{2} \div \dfrac{7x}{3}$

$= \dfrac{x}{2} \bullet \dfrac{3}{7x} = \dfrac{x \bullet 3}{2 \bullet 7 \bullet x} = \dfrac{\not{x} \bullet 3}{2 \bullet 7 \bullet \not{x}} = \dfrac{3}{14}$

The ratio of the opening of the cutting blades to the opening of the handles is $\frac{3}{14}$.

53. Total resistance $= \dfrac{1}{\frac{1}{R_1}+\frac{1}{R_2}}$

$= \dfrac{1}{\frac{1 \bullet R_2}{R_1 \bullet R_2}+\frac{1 \bullet R_1}{R_2 \bullet R_1}}$

$= \dfrac{1}{\frac{R_2+R_1}{R_1R_2}} = 1 \div \dfrac{R_2+R_1}{R_1R_2}$

$= 1 \bullet \dfrac{R_1R_2}{R_1+R_2} = \dfrac{R_1R_2}{R_1+R_2}$

REVIEW

57. $t^3 \bullet t^4 \bullet t^2 = t^{3+4+2} = t^9$

59. $-2r(r^3)^2 = -2r \bullet r^{3 \bullet 2} = -2r$[illegible]

$= -2r^{1+6} = -2r^7$

61. $\left(\dfrac{3r}{4r^3}\right)^4 = \dfrac{(3r)^4}{(4r^3)^4} = \dfrac{3^4r^4}{4^4r^{3 \bullet 4}}$

$= \dfrac{81r^4}{256r^{12}} = \dfrac{81}{256r^8}$

63. $\left(\dfrac{6r^{-2}}{2r^3}\right)^{-2} = \left(\dfrac{6}{2}r^{-2-3}\right)^{-2}$

$= (3r^{-5})^{-2} = 3^{-2}r^{-5 \bullet (-2)}$

$= 3^{-2}r^{10} = \dfrac{r^{10}}{3^2} = \dfrac{r^{10}}{9}$

8.7

$\frac{1}{2} = \frac{7}{2a}$

$2a\left(\frac{2}{a} + \frac{1}{2}\right) = 2a\left(\frac{7}{2a}\right)$

$\frac{4a}{a} + \frac{2a}{2} = \frac{14a}{2a}$
$4 + a = 7$
$4 + a - 4 = 7 - 4$
$a = 3$

PRACTICE

9. $\frac{x}{2} + 4 = \frac{3x}{2}$

$2\left(\frac{x}{2} + 4\right) = 2\left(\frac{3x}{2}\right)$

$\frac{2x}{2} + 8 = \frac{6x}{2}$
$x + 8 = 3x$
$x + 8 - x = 3x - x$
$8 = 2x$
$\frac{8}{2} = \frac{2x}{2}$
$4 = x$

11. $\frac{2y}{5} - 8 = \frac{4y}{5}$

$5\left(\frac{2y}{5} - 8\right) = 5\left(\frac{4y}{5}\right)$

$\frac{10y}{5} - 40 = \frac{20y}{5}$
$2y - 40 = 4y$
$2y - 40 - 2y = 4y - 2y$
$-40 = 2y$
$\frac{-40}{2} = \frac{2y}{2}$
$-20 = y$

13. $\frac{x}{3} + 1 = \frac{x}{2}$

$6\left(\frac{x}{3} + 1\right) = 6\left(\frac{x}{2}\right)$

$\frac{6x}{3} + 6 = \frac{6x}{2}$
$2x + 6 = 3x$
$2x + 6 - 2x = 3x - 2x$
$6 = x$

15. $\frac{x}{5} - \frac{x}{3} = -8$
$15\left(\frac{x}{5} - \frac{x}{3}\right) = 15(-8)$

$\frac{15x}{5} - \frac{15x}{3} = -120$
$3x - 5x = -120$
$-2x = -120$
$\frac{-2x}{-2} = \frac{-120}{-2}$
$x = 60$

17. $\frac{3a}{2} + \frac{a}{3} = -22$
$6\left(\frac{3a}{2} + \frac{a}{3}\right) = 6(-22)$
$\frac{18a}{2} + \frac{6a}{3} = -132$
$9a + 2a = -132$
$11a = -132$
$\frac{11a}{11} = \frac{-132}{11}$
$a = -12$

19. $\frac{x-3}{3} + 2x = -1$
$3\left(\frac{x-3}{3} + 2x\right) = 3(-1)$

$\frac{3(x-3)}{3} + 6x = -3$
$x - 3 + 6x = -3$
$7x - 3 = -3$
$7x - 3 + 3 = -3 + 3$
$7x = 0$
$\frac{7x}{7} = \frac{0}{7}$
$x = 0$

21. $\frac{z-3}{2} = z + 2$

$2\left(\frac{z-3}{2}\right) = 2(z + 2)$

$\frac{2(z-3)}{2} = 2z + 4$
$z - 3 = 2z + 4$
$z - 3 + 3 = 2z + 4 + 3$
$z = 2z + 7$
$z - 2z = 2z + 7 - 2z$
$-z = 7$
$(-1)(-z) = (-1)(7)$
$z = -7$

23. $\frac{5(x+1)}{8} = x + 1$

$8\left(\frac{5(x+1)}{8}\right) = 8(x + 1)$

$\frac{40(x+1)}{8} = 8x + 8$
$5(x + 1) = 8x + 8$
$5x + 5 = 8x + 8$
$5x + 5 - 5 = 8x + 8 - 5$
$5x = 8x + 3$
$5x - 8x = 8x + 3 - 8x$
$-3x = 3$
$\frac{-3x}{-3} = \frac{3}{-3}$
$x = -1$

25. $\frac{c-4}{4} = \frac{c+4}{8}$
LCD = 8
$8\left(\frac{c-4}{4}\right) = 8\left(\frac{c+4}{8}\right)$

$\frac{8(c-4)}{4} = \frac{8(c+4)}{8}$

$2(c - 4) = c + 4$
$2c - 8 = c + 4$
$2c - 8 + 8 = c + 4 + 8$
$2c = c + 12$
$2c - c = c + 12 - c$
$c = 12$

27. $\frac{x+1}{3} + \frac{x-1}{5} = \frac{2}{15}$
LCD = 15
$15\left(\frac{x+1}{3}\right) + 15\left(\frac{x-1}{5}\right) = 15\left(\frac{2}{15}\right)$

$\frac{15(x+1)}{3} + \frac{15(x-1)}{5} = \frac{30}{15}$

$5(x + 1) + 3(x - 1) = 2$
$5x + 5 + 3x - 3 = 2$
$8x + 2 = 2$
$8x + 2 - 2 = 2 - 2$
$8x = 0$
$\frac{8x}{8} = \frac{0}{8}$
$x = 0$

29. $\frac{3x-1}{6} - \frac{x+3}{2} = \frac{3x+4}{3}$
LCD = 6
$6\left(\frac{3x-1}{6}\right) - 6\left(\frac{x+3}{2}\right) = 6\left(\frac{3x+4}{3}\right)$

$\frac{6(3x-1)}{6} - \frac{6(x+3)}{2} = \frac{6(3x+4)}{3}$

$(3x - 1) - 3(x + 3) = 2(3x + 4)$
$3x - 1 - 3x - 9 = 6x + 8$
$-10 = 6x + 8$
$-10 - 8 = 6x + 8 - 8$
$-18 = 6x$
$\frac{-18}{6} = \frac{6x}{6}$
$-3 = x$

31. $\frac{3}{x} + 2 = 3$
$x\left(\frac{3}{x} + 2\right) = x(3)$
$\frac{3x}{x} + 2x = 3x$
$3 + 2x = 3x$
$3 + 2x - 2x = 3x - 2x$
$3 = x$
check:
$\frac{3}{(3)} + 2 = 3$
$1 + 2 = 3$
$3 = 3$

33. $\frac{5}{a} - \frac{4}{a} = 8 + \frac{1}{a}$

$a\left(\frac{5}{a} - \frac{4}{a}\right) = a\left(8 + \frac{1}{a}\right)$

$\frac{5a}{a} - \frac{4a}{a} = 8a + \frac{a}{a}$
$5 - 4 = 8a + 1$
$1 = 8a + 1$
$1 - 1 = 8a + 1 - 1$
$0 = 8a$
$\frac{0}{8} = \frac{8a}{8}$
$0 = a$
check:
$\frac{5}{(0)} - \frac{4}{(0)} = 8 + \frac{1}{(0)}$

Division by 0 is undefined, so no solution; 0 is extraneous.

35. $\frac{2}{y+1} + 5 = \frac{12}{y+1}$

$(y+1)\left(\frac{2}{y+1} + 5\right) = (y+1)\left(\frac{12}{y+1}\right)$

$\frac{2(y+1)}{y+1} + 5(y+1) = \frac{12(y+1)}{y+1}$

$2 + 5y + 5 = 12$
$5y + 7 = 12$
$5y + 7 - 7 = 12 - 7$
$5y = 5$
$\frac{5y}{5} = \frac{5}{5}$
$y = 1$
check:
$\frac{2}{(1)+1} + 5 = \frac{12}{(1)+1}$
$\frac{2}{2} + 5 = \frac{12}{2}$
$1 + 5 = 6$
$6 = 6$

37. $\frac{1}{x-1} + \frac{3}{x-1} = 1$

$(x-1)\left(\frac{1}{x-1} + \frac{3}{x-1}\right) = (x-1)(1)$

$\frac{x-1}{x-1} + \frac{3(x-1)}{x-1} = x - 1$

$1 + 3 = x - 1$
$4 = x - 1$
$4 + 1 = x - 1 + 1$
$5 = x$
check:
$\frac{1}{(5)-1} + \frac{3}{(5)-1} = 1$
$\frac{1}{4} + \frac{3}{4} = 1$
$\frac{1+3}{4} = 1$
$\frac{4}{4} = 1$
$1 = 1$

39. $\frac{a^2}{a+2} - \frac{4}{a+2} = a$

$(a+2)\left(\frac{a^2}{a+2} - \frac{4}{a+2}\right) = (a+2)(a)$

$\frac{a^2(a+2)}{a+2} - \frac{4(a+2)}{a+2} = a^2 + 2a$

$a^2 - 4 = a^2 + 2a$
$a^2 - 4 - a^2 = a^2 + 2a - a^2$
$-4 = 2a$
$\frac{-4}{2} = \frac{2a}{2}$
$-2 = a$
check:
$\frac{(-2)^2}{(-2)+2} - \frac{4}{(-2)+2} = (-2)$
$\frac{4}{0} - \frac{4}{0} = -2$
Division by 0 is undefined, so no solution; -2 is extraneous.

41. $\frac{x}{x-5} - \frac{5}{x-5} = 3$

$(x-5)\left(\frac{x}{x-5} - \frac{5}{x-5}\right) = (x-5)(3)$

$\frac{x(x-5)}{x-5} - \frac{5(x-5)}{x-5} = 3x - 15$

$x - 5 = 3x - 15$
$x - 5 + 15 = 3x - 15 + 15$
$x + 10 = 3x$
$x + 10 - x = 3x - x$
$10 = 2x$
$\frac{10}{2} = \frac{2x}{2}$
$5 = x$
check:
$\frac{(5)}{(5)-5} - \frac{5}{(5)-5} = 3$
$\frac{5}{0} - \frac{5}{0} = 3$
Division by 0 is undefined, so no solution; 5 is extraneous.

43. $\frac{3r}{2} - \frac{3}{r} = \frac{3r}{2} + 3$

$2r\left(\frac{3r}{2} - \frac{3}{r}\right) = 2r\left(\frac{3r}{2} + 3\right)$

$\frac{6r^2}{2} - \frac{6r}{r} = \frac{6r^2}{2} + 6r$

$3r^2 - 6 = 3r^2 + 6r$
$3r^2 - 6 = 3r^2 + 6r$
$3r^2 - 6 - 3r^2 = 3r^2 + 6r - 3r^2$
$-6 = 6r$
$\frac{-6}{6} = \frac{6r}{6}$
$-1 = r$
check:
$\frac{3(-1)}{2} - \frac{3}{(-1)} = \frac{3(-1)}{2} + 3$
$-\frac{3}{2} + \frac{3}{1} = -\frac{3}{2} + 3$
$-\frac{3}{2} + \frac{6}{2} = -\frac{3}{2} + \frac{6}{2}$
$\frac{3}{2} = \frac{3}{2}$

45. $\frac{1}{3} + \frac{2}{x-3} = 1$

$3(x-3)\left(\frac{1}{3} + \frac{2}{x-3}\right) = 3(x-3)(1)$

$\frac{3(x-3)}{3} + \frac{6(x-3)}{x-3} = 3x - 9$
$x - 3 + 6 = 3x - 9$
$x + 3 = 3x - 9$
$x + 3 - 3 = 3x - 9 - 3$
$x = 3x - 12$
$x - 3x = 3x - 12 - 3x$
$-2x = -12$
$\frac{-2x}{-2} = \frac{-12}{-2}$
$x = 6$
check:
$\frac{1}{3} + \frac{2}{(6)-3} = 1$
$\frac{1}{3} + \frac{2}{3} = 1$
$\frac{3}{3} = 1$
$1 = 1$

47. $\frac{u}{u-1} + \frac{1}{u} = \frac{u^2+1}{u^2-u}$

$\frac{u}{u-1} + \frac{1}{u} = \frac{u^2+1}{u(u-1)}$

$u(u-1)\left(\frac{u}{u-1} + \frac{1}{u}\right)$
$= u(u-1)\left(\frac{u^2+1}{u(u-1)}\right)$

$\frac{u^2(u-1)}{u-1} + \frac{u(u-1)}{u} = \frac{u(u-1)(u^2+1)}{u(u-1)}$
$u^2 + u - 1 = u^2 + 1$
$u^2 + u - 1 - u^2 = u^2 + 1 - u^2$
$u - 1 = 1$
$u - 1 + 1 = 1 + 1$
$u = 2$
check:
$\frac{(2)}{(2)-1} + \frac{1}{(2)} = \frac{(2)^2+1}{(2)^2-(2)}$
$\frac{2}{1} + \frac{1}{2} = \frac{4+1}{4-2}$
$\frac{4}{2} + \frac{1}{2} = \frac{5}{2}$
$\frac{4+1}{2} = \frac{5}{2}$
$\frac{5}{2} = \frac{5}{2}$

49. $\frac{3}{x-2}+\frac{1}{x}=\frac{2(3x+2)}{x^2-2x}$

$\frac{3}{x-2}+\frac{1}{x}=\frac{2(3x+2)}{x(x-2)}$

$x(x-2)\left(\frac{3}{x-2}+\frac{1}{x}\right)$
$=x(x-2)\left(\frac{2(3x+2)}{x(x-2)}\right)$

$\frac{3x(x-2)}{x-2}+\frac{x(x-2)}{x}$
$=\frac{2x(x-2)(3x+2)}{x(x-2)}$

$3x+x-2=2(3x+2)$
$4x-2=6x+4$
$4x-2+2=6x+4+2$
$4x=6x+6$
$4x-6x=6x+6-6x$
$-2x=6$
$\frac{-2x}{-2}=\frac{6}{-2}$
$x=-3$
check:
$\frac{3}{(-3)-2}+\frac{1}{(-3)}=\frac{2(3(-3)+2)}{(-3)^2-2(-3)}$
$\frac{3}{-5}-\frac{1}{3}=\frac{2(-9+2)}{9-(-6)}$
$-\frac{3}{5}-\frac{1}{3}=\frac{2(-7)}{9+6}$
$-\frac{3}{5}-\frac{1}{3}=\frac{-14}{15}$
$-\frac{9}{15}-\frac{5}{15}=-\frac{14}{15}$
$-\frac{14}{15}=-\frac{14}{15}$

51. $\frac{7}{q^2-q-2}+\frac{1}{q+1}=\frac{3}{q-2}$

$\frac{7}{(q+1)(q-2)}+\frac{1}{q+1}=\frac{3}{q-2}$

$(q+1)(q-2)\left(\frac{7}{(q+1)(q-2)}+\frac{1}{q+1}\right)$
$=(q+1)(q-2)\left(\frac{3}{q-2}\right)$

$\frac{7(q+1)(q-2)}{(q+1)(q-2)}+\frac{(q+1)(q-2)}{q+1}$
$=\frac{3(q+1)(q-2)}{q-2}$

$7+q-2=3(q+1)$
$5+q=3q+3$
$5+q-5=3q+3-5$
$q=3q-2$
$q-3q=3q-2-3q$
$-2q=-2$
$\frac{-2q}{-2}=\frac{-2}{-2}$
$q=1$
check:
$\frac{7}{(1)^2-(1)-2}+\frac{1}{(1)+1}=\frac{3}{(1)-2}$
$\frac{7}{1-1-2}+\frac{1}{2}=\frac{3}{-1}$
$\frac{7}{-2}+\frac{1}{2}=-3$
$\frac{-7+1}{2}=-3$
$\frac{-6}{2}=-3$
$-3=-3$

53. $\frac{3y}{3y-6}+\frac{8}{y^2-4}=\frac{2y}{2y+4}$

$\frac{3y}{3(y-2)}+\frac{8}{(y-2)(y+2)}=\frac{2y}{2(y+2)}$

$\frac{y}{(y-2)}+\frac{8}{(y-2)(y+2)}=\frac{y}{(y+2)}$

$(y+2)(y-2)\left(\frac{y}{(y-2)}+\frac{8}{(y-2)(y+2)}\right)$
$=(y+2)(y-2)\left(\frac{y}{(y+2)}\right)$

$\frac{y(y+2)(y-2)}{(y-2)}+\frac{8(y+2)(y-2)}{(y-2)(y+2)}$
$=\frac{y(y+2)(y-2)}{(y+2)}$

$y(y+2)+8=y(y-2)$
$y^2+2y+8=y^2-2y$
$y^2+2y+8-y^2=y^2-2y-y^2$
$2y+8=-2y$
$2y+8-2y=-2y-2y$
$8=-4y$
$\frac{8}{-4}=\frac{-4y}{-4}$
$-2=y$
check:
$\frac{3(-2)}{3(-2)-6}+\frac{8}{(-2)^2-4}=\frac{2(-2)}{2(-2)+4}$
$\frac{-6}{-6-6}+\frac{8}{4-4}=\frac{-4}{-4+4}$
$\frac{-6}{-12}+\frac{8}{0}=\frac{-4}{0}$

Division by 0 is undefined, so no solution; -2 is extraneous.

55. $y+\frac{2}{3}=\frac{2y-12}{3y-9}$

$y+\frac{2}{3}=\frac{2y-12}{3(y-3)}$

$3(y-3)\left(y+\frac{2}{3}\right)$
$=3(y-3)\left(\frac{2y-12}{3(y-3)}\right)$

$3(y-3)y+\frac{2\cdot 3(y-3)}{3}$
$=\frac{(2y-12)3(y-3)}{3(y-3)}$

$3y(y-3)+2(y-3)=2y-12$
$3y^2-9y+2y-6=2y-12$
$3y^2-7y-6=2y-12$
$3y^2-7y-6-2y=2y-12-2y$
$3y^2-9y-6=-12$
$3y^2-9y-6+12=-12+12$
$3y^2-9y+6=0$
$3(y^2-3y+2)=0$
$3(y-2)(y-1)=0$
$y-2=0 \qquad y-1=0$
$y=2 \qquad y=1$
check:
$(2)+\frac{2}{3}=\frac{2(2)-12}{3(2)-9}$
$\frac{6}{3}+\frac{2}{3}=\frac{4-12}{6-9}$
$\frac{8}{3}=\frac{-8}{-3}$
$\frac{8}{3}=\frac{8}{3}$

$(1)+\frac{2}{3}=\frac{2(1)-12}{3(1)-9}$
$\frac{3}{3}+\frac{2}{3}=\frac{2-12}{3-9}$
$\frac{5}{3}=\frac{-10}{-6}$
$\frac{5}{3}=\frac{5}{3}$

57. $\frac{5}{4y+12}-\frac{3}{4}=\frac{5}{4y+12}-\frac{y}{4}$

$\frac{5}{4(y+3)}-\frac{3}{4}=\frac{5}{4(y+3)}-\frac{y}{4}$

$4(y+3)\left(\frac{5}{4(y+3)}-\frac{3}{4}\right)$
$=4(y+3)\left(\frac{5}{4(y+3)}-\frac{y}{4}\right)$

$\frac{5\cdot 4(y+3)}{4(y+3)}-\frac{3\cdot 4(y+3)}{4}$
$=\frac{5\cdot 4(y+3)}{4(y+3)}-\frac{y\cdot 4(y+3)}{4}$

$5-3(y+3)=5-y(y+3)$
$5-3y-9=5-y^2-3y$
$-4-3y=5-y^2-3y$
$-4-3y+y^2=5-y^2-3y+y^2$
$-4-3y+y^2=5-3y$
$-4-3y+3y+y^2=5-3y+3y$
$-4+y^2=5$
$-4+y^2-5=5-5$
$y^2-9=0$
$(y-3)(y+3)=0$
$y-3=0 \qquad y+3=0$
$y=3 \qquad y=-3$
check:
$\frac{5}{4(3)+12}-\frac{3}{4}=\frac{5}{4(3)+12}-\frac{(3)}{4}$
$\frac{5}{12+12}-\frac{3}{4}=\frac{5}{12+12}-\frac{3}{4}$
$\frac{5}{24}-\frac{3}{4}=\frac{5}{24}-\frac{3}{4}$
$\frac{5}{24}-\frac{18}{24}=\frac{5}{24}-\frac{18}{24}$
$-\frac{13}{24}=-\frac{13}{24}$

$\frac{5}{4(-3)+12}-\frac{3}{4}=\frac{5}{4(-3)+12}-\frac{(-3)}{4}$
$\frac{5}{-12+12}-\frac{3}{4}=\frac{5}{-12+12}+\frac{3}{4}$
$\frac{5}{0}-\frac{3}{4}=\frac{5}{0}+\frac{3}{4}$
Division by 0 is undefined, so -3 is extraneous.

59. $\frac{x}{x-1} - \frac{12}{x^2-x} = \frac{-1}{x-1}$

$\frac{x}{x-1} - \frac{12}{x(x-1)} = \frac{-1}{(x-1)}$

$x(x-1)\left(\frac{x}{x-1} - \frac{12}{x(x-1)}\right)$
$= x(x-1)\left(\frac{-1}{x-1}\right)$

$\frac{x \cdot x(x-1)}{x-1} - \frac{12x(x-1)}{x(x-1)}$
$= \frac{-1x(x-1)}{x-1}$

$x^2 - 12 = -x$
$x^2 - 12 + x = -x + x$
$x^2 + x - 12 = 0$
$(x+4)(x-3) = 0$
$x + 4 = 0 \qquad x - 3 = 0$
$x = -4 \qquad x = 3$
check:
$\frac{(-4)}{(-4)-1} - \frac{12}{(-4)^2-(-4)} = \frac{-1}{(-4)-1}$
$\frac{-4}{-5} - \frac{12}{16-(-4)} = \frac{-1}{-5}$
$\frac{4}{5} - \frac{12}{20} = \frac{1}{5}$
$\frac{16}{20} - \frac{12}{20} = \frac{4}{20}$
$\frac{4}{20} = \frac{4}{20}$

$\frac{(3)}{(3)-1} - \frac{12}{(3)^2-(3)} = \frac{-1}{(3)-1}$
$\frac{3}{2} - \frac{12}{9-3} = \frac{-1}{2}$
$\frac{3}{2} - \frac{12}{6} = -\frac{1}{2}$
$\frac{9}{6} - \frac{12}{6} = -\frac{3}{6}$
$-\frac{3}{6} = -\frac{3}{6}$

61. $\frac{z-4}{z-3} = \frac{z+2}{z+1}$

$(z-3)(z+1)\left(\frac{z-4}{z-3}\right)$
$= (z-3)(z+1)\left(\frac{z+2}{z+1}\right)$

$\frac{(z-4)(z-3)(z+1)}{z-3}$
$= \frac{(z+2)(z-3)(z+1)}{z+1}$
$(z-4)(z+1) = (z+2)(z-3)$
$z^2 - 3z - 4 = z^2 - z - 6$
$z^2 - 3z - 4 + z^2 = z^2 - z - 6 + z^2$
$-3z - 4 = -z - 6$
$-3z - 4 + z = -z - 6 + z$
$-2z - 4 = -6$
$-2z - 4 + 4 = -6 + 4$
$-2z = -2$
$\frac{-2z}{-2} = \frac{-2}{-2}$
$z = 1$
check:
$\frac{(1)-4}{(1)-3} = \frac{(1)+2}{(1)+1}$
$\frac{-3}{-2} = \frac{3}{2}$
$\frac{3}{2} = \frac{3}{2}$

63. $\frac{n}{n^2-9} + \frac{n+8}{n+3} = \frac{n-8}{n-3}$

$\frac{n}{(n+3)(n-3)} + \frac{n+8}{n+3} = \frac{n-8}{n-3}$

$(n+3)(n-3)\left(\frac{n}{(n+3)(n-3)} + \frac{n+8}{n+3}\right)$
$= (n+3)(n-3)\left(\frac{n-8}{n-3}\right)$

$\frac{n(n+3)(n-3)}{(n+3)(n-3)} + \frac{(n+8)(n+3)(n-3)}{n+3}$
$= \frac{(n-8)(n+3)(n-3)}{n-3}$

$n + (n+8)(n-3) = (n+8)(n+3)$
$n + n^2 + 5n - 24 = n^2 + 11n - 24$
$n^2 + 6n - 24 = n^2 + 11n - 24$
$n^2 + 6n - 24 - n^2 = n^2 + 11n - 24 - n^2$
$6n - 24 = 11n - 24$
$6n - 24 - 11n = 11n - 24 - 11n$
$-5n - 24 = -24$
$-5n - 24 + 24 = -24 + 24$
$-5n = 0$
$\frac{-5n}{-5} = \frac{0}{-5}$
$n = 0$
check:
$\frac{(0)}{(0)^2-9} + \frac{(0)+8}{(0)+3} = \frac{(0)-8}{(0)-3}$
$\frac{0}{0-9} + \frac{8}{3} = \frac{-8}{-3}$
$\frac{0}{-9} + \frac{8}{3} = \frac{8}{3}$
$0 + \frac{8}{3} = \frac{8}{3}$
$\frac{8}{3} = \frac{8}{3}$

65. $\frac{b+2}{b+3}+1=\frac{-7}{b-5}$

$$(b+3)(b-5)\left(\frac{b+2}{b+3}+1\right)$$
$$=(b+3)(b-5)\left(\frac{-7}{b-5}\right)$$

$$\frac{(b+2)(b+3)(b-5)}{b+3}+1(b+3)(b-5)$$
$$=\frac{-7(b+3)(b-5)}{b-5}$$

$$(b+2)(b-5)+(b+3)(b-5)$$
$$=-7(b+3)$$

$$b^2-3b-10+b^2-2b-15$$
$$=-7b-21$$

$$2b^2-5b-25=-7b-21$$
$$2b^2-5b-25+7b$$
$$=-7b-21+7b$$

$$2b^2+2b-25=-21$$
$$2b^2+2b-25+21=-21+21$$
$$2b^2+2b-4=0$$
$$2(b^2+b-2)=0$$
$$2(b+2)(b-1)=0$$
$$b+2=0 \qquad b-1=0$$
$$b=-2 \qquad b=1$$

check:

$$\frac{(-2)+2}{(-2)+3}+1=\frac{-7}{(-2)-5}$$
$$\frac{0}{1}+1=\frac{-7}{-7}$$
$$0+1=1$$
$$1=1$$

$$\frac{(1)+2}{(1)+3}+1=\frac{-7}{(1)-5}$$
$$\frac{3}{4}+1=\frac{-7}{-4}$$
$$\frac{3}{4}+\frac{4}{4}=\frac{7}{4}$$
$$\frac{7}{4}=\frac{7}{4}$$

67. $\frac{1}{a}+\frac{1}{b}=1$

$$ab\left(\frac{1}{a}+\frac{1}{b}\right)=ab(1)$$

$$\frac{ab}{a}+\frac{ab}{b}=ab$$
$$b+a=ab$$
$$b+a-b=ab-b$$
$$a=ab-b$$
$$a-ab=ab-b-ab$$
$$a-ab=-b$$
$$a(1-b)=-b$$
$$\frac{a(1-b)}{1-b}=\frac{-b}{1-b}$$
$$a=\frac{-b}{1-b}$$
$$a=\frac{(-1)b}{(-1)(-1+b)}$$

$$a=\frac{b}{(-1+b)}$$

$$a=\frac{b}{(b-1)}$$

check:

$$\frac{1}{\left(\frac{b}{(b-1)}\right)}+\frac{1}{b}=1$$
$$1\bullet\frac{b-1}{b}+\frac{1}{b}=1$$
$$\frac{b-1}{b}+\frac{1}{b}=1$$

$$\frac{b-1+1}{b}=1$$
$$\frac{b}{b}=1$$
$$1=1$$

69. $\frac{1}{f}=\frac{1}{d_1}+\frac{1}{d_2}$

$$fd_1d_2\left(\frac{1}{f}\right)=fd_1d_2\left(\frac{1}{d_1}+\frac{1}{d_2}\right)$$

$$\frac{fd_1d_2}{f}=\frac{fd_1d_2}{d_1}+\frac{fd_1d_2}{d_2}$$
$$d_1d_2=fd_2+fd_1$$
$$d_1d_2=f(d_2+d_1)$$
$$\frac{d_1d_2}{(d_2+d_1)}=\frac{f(d_2+d_1)}{(d_2+d_1)}$$
$$\frac{d_1d_2}{(d_2+d_1)}=f$$

71. $H = \frac{RB}{R+B}$

$$(R+B)(H) = (R+B)\left(\frac{RB}{R+B}\right)$$
$$RH + BH = \frac{RB(R+B)}{R+B}$$
$$RH + BH = RB$$
$$RH + BH - RB = RB - RB$$
$$RH + BH - RB = 0$$
$$RH + BH - RB - BH = 0 - BH$$
$$RH - RB = -BH$$
$$R(H - B) = -BH$$
$$\frac{R(H-B)}{H-B} = \frac{-BH}{H-B}$$
$$R = \frac{BH}{-1(H-B)} = \frac{BH}{-H+B} = \frac{BH}{B+H}$$

73. To find the constant of variation, substitute 3 for t and 50 for r in the formula:

$$t = \frac{k}{r}$$
$$3 = \frac{k}{50}$$
$$50(3) = 50\left(\frac{k}{50}\right)$$
$$150 = k$$

To find the time it takes driving 60 mph, substitute 150 for k and 60 for r in

$$t = \frac{k}{r}$$
$$t = \frac{150}{60}$$
$$t = \frac{10 \cdot 5 \cdot 3}{10 \cdot 2 \cdot 3}$$
$$t = \frac{\cancel{10} \cdot 5 \cdot \cancel{3}}{\cancel{10} \cdot 2 \cdot \cancel{3}}$$
$$t = \frac{5}{2} = 2\frac{1}{2}$$

The trip will take $2\frac{1}{2}$ hours.

75. To find the constant of variation, substitute 40 for V and 8 for p in the formula:

$$V = \frac{k}{p}$$
$$40 = \frac{k}{8}$$
$$8(40) = 8\left(\frac{k}{8}\right)$$
$$320 = k$$

To find the volume when the pressure is 6 atmospheres, substitute 320 for k and 6 for p in the formula:

$$V = \frac{k}{p}$$
$$V = \frac{320}{6}$$
$$V = \frac{2 \cdot 160}{2 \cdot 3}$$
$$V = \frac{\cancel{2} \cdot 160}{\cancel{2} \cdot 3}$$
$$V = \frac{160}{3} = 53\frac{1}{3}$$

The volume will be $53\frac{1}{3}$ m^3.

REVIEW

79. $x^2 + 4x = x(x+4)$

81. $2x^2 + x - 3 = (2x+3)(x-1)$

83. $x^4 - 16 = (x^2 - 4)(x^2 + 4)$
$= (x-2)(x+2)(x^2+4)$

STUDY SET Section 8.8

VOCABULARY

1. interest; principal; rate

3. clear; LCD

CONCEPTS

5. analyze, form, solve, state, & check

7. a) $d = rt$

$\frac{d}{t} = \frac{rt}{t}$

$\frac{d}{t} = r$

b) $d = rt$

$\frac{d}{r} = \frac{rt}{r}$

$\frac{d}{r} = t$

9. a) Amount of shed assembled in 1 hr by Marvin $= \frac{1}{6}$.

Amount of shed assembled in 1 hr by Kyla $= \frac{1}{5}$.

b) Amount of shed assembled in 1 hr by both $= \frac{1}{6} + \frac{1}{5}$

$= \frac{1 \cdot 5}{6 \cdot 5} + \frac{1 \cdot 6}{5 \cdot 6}$

$= \frac{5+6}{30}$

$= \frac{11}{30}$

11. Since the 2 ice machines can fill a supermarkets' order in x hours, then they can fill $\frac{1}{x}$ of the order in 1 hr.

PRACTICE

13. **Analyze:** We are asked to find a number. If we add it to the denominator of $\frac{3}{4}$ and multiply the numerator by 2, the result is 1.

Form: Let n represent the unknown number.

The denominator will be $4 + n$.

The numerator will be $3 \cdot 2$.

$\frac{3 \cdot 2}{n+4} = 1$

Solve:

$\frac{6}{n+4} = 1$

$(n+4)\left(\frac{6}{n+4}\right) = (n+4)(1)$

$\frac{6(n+4)}{n+4} = n + 4$

$6 = n + 4$

$6 - 4 = n + 4 - 4$

$2 = n$

State: The number is 2.

Check:

$\frac{3 \cdot 2}{(2)+4} = 1$

$\frac{6}{6} = 1$

$1 = 1$

15. **Analyze:** We are asked to find a number. If we add it to the numerator of $\frac{3}{4}$ and add 2 times it to the denominator, the result is $\frac{4}{7}$.

Form: Let n represent the unknown number.

The numerator will be $3 + n$.

The denominator will be $4 + 2n$.

$\frac{3+n}{4+2n} = \frac{4}{7}$

Solve:

$\frac{3+n}{4+2n} = \frac{4}{7}$

$7(4+2n)\left(\frac{3+n}{4+2n}\right)$

$= 7(4+2n)(\frac{4}{7})$

$\frac{7(4+2n)(3+n)}{4+2n} = \frac{28(4+2n)}{7}$

$7(3+n) = 4(4+2n)$

$21 + 7n = 16 + 8n$

$21 + 7n - 8n = 16 + 8n - 8n$

$21 - n = 16$

$21 - n - 21 = 16 - 21$

$-n = -5$

$(-1)(-n) = (-1)(-5)$

$n = 5$

State: The number is 5.

Check:

$\frac{3+(5)}{4+2(5)} = \frac{4}{7}$

$\frac{8}{4+10} = \frac{4}{7}$

$\frac{8}{14} = \frac{4}{7}$

$\frac{\cancel{2} \cdot 4}{\cancel{2} \cdot 7} = \frac{4}{7}$

$\frac{4}{7} = \frac{4}{7}$

17. Analyze: We are asked to find a number. If we add it to its reciprocal, the result is $\frac{13}{6}$.

Form: Let n represent the unknown number.

The reciprocal is $\frac{1}{n}$.

$$n + \frac{1}{n} = \frac{13}{6}$$

Solve:

$$n + \frac{1}{n} = \frac{13}{6}$$
$$6n\left(n + \frac{1}{n}\right) = 6n\left(\frac{13}{6}\right)$$
$$6n^2 + \frac{6n}{n} = \frac{108n}{6}$$
$$6n^2 + 6 = 13n$$
$$6n^2 + 6 - 13n = 13n - 13n$$
$$6n^2 - 13n + 6 = 0$$
$$(3n - 2)(2n - 3) = 0$$
$$3n - 2 = 0 \qquad 2n - 3 = 0$$
$$3n = 2 \qquad 2n = 3$$
$$n = \frac{2}{3} \qquad n = \frac{3}{2}$$

State: The number is $\frac{2}{3}$ or $\frac{3}{2}$.

Check:

$$\left(\frac{2}{3}\right) + \frac{1}{\left(\frac{2}{3}\right)} = \frac{13}{6}$$
$$\frac{2}{3} + 1 \cdot \frac{3}{2} = \frac{13}{6}$$
$$\frac{2}{3} + \frac{3}{2} = \frac{13}{6}$$
$$\frac{2 \cdot 2}{3 \cdot 2} + \frac{3 \cdot 3}{2 \cdot 3} = \frac{13}{6}$$
$$\frac{4}{6} + \frac{9}{6} = \frac{13}{6}$$
$$\frac{13}{6} = \frac{13}{6}$$

$$\left(\frac{3}{2}\right) + \frac{1}{\left(\frac{3}{2}\right)} = \frac{13}{6}$$
$$\frac{3}{2} + 1 \cdot \frac{2}{3} = \frac{13}{6}$$
$$\frac{3}{2} + \frac{2}{3} = \frac{13}{6}$$
$$\frac{3 \cdot 3}{2 \cdot 3} + \frac{2 \cdot 2}{3 \cdot 2} = \frac{13}{6}$$
$$\frac{9}{6} + \frac{4}{6} = \frac{13}{6}$$
$$\frac{13}{6} = \frac{13}{6}$$

19. Analyze: If the first inlet pipe can fill the pool in 5 hours, then it can fill $\frac{1}{5}$ of the pool in 1 hour. If the other inlet pipe can fill the pool in 4 hours, then it can fill $\frac{1}{4}$ of the pool in 1 hour. If it takes x hours for both pipes to fill the pool, then they can fill $\frac{1}{x}$ of the pool in 1 hour.

Form: Let x represent the number of hours it will take to fill the pool if both pipes are used. Then,

$$\frac{1}{5} + \frac{1}{4} = \frac{1}{x}$$

Solve:

$$\frac{1}{5} + \frac{1}{4} = \frac{1}{x}$$
$$20x\left(\frac{1}{5} + \frac{1}{4}\right) = 20x\left(\frac{1}{x}\right)$$
$$\frac{20x}{5} + \frac{20x}{4} = \frac{20x}{x}$$
$$4x + 5x = 20$$
$$9x = 20$$
$$\frac{9x}{9} = \frac{20}{9}$$
$$x = \frac{20}{9} = 2\frac{2}{9}$$

State: It will take $\frac{20}{9}$ or $2\frac{2}{9}$ hrs. for both inlet pipes to fill the pool.

Check:

$$\frac{1}{5} + \frac{1}{4} = \frac{1}{\left(\frac{20}{9}\right)}$$
$$\frac{1}{5} + \frac{1}{4} = 1 \cdot \frac{9}{20}$$
$$\frac{1}{5} + \frac{1}{4} = \frac{9}{20}$$
$$\frac{1 \cdot 4}{5 \cdot 4} + \frac{1 \cdot 5}{4 \cdot 5} = \frac{9}{20}$$
$$\frac{4}{20} + \frac{5}{20} = \frac{9}{20}$$
$$\frac{9}{20} = \frac{9}{20}$$

21. Analyze: If the homeowner can roof his house in 7 days, then he can roof $\frac{1}{7}$ of the house in 1 day. If the professional can roof the house 4 days, then he can roof $\frac{1}{4}$ of the house in 1 day. If it takes x days for both people to roof the house, then they can roof $\frac{1}{x}$ of house in 1 day.

Form: Let x represent the number of days it will take the homeowner and the professional to roof the house.

Then,

$$\frac{1}{7} + \frac{1}{4} = \frac{1}{x}$$

Solve:

$$\frac{1}{7} + \frac{1}{4} = \frac{1}{x}$$
$$28x\left(\frac{1}{7} + \frac{1}{4}\right) = 28x\left(\frac{1}{x}\right)$$

$\frac{28x}{7}+\frac{28x}{4}=\frac{28x}{x}$
$4x+7x=28$
$11x=28$
$\frac{11x}{11}=\frac{28}{11}$
$x=\frac{28}{11}=2\frac{6}{11}$

State: It will take $\frac{28}{11}$ or $2\frac{6}{11}$ hrs. for the homeowner and the professional to roof the house.

Check:

$\frac{1}{7}+\frac{1}{4}=\frac{1}{\left(\frac{28}{11}\right)}$
$\frac{1}{7}+\frac{1}{4}=1\bullet\frac{11}{28}$
$\frac{1}{7}+\frac{1}{4}=\frac{11}{28}$
$\frac{1\bullet 4}{7\bullet 4}+\frac{1\bullet 7}{4\bullet 7}=\frac{11}{28}$
$\frac{4}{28}+\frac{7}{28}=\frac{11}{28}$
$\frac{11}{28}=\frac{11}{28}$

23. **Analyze:** We can use the formula $d=rt$. We solve this formula for t :

$d=rt$
$\frac{d}{r}=\frac{rt}{r}$
$\frac{d}{r}=t$

If the man walks 8 miles at some unknown rate r mph, it will take $\frac{8}{r}$ hours. If he bicycles 28 miles at some unknown rate of $r+10$, it will take $\frac{28}{r+10}$ hours. We will find the rate at which the man can walk, then use that to find the time it will take him to walk 30 miles.

Form: The man can bicycle 29 miles in the same time as it takes him to walk 8 miles. Therefore,

$\frac{8}{r}=\frac{28}{r+10}$

Solve:

$\frac{8}{r}=\frac{28}{r+10}$
$r(r+10)\left(\frac{8}{r}\right)$
$=r(r+10)\left(\frac{28}{r+10}\right)$
$\frac{8r(r+10)}{r}=\frac{28r(r+10)}{r+10}$
$8(r+10)=28r$
$8r+80=28r$
$8r+80-8r=28r-8r$
$80=20r$
$\frac{80}{20}=\frac{20r}{20}$
$4=r$

State: The man can walk at a rate of 4 mph, so it will take $\frac{d}{r}=\frac{30}{4}=7\frac{1}{2}$ hrs. to walk 30 miles.

Check:

$\frac{8}{(4)}=\frac{28}{(4)+10}$
$\frac{8}{4}=\frac{28}{14}$
$2=2$

25. **Analyze:** We can use the formula $d=rt$. We solve this formula for t :

$d=rt$
$\frac{d}{r}=\frac{rt}{r}$
$\frac{d}{r}=t$

When the boat travels upstream, it is traveling against the current and when it travels downstream, it is traveling with the current. If the speed of the current is some unknown rate r, then the speed of the boat going downstream will be $18+r$ mph. the boat can travel 22 miles downstream at a rate of $18+r$ mph, so it will take $\frac{22}{18+r}$ hours for the boat to travel downstream. The speed of the boat going upstream will be $18-r$ mph, so it will take $\frac{14}{18-r}$ hours.

Form: The boat can travel 22 miles downstream in the same amount of time it takes to travel 14 miles upstream. Therefore,

$\frac{22}{18+r}=\frac{14}{18-r}$

Solve:

$\frac{22}{18+r}=\frac{14}{18-r}$
$(18+r)(18-r)\left(\frac{22}{18+r}\right)$
$=(18+r)(18-r)\left(\frac{14}{18-r}\right)$
$\frac{22(18+r)(18-r)}{18+r}$
$=\frac{14(18+r)(18-r)}{18-r}$
$22(18-r)=14(18+r)$
$396-22r=252+14r$

$396 - 22r - 396$

$= 252 + 14r - 396$

$-22r = -144 + 14r$

$-22r - 14r$

$= -144 + 14r - 14r$

$-36r = -144$

$\frac{-36r}{-36} = \frac{-144}{-36}$

$r = 4$

State: The speed of the current is 4 mph.

Check:

$\frac{22}{18+(4)} = \frac{14}{18-(4)}$

$\frac{22}{22} = \frac{14}{14}$

$1 = 1$

27. Analyze: We can use the formula $i = pr$. We solve this formula for p :

$i = pr$

$\frac{i}{p} = \frac{pr}{p}$

$\frac{i}{p} = i$

Let r represent the first CD's rate, then $r + 0.01$ is the second CD's rate. If the first CD earns $r\%$ interest, the principal invested was $\frac{175}{r}$. If the second CD earns $r + 0.01\%$, the principal invested was $\frac{200}{r+0.01}$ hours.

Form: The same principal is invested in each CD, so

$\frac{175}{r} = \frac{200}{r+0.01}$

Solve:

$\frac{175}{r} = \frac{200}{r+0.01}$

$r(r + 0.01)\left(\frac{175}{r}\right)$

$= r(r + 0.01)\left(\frac{200}{r+0.01}\right)$

$\frac{175r(r+0.01)}{r}$

$= \frac{200r(r+0.01)}{r+0.01}$

$175(r + 0.01) = 200r$

$175r + 1.75 = 200r$

$175r + 1.75 - 175r = 200r - 175r$

$1.75 = 225r$

$\frac{1.75}{225} = \frac{225r}{225}$

$0.07 = r$

State: The first CD's interest rate is 0.07 or 7%. The second CD's rate is 1% greater, or 8%.

Check:

$\frac{175}{(0.07)} = \frac{200}{(0.07)+0.01}$

$\frac{175}{0.07} = \frac{200}{0.08}$

$2500 = 2500$

29. Analyze: If we let x represent the number of workers to contributed to the gift, then the cost for each worker is $\frac{35}{x}$. If 2 more workers contribute, then the cost for each is reduced by \$2. thus, the original cost for each worker is $\frac{35}{x+2} + 2$.

Form: Then,

$\frac{35}{x} = \frac{35}{x+2} + 2$

Solve:

$\frac{35}{x} = \frac{35}{x+2} + 2$

$x(x + 2)\left(\frac{35}{x}\right)$

$= x(x + 2)\left(\frac{35}{x+2} + 2\right)$

$\frac{35x(x+2)}{x}$

$= \frac{35x(x+2)}{x+2} + 2x(x + 2)$

$35(x + 2) = 35x + 2x(x + 2)$

$35x + 70 = 35x + 2x^2 + 4x$

$35x + 70 = 39x + 2x^2$

$35x + 70 - 35x$

$= 39x + 2x^2 - 35x$

$70 = 4x + 2x^2$

$70 - 70 = 4x + 2x^2 - 70$

$0 = 2x^2 + 4x - 70$

$0 = 2(x^2 + 2x - 35)$

$0 = 2(x - 5)(x + 7)$

$x - 5 = 0 \qquad x + 7 = 0$

$x = 5 \qquad x = -7$

State: Since the number of workers must be positive, then 5 workers contributed to the gift.

Check:

$\frac{35}{(5)} = \frac{35}{(5)+2} + 2$

$\frac{35}{5} = \frac{35}{7} + 2$

$7 = 5 + 2$

$7 = 7$

31. **Analyze:** If we let x represent the number of calculators purchased at the regular price, then the cost of each calculator is $\frac{120}{x}$.

If each calculator costs \$1 less then the original cost of each calculator is $\frac{120}{x+10} + 1$.

Form: Then,

$$\frac{120}{x} = \frac{120}{x+10} + 1$$

Solve:

$$\frac{120}{x} = \frac{120}{x+10} + 1$$
$$x(x+10)\left(\frac{120}{x}\right) = x(x+10)\left(\frac{120}{x+10} + 1\right)$$
$$\frac{120x(x+10)}{x} = \frac{120x(x+10)}{x+10} + x(x+10)$$
$$120(x+10) = 120x + x(x+10)$$
$$120x + 1200 = 120x + x^2 + 10x$$
$$120x + 1200 = 130x + x^2$$
$$120x + 1200 - 120x = 130x + x^2 - 120x$$
$$1200 = 10x + x^2$$
$$1200 - 1200 = 10x + x^2 - 1200$$
$$0 = x^2 + 10x - 1200$$
$$0 = (x-30)(x+40)$$
$$x - 30 = 0 \qquad x + 40 = 0$$
$$x = 30 \qquad x = -40$$

State: Since the number of calculators must be positive, the bookstore an purchase 30 calculators at the regular price.

Check:

$$\frac{120}{(30)} = \frac{120}{(30)+10} + 1$$
$$\frac{120}{30} = \frac{120}{40} + 1$$
$$4 = 3 + 1$$
$$4 = 4$$

33. **Analyze:** We can use the formula $d = rt$. We solve this formula for t :

$$d = rt$$
$$\frac{d}{r} = \frac{rt}{r}$$
$$\frac{d}{r} = t$$

If the speed of the current is some unknown rate, r, then the speed of the boat going upstream is $r - 5$ mph and going downstream is $r + 5$ mph. The boat goes 60 miles upstream at $r - 5$ mph so it takes $\frac{60}{r-5}$ hours. It then goes 60 miles downstream at $r + 5$ mph so it takes $\frac{60}{r+5}$ hours.

Form: The total trip is to take 5 hours, so

$$\frac{60}{r-5} + \frac{60}{r+5} = 5$$

Solve:

$$\frac{60}{r-5} + \frac{60}{r+5} = 5$$
$$(r-5)(r+5)\left(\frac{60}{r-5} + \frac{60}{r+5}\right) = (r-5)(r+5)(5)$$
$$\frac{60(r-5)(r+5)}{r-5} + \frac{60(r-5)(r+5)}{r+5} = 5(r-5)(r+5)$$
$$60(r+5) + 60(r-5) = 5(r^2 - 25)$$
$$60r + 300 + 60r - 300 = 5r^2 - 125$$
$$120r = 5r^2 - 125$$
$$120r - 120r = 5r^2 - 125 - 120r$$
$$0 = 5r^2 - 120r - 125$$
$$0 = 5(r^2 - 24r - 25)$$
$$0 = 5(r-25)(r+1)$$
$$r - 25 = 0 \qquad r + 1 = 0$$
$$r = 25 \qquad r = -1$$

State: The still-water speed should be positive, so the still-water speed is 25 mph.

Check:

$$\frac{60}{(25)-5} + \frac{60}{(25)+5} = 5$$
$$\frac{60}{20} + \frac{60}{30} = 5$$
$$3 + 2 = 5$$
$$5 = 5$$

REVIEW

37. $x^2 - 5x - 6 = 0$

$$(x+6)(x+1) = 0$$
$$x - 6 = 0 \qquad x + 1 = 0$$
$$x = 6 \qquad x = -1$$

39. $(t+2)(t^2+7t+12=0$
$(t+2)(t+4)(t+3)=0$
$t+2=0 \quad t+4=0 \quad t+3=0$
$t=-2 \quad t=-4 \quad t=-3$

41. $y^3-y^2=0$
$y^2(y-1)=0$
$y^2=0 \quad y-1=0$
$y=0 \quad y=1$

43. $(x^2-1)(x^2-4)=0$
$(x-1)(x+1)(x-2)(x+2)=0$
$x-1=0 \quad x+1=0$
$x=1 \quad x=-1$

$x-2=0 \quad x+2=0$
$x=2 \quad x=-2$

CHAPTER 8 REVIEW

1. a) 3 to 6 :

$$\frac{3}{6} = \frac{1\cdot 3}{2\cdot 3} = \frac{1\cdot \not{3}}{2\cdot \not{3}} = \frac{1}{2}$$

b) $12x$ to $15x$:

$$\frac{12x}{15x} = \frac{3\cdot 4\cdot x}{3\cdot 5\cdot x} = \frac{\not{3}\cdot 4\cdot \not{x}}{\not{3}\cdot 5\cdot \not{x}} = \frac{4}{5}$$

c) 2 feet to 1 yard :

$$\frac{2\text{ ft}}{1\text{ yd}} = \frac{2\text{ ft}}{3\text{ ft}} = \frac{2}{3}$$

d) 5 pints to 3 quarts :

$$\frac{5\text{ pints}}{3\text{ quarts}} = \frac{5\text{ pints}}{3(2)\text{ pints}} = \frac{5}{6}$$

3. Let c represent the unit cost. Then,

$$\frac{\$8.79}{3\text{ lbs}} = \frac{x}{1\text{ lb.}}$$
$$\$8.79 = 3x$$
$$\frac{\$8.79}{3} = \frac{3x}{3}$$
$$\$2.93 = x$$

Therefore, 1 lb. of coffee costs \$2.93.

5. **Analyze:** We are given the annual salary of a U.S. Supreme Court Justice. We will use the fact that there are 52 weeks in 1 year. We are to find the weekly rate of pay.

Form: Let w represent the weekly rate of pay.

$$\frac{\$164{,}100}{52\text{ weeks}} = \frac{w}{1\text{ week}}$$

Solve:

$$\frac{164{,}100}{52} = \frac{w}{1}$$
$$164,100(1) = 52(w)$$
$$164,1000 = 52w$$
$$\frac{164{,}100}{52} = \frac{52w}{52}$$
$$3155.77 = w$$
$$w \approx 3156$$

State: The weekly rate of pay is \$3156.

Check:

$$\frac{164{,}100}{52} = \frac{(3156)}{1}$$
$$3155.77 \approx 3156$$

7. a) $\frac{3}{x} = \frac{6}{9}$

$$6x = 27$$
$$\frac{6x}{6} = \frac{27}{6}$$
$$x = \frac{3\cdot 9}{2\cdot 3} = \frac{\not{3}\cdot 9}{2\cdot \not{3}} = \frac{9}{2}$$

b) $\frac{x}{3} = \frac{x}{5}$

$$5x = 3x$$
$$5x - 3x = 3x - 3x$$
$$2x = 0$$
$$\frac{2x}{2} = \frac{0}{2}$$
$$x = 0$$

c) $\frac{x-2}{5} = \frac{x}{7}$

$$7(x-2) = 5x$$
$$7x - 14 = 5x$$
$$7x - 14 - 7x = 5x - 7x$$
$$-14 = -2x$$
$$\frac{-14}{-2} = \frac{-2x}{-2}$$
$$7 = x$$

d) $\frac{4x-1}{18} = \frac{x}{6}$

$$6(4x-1) = 18x$$
$$24x - 6 = 18x$$
$$24x - 6 - 24x = 18x - 24x$$
$$-6 = -6x$$
$$\frac{-6}{-6} = \frac{-6x}{-6}$$
$$1 = x$$

9. **Analyze:** According to the illustration, 3 out of 4 people will develop gum disease. We are to find how many people out of 340 will develop gum disease.

Form: Let n represent the number of people who will develop gum disease.

$$\frac{3}{4} = \frac{n}{340}$$

Solve:

$$3(340) = 4n$$
$$1020 = 4n$$
$$\frac{1020}{4} = \frac{4n}{4}$$
$$255 = n$$

State: 255 out of 340 people will develop gum disease.

Check:

$$\frac{3}{4} = \frac{(255)}{340}$$
$$\frac{3}{4} = \frac{3•85}{4•85}$$
$$\frac{3}{4} = \frac{3}{4}$$

11. Because of the addition in the numerator, x is not a common factor of the numerator and the denominator

13. a) $\frac{3x^2}{5x^2y} \div \frac{6x}{15xy^2}$

$$= \frac{3x^2}{5x^2y} • \frac{15xy^2}{6x}$$
$$= \frac{3x^2•15xy^2}{5x^2y•6x}$$
$$= \frac{3•x•x•3•5•x•y•y}{5•x•x•y•2•3•x}$$
$$= \frac{\not{3}•\not{x}•\not{x}•3•\not{5}•\not{x}•\not{y}•y}{\not{5}•\not{x}•\not{x}•\not{y}•2•\not{3}•\not{x}}$$
$$= \frac{3•y}{2}$$
$$= \frac{3y}{2}$$

b) $\frac{x^2+5x}{x^2+4x-5} \div \frac{x^2}{x-1}$

$$= \frac{x^2+5x}{x^2+4x-5} • \frac{x-1}{x^2}$$
$$= \frac{(x^2+5x)•(x-1)}{(x^2+4x-5)•x^2}$$
$$= \frac{x(x+5)•(x-1)}{(x+5)(x-1)•x•x}$$
$$= \frac{1}{x}$$

c) $\frac{x^2-x-6}{2x-1} \div \frac{x^2-2x-3}{2x^2+x-1}$

$$= \frac{x^2-x-6}{2x-1} • \frac{2x^2+x-1}{x^2-2x-3}$$
$$= \frac{(x^2-x-6)•(2x^2+x-1)}{(2x-1)•(x^2-2x-3)}$$
$$= \frac{(x-3)(x+2)•(2x-1)(x+1)}{(2x-1)•(x-3)(x+1)}$$
$$= x + 2$$

d) $\frac{x^2-3x}{x^2-x-6} \div \frac{x^2-x}{x^2+x-2}$

$$= \frac{x^2-3x}{x^2-x-6} • \frac{x^2+x-2}{x^2-x}$$
$$= \frac{(x^2-3x)•(x^2+x-2)}{(x^2-x-6)•(x^2-x)}$$
$$= \frac{x(x-3)•(x-1)(x+2)}{(x-3)(x+2)•x(x-1)}$$
$$= 1$$

e) $\frac{x^2+4x+4}{x^2+x-6}\left(\frac{x-2}{x-1} \div \frac{x+2}{x^2+2x-3}\right)$

$$= \frac{x^2+4x+4}{x^2+x-6}\left(\frac{x-2}{x-1} • \frac{x^2+2x-3}{x+2}\right)$$
$$= \frac{x^2+4x+4}{x^2+x-6} • \frac{x-2}{x-1} • \frac{x^2+2x-3}{x+2}$$
$$= \frac{(x^2+4x+4)(x-2)(x^2+2x-3)}{(x^2+x-6)(x-1)(x+2)}$$
$$= \frac{(x+2)(x+2)(x-2)(x+3)(x-1)}{(x+3)(x-2)(x-1)(x+2)}$$
$$= x + 2$$

15. a) $2x^2$, $4x$

$2x^2 = 2•x•x$

$4x = 2•2•x$

LCD $= 2•2•x•x = 4x^2$

b) $3y^2$, $9x$, $6x$, $18xy^2$

$3y^2 = 3•y•y$

$9x = 3•3•x$

$6x = 2•3•x$

$18xy^2 = 2•3•3•x•y•y$

LCD $= 3•3•y•y•x•2 = 18xy^2$

c) $x + 1$, $x - 1$

$x + 1 = x + 1$

$x - 1 = x - 1$

LCD $= (x + 1)(x - 1)$

d) $y^2 - 25$, $y - 5$

$y^2 - 25 = (y - 5)(y + 5)$

$y - 5 = y - 5$

LCD $= (y - 5)(y + 5)$

17. a) $\frac{\frac{3}{2}}{\frac{2}{3}} = \frac{3}{2} • \frac{3}{2} = \frac{9•9}{2•2} = \frac{81}{4}$

b) $\frac{\frac{3}{2}+1}{\frac{2}{3}+1} = \frac{\frac{3}{2}+\frac{2}{2}}{\frac{2}{3}+\frac{3}{3}} = \frac{\frac{5}{2}}{\frac{5}{3}} = \frac{5}{2} • \frac{3}{5}$

$$= \frac{\not{5}•3}{2•\not{5}} = \frac{3}{2}$$

c) $\frac{\frac{1}{x}+1}{\frac{1}{x}-1} = \frac{x\left(\frac{1}{x}+1\right)}{x\left(\frac{1}{x}-1\right)} = \frac{\frac{x}{x}+x}{\frac{x}{x}-x}$

$$= \frac{1+x}{1-x}$$

d) $\frac{1+\frac{3}{x}}{2-\frac{1}{x^2}} = \frac{x^2\left(1+\frac{3}{x}\right)}{x^2\left(2-\frac{1}{x^2}\right)}$

$= \frac{x^2+\frac{3x^2}{x}}{2x^2-\frac{x^2}{x^2}}$

$= \frac{x^2+3x}{2x^2-1} = \frac{x(x+3)}{2x^2-1}$

e) $\frac{\frac{2}{x-1}+\frac{x-1}{x+1}}{\frac{1}{x^2-1}} = \frac{\frac{2}{x-1}+\frac{x-1}{x+1}}{\frac{1}{(x-1)(x+1)}}$

$= \frac{(x-1)(x+1)\left(\frac{2}{x-1}+\frac{x-1}{x+1}\right)}{(x-1)(x+1)\left(\frac{1}{(x-1)(x+1)}\right)}$

$= \frac{\frac{2(x-1)(x+1)}{x-1}+\frac{(x-1)(x-1)(x+1)}{x+1}}{\frac{(x-1)(x+1)}{(x-1)(x+1)}}$

$= \frac{2(x+1)+(x-1)(x-1)}{1}$

$= 2(x+1)+(x-1)(x-1)$
$= 2x+2+x^2-2x+1$
$= x^2 \mid 3$

f) $\frac{\frac{a}{b}+c}{\frac{b}{a}+c} = \frac{ab\left(\frac{a}{b}+c\right)}{ab\left(\frac{b}{a}+c\right)}$

$= \frac{\frac{a(ab)}{b}+abc}{\frac{b(ab)}{a}+abc}$

$= \frac{a^2+abc}{b^2+abc}$

$= \frac{a(a+bc)}{b(b+ac)}$

g) $\frac{\frac{1}{y^4}}{c^{-2}} = \frac{\frac{1}{y^4}}{\frac{1}{c^2}} = \frac{1}{y^4}\bullet\frac{c^2}{1} = \frac{c^2}{y^4}$

h) $\frac{x^{-2}+1}{x^{-2}-1} = \frac{\frac{1}{x^2}+1}{\frac{1}{x^2}-1}$

$= \frac{x^2\left(\frac{1}{x^2}+1\right)}{x^2\left(\frac{1}{x^2}-1\right)}$

$= \frac{\frac{x^2}{x^2}+x^2}{\frac{x^2}{x^2}-x^2}$

$= \frac{1+x^2}{1-x^2}$

19. $E = 1-\frac{T_2}{T_1}$
$T_1(E) = T_1\left(1-\frac{T_2}{T_1}\right)$
$T_1E = T_1 - \frac{T_1T_2}{T_1}$
$T_1E = T_1 - T_2$
$T_1E - T_1 = T_1 - T_2 - T_1$
$T_1E - T_1 = -T_2$
$T_1(E-1) = -T_2$
$\frac{T_1(E-1)}{(E-1)} = \frac{-T_2}{(E-1)}$
$T_1 = \frac{T_2}{-(E-1)}$
$T_1 = \frac{T_2}{-E+1}$
$T_1 = \frac{T_2}{1-E}$

21. $l = 30,\ w = 20$
$l = \frac{k}{w}$
$30 = \frac{k}{20}$
$20(30) = 20\left(\frac{k}{20}\right)$
$600 = \frac{20k}{20}$
$600 = k$

23. Since the maid can clean the whole house in 4 hours, then she can clean $\frac{1}{4}$ of the house in 1 hour.

25. **Analyze:** We can use the formula $i = pr$. We solve this formula for p :

$$i = pr$$
$$\frac{i}{r} = \frac{pr}{r}$$
$$\frac{i}{r} = p$$

Let r represent the savings and loan interest rate, then $r + 0.01$ is the credit union's interest rate. If the student earned \$100 at the savings and loan then the principal invested was $\frac{100}{r}$. If she had deposited her money into the credit union, she would have earned \$120. So, the principal invested was $\frac{120}{r+0.01}$ hours.

Form: The same principal is invested, so $\frac{100}{r} = \frac{120}{r+0.01}$

Solve:

$$\frac{100}{r} = \frac{120}{r+0.01}$$
$$100(r + 0.01) = 120r$$
$$100r + 1 = 120r$$
$$100r + 1 - 100r = 120r - 100r$$
$$1 = 20r$$
$$\frac{1}{20} = \frac{20r}{20}$$
$$0.05 = r$$

State: The student earned 5% interest at the saving and loan.

Check:

$$\frac{100}{(0.05)} = \frac{120}{(0.05)+0.01}$$
$$\frac{100}{0.05} = \frac{120}{0.06}$$
$$2000 = 2000$$

27. **Analyze:** We can use the formula $d = rt$. We solve this formula for t :

$$d = rt$$
$$\frac{d}{r} = \frac{rt}{r}$$
$$\frac{d}{r} = t$$

If the velocity of the wind is the some unknown rate r, then the velocity of the plane going downwind is $360 + r$ mph. The velocity of the plane going upwind is $360 - r$ mph. The plane can travel 400 miles downwind at a rate of $360 + r$ mph, so it will take $\frac{400}{360+r}$ hours. The plane can travel 320 miles upwind at a rate of $360 - r$ mph, so it will take $\frac{320}{360-r}$ hours.

Form: The plane can travel 400 miles downwind in the same amount of time it takes to travel 320 miles upwind. Thus,

$$\frac{400}{360+r} = \frac{320}{360-r}$$

Solve:

$$\frac{400}{360+r} = \frac{320}{360-r}$$
$$400(360 - r) = (360 + r)320$$
$$144,000 - 400r = 115,200 + 320r$$
$$144,000 - 400r + 400r$$
$$= 115,200 + 320r + 400r$$
$$144,000 = 115,200 + 720r$$
$$144,000 - 115,200$$
$$= 115,200 + 720r - 115,200$$
$$28,800 = 720r$$
$$\frac{28,800}{720} = \frac{720r}{720}$$
$$40 = r$$

State: The velocity of the wind is 40 mph.

Check:

$$\frac{400}{360+(40)} = \frac{320}{360-(40)}$$
$$\frac{400}{400} = \frac{320}{320}$$
$$1 = 1$$

Turn the graphing calculator into a powerful tool for your success!

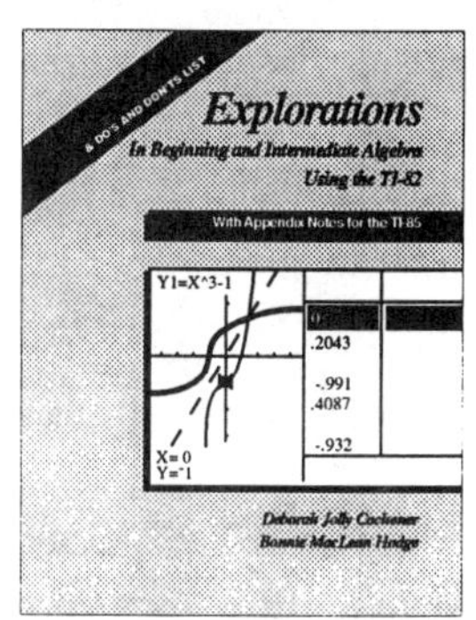

Explorations in Beginning and Intermediate Algebra Using the TI-82 With Appendix Notes for the TI-85, CASIO fx-7700GE, and HP48G
by Deborah J. Cochener and Bonnie M. Hodge,
both of Austin Peay State University

$25.95 Single Copy Price. 314 pages. Spiral bound. 8 1/2 x 11.
ISBN: 0-534-34091-1. © 1996. Published by Brooks/Cole.

Learn to use the graphing calculator to develop problem-solving and critical-thinking skills that will help improve your performance in your beginning or intermediate algebra course!

Designed to help you succeed in your beginning or intermediate algebra course, this unique and student-friendly workbook improves both your understanding and retention of algebra concepts—using the graphing calculator. By integrating technology into mathematics, the authors help you develop problem-solving and critical-thinking skills.

To guide you in your explorations, you'll find:

- hands-on applications with solutions
- correlation charts that relate course topics to the workbook units
- a key chart that shows which units introduce keys on the calculator
- a Troubleshooting Section to help you avoid common errors

I found this workbook to be very well written and thought out and to be quite different from calculator supplements on the market. The integration of concept development and applications makes this usable as an ancillary textbook rather than just a calculator supplement. . . . The students are not taught to rely on the calculator, but to use it as another problem solving tool and as reinforcement of algebraic principles."
Mary Lou Hammond, Spokane Community College

The authors are clearly aware of important issues in mathematics education today, and have attempted to respond with a constructive, active-oriented approach. . . . A surprisingly valuable introduction to some not so well understood arithmetic concepts.
Don Shriner, Frostburg State University

Topics

This text contains 35 units divided into the following subsections:

Basic Calculator Operations
Graphically Solving Equations and Inequalities
Graphing and Applications of Equations in Two Variables
Stat Plots

Also included are:

Concept Correlation Charts
Key Introduction Charts
Troubleshooting

Order your copy today!

To receive your copy of ***Explorations in Beginning and Intermediate Algebra Using the TI-82: With Appendix Notes for the TI-85, CASIO fx-7700GE, and HP48G,*** simply mail in the order form attached.

ORDER FORM

To order, simply fill out this coupon and return it to Brooks/Cole along with your check, money order, or credit card information.

______Yes! I would like to order ***Explorations in Beginning and Intermediate Algebra Using the TI-82: With Appendix Notes for the TI-85, CASIO fx-7700 GE, and HP48G***, by Deborah J. Cochener and Bonnie M. Hodge, ISBN: 0-534-34227-2 for $25.95.

Residents of: AL, AZ, CA, CT, CO, FL, GA, IL, IN, KS, KY, LA, MA, MD, MI, MN, MO, NC, NJ, NY, OH, PA, RI, SC, TN, TX, UT, VA, WA, WI must add appropriate state sales tax.

Subtotal __________
Tax ____________
Handling ___$4.00___
Total ____________

Payment Options

_______ Check or money order enclosed

or bill my ____VISA ____MasterCard ____American Express

Card Number: ____

Expiration Date: ______________________________________

Signature: ____

Note: Credit card billing and shipping addresses must be the same.

Please ship my order to: (Please print.)

Name ____

Street Address_______________________________________

City ____________________ State __________ Zip+4________________

Telephone (   )______________________ e-mail __________________

Mail to:

Brooks/Cole Publishing Company
Dept. 9BCMA008
511 Forest Lodge Road, Pacific Grove, California 93950-5098
Phone: (408) 373-0728; Fax: (408) 375-6414
Internet: http://www.brookscole.com

ITP **International Thomson Publishing Education Group**

P.S. If this book is appropriate for a course that you teach, please send your request for a complimentary review copy, on department letterhead, to the address listed above. Prices subject to change without notice.